Genomics and Bioethics:
Interdisciplinary Perspectives, Technologies and Advancements

Soraj Hongladarom
Chulalongkorn University, Thailand

Medical Information Science Reference

MEDICAL INFORMATION SCIENCE REFERENCE

Hershey · New York

Director of Editorial Content: Kristin Klinger
Director of Book Publications: Julia Mosemann
Acquisitions Editor: Lindsay Johnston
Development Editor: Mike Killian
Publishing Assistant: Natalie Pronio
Typesetters: Natalie Pronio and Casey Conapitski
Production Editor: Jamie Snavely
Cover Design: Lisa Tosheff

Published in the United States of America by
Medical Information Science Reference (an imprint of IGI Global)
701 E. Chocolate Avenue
Hershey PA 17033
Tel: 717-533-8845
Fax: 717-533-8661
E-mail: cust@igi-global.com
Web site: http://www.igi-global.com

Library of Congress Cataloging-in-Publication Data

Genomics and bioethics : interdisciplinary perspectives, technologies, and advancements / Soraj Hongladarom, editor.
p. ; cm.
Includes bibliographical references and index.
Summary: "This book focuses on ethical, social, cultural, and legal implications of genetics, genomics and genetic databanking as they relate to concrete cultural and historical traditions"--Provided by publisher.
ISBN 978-1-61692-883-4 (h/c) -- ISBN 978-1-61692-885-8 (eISBN) 1. Genomics--Moral and ethical aspects. 2. Genetics--Moral and ethical aspects. 3. Human genetics--Moral and ethical aspects. I. Soraj Hongladarom, 1962-
[DNLM: 1. Genomics--ethics. 2. Bioethical Issues. 3. Genetic Research--ethics. QU 58.5 G33515 2011]
QH438.7.G4615 2011
174'.957--dc22

2010021004

British Cataloguing in Publication Data
A Cataloguing in Publication record for this book is available from the British Library.

Editorial Advisory Board

Table of Contents

Detailed Table of Contents

Section 1
Theoretical Concerns

Nothing is unnatural in Buddhism. That is to say, everything that happens does happen according to natural law; even acts perpetrated and designed by human beings are natural as they would not be possible in the first place if natural law is violated. A consequence of this is that acts such as cloning and manipulation of genetic coding, which are something criticized as being 'unnatural,' are accepted as natural in Buddhism, and thus Buddhism does not have anything in principle against such acts, or indeed any acts at all so long as they are natural. Hence, if there is to be anything wrong in these acts the wrongness must come from somewhere else. Buddhism regards the wrongness or rightness of action as arising from their consequences.

The contrast between public health measures and medical intervention is discussed. In some cases the former are more effective than the latter in preventing infectious diseases and in raising the overall health standard of a population. If this is the case, then the promises of genomics would appear to be vacuous because simple public health or epidemiological measures would be equally good. Since genomics promises to individualize care, it is at odds with public health and epidemiology. Genomics may prove to be valuable in individual care, but its role in how the average health standard in population at large appears to be contested. As long as population health factors are not taken into account, genomics would not realize its full potential.

The role of international bodies as the place where international debates on bioethical issues take place is investigated. In theory these international venues are supposed to be neutral in that they do not favor any particular traditions or belief systems. As a neutral venue for debates, these international arenas should recede to the background, so to speak, and let all the voices from among all parties involved in the debate and discussion be heard. However, in practice such a scenario scarcely happens, as these fora and venues are often criticized as being dominated by the West and their claim of universal values. In order for the international venues to be a really just place where all the voices are given due prominence, the roles and arrangements of these international organizations need to be scrutinized.

The chapter deals with moral deliberation over the status of the embryo, observed from two perspectives, namely the inner context of physical and biological composition including the argument of potentiality as a driving force of development, and the external context within lived and experienced practices in which an embryo is inevitably embedded. Both components are integral parts of what constitutes the life of the embryo, and therefore any separated observation is biased and does not fulfill the demands of the universal truth. Hence, the usual argument that focuses exclusively on the embryo itself, whether the embryo deserves moral right as a result of its potentiality for autonomy, is misguided.

The chapter discusses both the benefits and ethical together with other concerns that arise from direct-to-consumer genetic testing. On the one hand, it is the right of an individual to know what their genetic constitution is like and what could result to their body afterwards. But on the other hand there are a number of concerns, such as clinical relevance of the consumer testing service, the role of counseling, sensitivity and specificity of the testing, and the danger of genetic determinism. The author points out that public understanding and education on basic relations between genetics and susceptibility for diseases is very important and should play a larger role. The role of incidentaloma, where testing intended for one purposes turns out to yield another kind of information not originally intended, is also discussed.

The chapter discusses the emerging trend where genetic knowledge is used in nutritional sciences resulting in tailor made dietary recommendations for specific individuals. A number of ethical concerns with such new technologies are raised. The idea of personalized dietary advice is that, in order to help combat diseases, food intake should be adapted so as to fight diseases most effectively through knowledge obtained from the individual's genetic makeup. However, people are much more attached to food, viewing food as part and parcel of who they are, than they are to drugs. This raises a salient ethical point.

A new model of obtained informed consent for research involving genetic constitution of population is proposed. The mainstream method of emphasizing individual autonomy and individual consent alone might not be adequate in furthering and facilitating research that could translate into a lot of benefits for humankind. Instead of the idea of individual autonomy, the information in question should be subject to multiple ownership or custodianship that call for multiple loci of consent. If the human genome is the common heritage of humanity, a model of collective guardianship of genetic information might be more appropriate.

A common argument against research on genetics is that human beings are "playing God." What this means is that it is unethical for humans to emulate God; whatever is done by human beings when they play God would be utterly unnatural and will bring upon dire consequences. The question is whether advanced genetic research is 'natural.' The chapter traces the origin on the playing God argument, one that chastises humans for transgressing the boundary between the mortal and the divine, and found that it stems from two sources, namely Augustine and Aquinas. The conclusion is that playing God arguments are equivocal and hence do not have much force.

The chapter criticizes arguments purporting to show that the human body could be made available in the market as property and those arguing that the concept of property could be applicable to the human bodily parts or human DNA. The author argues that the genetic information contained in matter such as DNA cannot be classified as property. There are three reasons: DNA is too personal to be commodified; DNA is of familial nature; and commercialization of DNA runs the risk of exploitation of the disadvantaged.

Section 2
Country Experiences

The chapter deals with ethical aspects of patent law and how the global patent regime helps or hinders the development of a developing country such as Thailand. More specifically, Article 27.3 of the World Trade Organization (WTO) Agreement on Trade-Related Aspects of Intellectual Property Rights (TRIPS) is analyzed; the Article states that countries may exclude plants and animals (but not microorganisms) from patent protection, thus it allows for a leeway where developing countries could leverage against the multinational corporations from rich developed countries.

The chapter reports on the experiences of both experts and lay people on the level of knowledge and social representations of genomic research and its applications in a number Latin American developing countries. Issues discussed include access to prevention and therapeutic methods related to genomic medicine in Latin America, risks associated to genetic modifications in humans, lack of equity in the access to health benefits, control by biotechnological companies, commercialization of gene sequences through patents which leads to commercial exploitation of underdeveloped countries, among others.

The authors focus their attention on the structure of Indonesian law and policies on biotechnology issues; they also address some issues related to bioethics and research activities and economic activities, such as the issue of bioprospecting and biopiracy, and how the legal and governance structure within Indonesia are designed to cope with this issue. An issue that looms large is about intellectual property rights.

The chapter presents a careful comparative study on ethical and legal aspects of human biobanks both in Europe and elsewhere. The rapid expansion of human DNA sampling and data collection has taken place in the last few years, but the legal and ethical perception of this situation looks very different in European countries and beyond. The author focuses her attention on the European Union, especially in Estonia, where a population wide gene back has been established; moreover, she also discusses what is happening in Macedonia, a relatively neglected country in eastern Europe, as well as Australia, India and Israel.

Singapore has a well developed committee (namely the Bioethics Advisory Committee) that conducts studies and provides advice on bioethical issues. A particular focus is put on the tension between privacy concerns and the imperatives of access for biomedical research, which has been championed by the Singaporean government as one of the future leading sectors of its economy. The notion of "genetic exceptionalism" is discussed. The author thinks that the exception is not, or should not, be as strong as what its proponents take it to be. This weakening of genetic exceptionalism informs many decisions of the Singaporean Bioethics Advisory Committee.

The chapter presents a detailed account of the available legal mechanisms on biotechnological issues in her country. Among the issues discussed are biosafety regulations, especially ones concerning the Cartagena Protocol as well as the governance structure in Tajikistan on these issues.

The chapter focuses on a comparison of legal mechanisms and regulatory frameworks on genetic testing and privacy in Europe and Australia. The rationale for examining genetic issues in the European context is that the framework for personal data protection has undergone much greater scrutiny at the policy level, thus providing one possible model that could be adopted much more broadly. In addition, the Australian perspective is also interesting because of the considerable attention given to the protection of genetic information in this country in recent years. These recent developments thus represent a useful counterpoint for analysis of the issues with regard to protecting genetic data.

Preface

This book grew out of an international workshop on "Technology and Culture," which was held in Bangkok, Thailand on March 27, 2007. The workshop was supported by a generous grant from the International Institute for Asian Studies (IIAS), a research organization based in Leiden, the Netherlands focusing on supporting research mostly in social scientific and cultural studies of Asia. The grant was part of the series of grants made available under the joint effort between the IIAS, the European Alliance for Asian Studies, and the Asia-Europe Foundation. This grant, entitled "The Asia-Europe Workshop Series" aims at fostering closer ties between academics in Asia and Europe so as to create new knowledge and understanding on issues of common interest between the two continents.

The workshop, whose full title is "Technology and Culture: Genetics and its Social and Ethical Implications in Asia and Europe," was held in conjunction with the Eighth Asian Bioethics Conference, which was a rather large event for those who are interested in issues related to ethical, social and legal aspects of biomedical sciences and biotechnology, especially in the context of Asia and Asian cultures. As befitting the overall theme of the Asian Bioethics Conference, the workshop on "Genomics and Culture" looks at how the cultures and traditions interact with issues that are quickly emerging due to the rapid advancements in genetic sciences. However, not all chapters that were presented during the workshop are published here, and in fact several chapters in this book were not presented during the workshop. So this book is not a mere collection of chapters presented during the workshop, but an original work that only has its starting point in that 2007 workshop.

It is undeniable that the world today is one where science and technology play an essential role in almost all areas of life. Two technologies that permeate people's lives the most are those related to information and communication on the one hand and the manipulation of the basic constituents of life itself on the other. Information and communication technologies are making themselves present ever increasingly, so much so that the lives of many people today is almost saturated by the constant demands these technologies are making on them. It might be common place a few years ago to say that the Internet has become more prevalent in people's lives, but today the Internet itself has become more like the environment itself, something not quite unlike the air we breathe or the water we drink and bathe with. So the Internet is not only prevalent, it has become totally saturated. Even in the developing world, more and more people are using the mobile phones. Many of these mobile phones, furthermore, are internet capable, meaning that people can carry internet connection everywhere without being tied down to a desktop computer. The impacts and implications of this massive transformation in people's lives everywhere on the globe are indeed staggering.

The other massively transformative technology is biotechnology. As the Internet is deeply transformative in how we think and how we relate to one another in the social world, biotechnology and genetic sciences are poised to transform the very constitution of our own physical bodies themselves. The new era in life sciences and biotechnology is quickly unfolding itself before our very eyes. The era started in earnest with the success in cloning mammals by the team in Scotland a few years ago, and this beginning was strongly reaffirmed through the success of sequencing of the entire human genome a short while later. All these, together with the explosion of the Internet and mobile computing, make the time we are living in a very exciting one. New possibilities and vistas are opening up in a way that was not even conceivable only within the memories of many living nowadays.

What is more exciting, perhaps, is that the information and communication technologies and the technologies in biomedical sciences are in fact merging together. Computers were used in computing the huge quantity of data available in the human genome. The genome itself has been 'digitized,' meaning that the information contained in the genome can be directly manipulated and processed through computing machinery. Many groups are doing research on the brain-computer interaction, with many exciting applications. The merging of the two technologies could point to a time when the physical bodies, made of course of flesh and blood, and the bodies of computer hardware are fused together. Not only did computing technologies make inroad into biotechnology, but the other way around is also the case. Functions of certain aspects of biology are being used to help increase the power of computers. Designs of hugely powerful computers are being modeled on the brain. Studies are being conducted to search for ways to store digital numbers on biological material, which would make computers much smaller and more powerful. Moreover, the way in which the synapses arrange themselves in the brain is becoming a model for computers. Even how different brains (different persons) interact and collaborate in accomplishing certain tasks could be a model on how computers themselves collaborate. The line between the biological and the metallurgical or artificial are becoming fuzzier by the day.

However, we do not live in a purely scientific and technological world. That is to say, if everything in the world functions as does an idealized computer or a biological laboratory, then things would be much tidier and much more tractable than they are now. But we do not live in the idealized world. Scientific research and technological innovation are always bound to have repercussions when they are introduced to the public at large. Since science and technology do not exist in the vacuum in idealized conditions, they need to factor in these surrounding contexts where people's goals, beliefs, desires, values, traditions, histories, cultures are taken into account. The overall goal of the 2007 workshop on Genomics and Culture was precisely to investigate how the ethical implications of genomics and related sciences and technologies could be investigated through the lenses of the world's major cultural traditions, that of East and West. This overall goal also informs the basic orientation of this book.

The intersection and the meshing of the activities, procedures and end results of science and technology with the people's goals and values discussed above present themselves in ethical dilemmas as well as social and cultural implications of science and technology. For instance, there is the technical problem of how to clone a mammal. What Wilmut discovered was a technique of taking out the nucleus of a certain cell, injecting it with another nucleus taken from another organism and somehow coaxing the mixture to grow. This in itself is the scientific breakthrough. But when the activity of cloning is considered in the wider perspective, we find that it gives rise to a large number of problems, many of which are familiar. First of all, is it ethical to do the cloning? Here the question does not address the technical aspect of cloning; instead it demands an answer as to the value of the act itself. And how are we to know those values? Is it enough to search down the depth of your soul, so to speak, in order to find the answer whether cloning is ethical or not? Or should we conduct a survey to sound out the public's opinion on the issue, and if the majority finds that it is ethical (or unethical) then could we pass the judgment that it is really ethical (or unethical)? Or do we wait and see what happens as a result of introducing the cloning? Or do we take as an assumption that reason and argument will always lead us to eventual agreement and perhaps the truth, including truths about moral values?

These are of course very intractable problems, stuff that philosophers and ethicists have been tackling for millennia. The intractability has led many to say that these kinds of problems do not have real solutions and it is actually a waste of time to find solutions to them. But that is a grave mistake. The difficulty in learning the ethical value of an act does not mean that the act does not have objective values. Ethics is not a calibration of different opinions and viewpoints; it is a serious undertaking aiming to go into the depth of things and of action. At any rate not bothering with thinking about ethics would certainly result in a much worse off situation than thinking about it. A society that does not reflect at all on the ethical value or implications of their actions would be a dry and meaningless one indeed.

So this is what the chapters in this book are trying to offer. As I mentioned, some of the authors were present during the 2007 meeting, and their chapters have been much revised during the course of the two to three years from that meeting to the publication of this book. Those who were present at the meeting whose chapters are available here are: Somparn Promta, Chee-Khoon Chan, Leonardo de Castro, Ole Döring, Jakkrit Kuanpoj, Brigitte Jansen, Terry Kaan, and Jürgen Simon. As the 2007 workshop was held under the auspice of the Asia-Europe Foundation, it was natural that scholars and scientists from Europe and Asia came and interacted during the workshop. Somparn and Jakkrit are from Thailand. Chan is from Malaysia. Leo de Castro was from the Philippines, and Terry Kaan was from Singapore. From the Europe side, Ole Döring, Brigitte Jansen and Jürgen Simon are all from Germany.

Furthermore, we have a number of authors who were not present at the workshop but nonetheless responded to our general call for chapters and presented their works for consideration. Their works significantly broadened the scope and increased the depth of the book. Instead of only Asia and Europe these authors have made the book really global. Chamundeeswari Kuppuswamy is from India, but now is working in the UK. Elena Ignovska is from Macedonia. Richard Stein is from the US. Varghese Daniel is from Australia. Eduardo Rodriguez is from Chile. Theofrantus Litaay is from Indonesia. Firuza Nasyrova is from Tajikistan, and, last but not least, Sergio Malanda is from Spain and currently working in Australia. These authors show the breadth and the international character of the chapters in the book very well.

The topic and the academic disciplines represented by these authors also show the depth and breadth of the collection. Several chapters deal with the legal aspects of genetics and biotechnology in the authors' own countries. A few deal with theoretical and philosophical formulations, while more discuss the roles of particular religions, some of which are rather seldom discussed in the literature, in bioethics.

The book is divided into two sections. Section 1 contains chapters that are of more philosophical and normative nature, and Section 2 deals with more specific topics, namely the legal situations in certain countries and discussions as to the ethical implications of these legal mechanisms. The first chapter, *Buddhism and Human Genetic Research*, by Somparn Promta, is an attempt to find out how Buddhism responds to the advancements in biotechnology. What is notable in his presentation is that Somparn argues that in Buddhism nothing is unnatural. That is, everything that happens does happen according to natural law; thus even acts perpetrated and designed by human beings are natural as they would not be possible in the first place if natural law is violated. The implication of this is that acts such as cloning and manipulation of genetic coding, which are something criticized as being 'unnatural' are accepted as natural in Buddhism, and thus Buddhism does not have anything in principle against such acts, or indeed any acts at all so long as they are natural. Hence, if there is to be anything wrong in these acts, the wrongness must come from somewhere else, and Somparn argues that for Buddhism the wrongness or rightness of action comes from their consequences.

The second chapter, *Genomics and Population Health: A Social Epidemiology Perspective*, by Chee-Khoon Chan deals with the contrast between public health measures and medical intervention, where he argues that in some cases the former is more effective than the latter in preventing infectious diseases and in raising the overall health standard of a population. Chan argues that if this is the case (and he presents evidence and reasons showing why it is really the case), then the promises of genomics would appear to be vacuous because simple public health or epidemiological measures would be equally good. Since genomics promises to individualize care, it is at odds with public health and epidemiology. Genomics may prove to be valuable in individual care, but its role in how the average health standard in the population at large appears to be contested. As long as population health factors are not taken into account, genomics, so he argues, would not realize its full potential.

Moving from epidemiology to the realm of international law, Chamu Kappuswamy, in her *International Organizations as Fora for International Bioethical Debate: Towards a Just International Bioethical Law?* looks at the role of international bodies as the place where international debates on bioethical issues take place. In theory these international venues are supposed to be neutral in that they do not favor any particular traditions or belief systems. As a neutral venue for debates, these international arenas should recede to the background, so to speak, and let all the voices from among all parties involved in the debate and discussion be heard. However, in practice such a scenario scarcely happens, as these fora and venues are often criticized as being dominated by the West and their claim of universal values. In order for the international venues to be a really just place where all the voices are given due prominence, the roles and arrangements of these international organizations need to be scrutinized. For Kappuswamy, the peoples of the world need to utilize their various firepower coming from their respective cultural traditions to tackle the difficult dilemmas arising from the use of genomics and biotechnology. She is still optimistic in the role of the international organizations as a meeting place for people coming from very different backgrounds. This will ensure that genuine international agreement and perhaps legal mechanism is possible.

The next chapter is by Elena Ignovska *The Value of Life of the Embryo Observed from Two Different Lenses: From its Own Potential to Develop, and from the Context in which It is Embedded.* Here the chapter deals with moral deliberation over the status of the embryo, observed from two perspectives, namely the inner context of physical and biological composition including the argument of potentiality as a driving force of development, and the external context within lived and experienced practices in which an embryo is inevitably embedded. For Ignovska both components are integral parts of what constitutes the life of the embryo, and therefore any separated observation is biased and does not fulfill the demands of the universal truth. Hence, the usual argument that focuses exclusively on the embryo itself, whether the embryo deserves moral right as a result of its potentiality for autonomy, is misguided. Instead of focusing only on the embryo, Ignovska invites us to have a look at the context wherein the embryo itself is embedded. Not only the right of the embryo is in question, but the right of the parents and others related to the embryo needs to be considered too. For instance, she looks at embryos created in vitro within the context of a divorce dispute in which progenitors are confronting different interests for the future of the embryo itself.

In the next chapter, Richard Stein in *Direct-to-Consumer Genetic Testing* discusses the emerging trend of direct-to-consumer genetic testing and identifies a number of potential pitfalls that emerge from this practice. After providing some very useful details and background information about the Human

Genome Project and genetic testing in general, Stein outlines both the benefits and the ethical and other concerns that arise from it. On the one hand, it is the right of an individual to know what their genetic constitution is like and what conditions they are predisposed to (some individuals, however, may prefer to have the right not to know so that they don't have to bother with information that is potentially challenging psychologically – in any case, it is the individual himself or herself who chooses). But, on the other hand, there are a number of concerns, such as the clinical relevance of the consumer testing service, the role of counseling, the sensitivity and specificity of the testing, and the danger of genetic determinism (the belief that someone's genetic makeup accounts for all their physical and personal being). Stein points out that public understanding and education on basic relations between genetics and disease susceptibility is very important and should play a larger role in society. He also discusses the role of incidentaloma, where testing intended for one purpose turns out to yield another kind of information, which not originally sought. This can have disruptive effects on the individual and, hence, the ethical considerations of the direct-to-consumer genetic testing need to be fully taken into account.

Minakshi Bhardwaj also stays in the same general area when she discusses what is called the "omics" technologies in *The Applications of Omics Technologies and the Challenges of Ethics in Nutritional Sciences*. Here she discusses another emerging trend where genetic knowledge is used in nutritional sciences resulting in tailored dietary recommendations for specific individuals. This practice is very close to the direct-to-consumer genetic testing discussed by Stein in the previous chapter. In Bhardwaj's words, "several ... technologies are developed in the last decade that take an 'omics' approach i.e. an integrated approach in the study of cell function. It is hoped that the applied integrative omics approaches may be helpful in establishing cause and effect relationships between genotype and phenotype. These 'omics' approaches include the integration of genomics, proteomics, transcriptomics, metabolomics and other omic technologies to do the non-targeted studies of biomolecules involved in the proper functioning of the cells and their responses to environmental changes." And she raises a number of ethical concerns with such new technologies. What is interesting is her discussion of personalized dietary advice. The idea is that, in order to help combat diseases, food intake should be adapted so as to fight diseases most effectively through knowledge obtained from the individual's genetic makeup. But according to Bhardwaj, food is not like medicine. People are much more attached to food, viewing food as part and parcel of who they are, than they are to drugs. This raises a salient ethical point. Furthermore, Bhardwaj also discusses the impact that these new technologies could have on the developing countries and the whole issues of ethics of food and global justice.

Leaping from genetic testing to the practice of conducting biomedical research on human subjects, Leonardo de Castro and Chin Leong Teoh, in *Beyond Informed Consent: A Model of Collective Guardianship for Ethical Genetic Research*, propose a new model of obtained informed consent for research involving the genetic constitution of population. De Castro and Teoh present a convincing case that the mainstream method of emphasizing individual autonomy and individual consent alone might not be adequate in furthering and facilitating research that could translate into a lot of benefits for humankind. Instead of the idea of individual autonomy, where the lone individual possesses and controls his or her genetic information, the information in question should be subject to multiple ownerships or custodianships that call for multiple loci of consent. If the human genome is the common heritage of humanity, de Castro and Teoh believe that a model of collective guardianship of genetic information may be more appropriate. In such a case, guidelines which combine the principle of informed consent with the institutionalization of the oversight role of ethics committees in the management of genetic databases, in their use for secondary research, in the crossmatching of databases, in the decoding or unlinking of samples or data, and so on, can be drafted so that important work in population genetic research can proceed whilst still respecting the rights and interests of human subjects.

Moving from informed consent to theology, we find Varghese M. Daniel discussing genetics and theology in the next chapter, *Genomics and Genetic Engineering: Playing God?*. A common argument against research on genetics is that human beings are "playing God." What this means is that it is unethical for humans to emulate God; whatever is done by human beings when they play God would be utterly unnatural and will bring upon dire consequences. Here Daniel touches upon the question of whether advanced genetic research is 'natural,' a topic which is also taken up in detail by Somparn Promta in the first chapter. Varghese traces the origin on the playing God argument, one that chastises humans for transgressing the boundary between the mortal and the divine, and finds that there are two sources: the Augustinian view on 'original sin' and the Aristotelian-Aquinas thought on natural law. (There is also another strain of 'playing God' which is positive. Humans, being created in the image of God, have the duty for stewardship, thus taking over the role that God plays.) Here Daniel argues that such negative views on humans playing God do not have either a negative or positive role in Eastern (namely, Orthodox) Christianity. Here Daniel agrees with Somparn in that both do not view advanced research in genetic or other branches of science to be unnatural. Such research works cannot be unnatural because everything falls under the realm of cosmic law for Buddhism, or because in the eyes of God everything done by humans are only "human plays." Daniel then concludes that playing God arguments are equivocal and hence do not have much force.

The next chapter, *A Philosophical Exploration of the Concept of 'Property' in Genetics and Databanking: Challenges for Bioethics in Asia and Europe*, by Ole Döring is the last chapter in Section One, and is a strong criticism of any attempt to argue that the human body, including the human genetic information, could be made available in the market as property or even to argue that the concept of property could be applicable to the human bodily parts or human DNA. Following Williams-Jones, who analyzes the tendency to materialize and commodify human bodily parts as arising from three factors, namely Cartesian dualism, materialist conception of the person, and the principle of self-determination or autonomy, all of which contribute to the idea that the human body is little more than an a clump of inert matter, Döring argues that the genetic information contained in matter such as DNA cannot be classified as property. He gives three reasons: DNA is too personal to be commodified; DNA is of familial nature; and commercialization of DNA runs the risk of exploitation of the disadvantaged. True to the German tradition of stressing the uncompromising importance of individual rights, Döring writes: "The true challenge for bioethics is not to organize the dissent that exists among those who already accept the foregoing cultural grammar of a qualified dualism of the human being that renders one part, quality or property of the human disposable (*verfügbar*) in principle. Ethics rather seeks to provide strong grounds to explain and argue for the unavailability (*Unverfügbarkeit*) of the human, for any reason or purpose."

Chapters in Section 2 deal with more specific issues of legal situations in many countries, and they together provide a broad panorama where we can look at how different countries respond to the challenges posed by genetic sciences. Jakkrit Kuanpoth discusses the Thai legal mechanisms dealing with biotechnology in his chapter, *Biotechnological Patents and Morality: A Critical View from a Developing Country*. He concentrates on the ethical aspect of patent law and how the global patent regime helps or hinders the development of a developing country such as Thailand. More specifically, Kuanpoth focuses his analysis on Article 27.3 of the World Trade Organization (WTO) Agreement on Trade-Related Aspects of Intellectual Property Rights (TRIPS), which states that countries may exclude plants and animals (but not micro-organisms) from patent protection, thus allow a leeway where developing countries could leverage against the multinational corporations from rich developed countries and protect the plant and animal varieties within their domains. Kuanpoth discusses the ongoing debates and the intricate interpretations being offered on this and other international regulations.

Still on the topic of genetics and the developing countries, the next chapter by Eduardo Rodriguez and Fernando Lolas, *Social Issues Related To Gene Patenting At Latin America: A Bioethical Reflection* reports the authors' research experience about the level of knowledge and social representations of genomic research and its applications of experts (legal, biomedical researches) and lay people in relation to genomic medicine, the possibility of genetic manipulation and legal protections at Latin American developing countries, specifically Argentina, Chile, Mexico and Peru. The authors discuss issues such as little access to prevention and therapeutic methods related to genomic medicine in Latin America, risks associated to genetic modifications in humans, lack of equity in the access to health benefits, control by biotechnological companies, commercialization of gene sequences through patents which leads to commercial exploitation of underdeveloped countries, little participation of indigenous communities in the studies done on their DNA sometimes without proper informed consent, and the necessity of legal regulation to prevent the pathway towards enhancement genetic modifications or reproductive human cloning and of regulating access to genetic information.

Coming back from Latin America back to Asia again, the chapter by Theofrantus Litaay, Dyah Hapsari Prananingrum, Yakub Adi Krisanto, *Indonesian Legal Perspectives on Biotechnology and Intellectual Property Rights* looks at the situation in Indonesia. Many concerns that are voiced by Kuanpoth and Rodriguez and Lolas are also raised by Litaay and his colleagues in the case of Indonesia. More specifically, the authors focus their attention on the structure of Indonesian law and policies on biotechnology issues; they also address some issues related to bioethics and research activities and economic activities, such as the issue of bioprospecting and biopiracy, and how the legal and governance structure within Indonesia are designed to cope with this issue. It is not surprising that intellectual property issues loom large in these chapters, as this is a very contentious issue between the developing and the developed world, an issue that merits a much closer look and more careful research.

Brigitte Jansen's chapter, *Human Biobanks: Selected Examples from and beyond Europe* is a careful comparative study on ethical and legal aspects of human biobanks both in Europe and elsewhere. According to her, a rapid expansion of human DNA sampling and data collection in order to exploit and study the genetic information has taken place in the last few years, but the legal and ethical perception of this situation looks very different in European countries and beyond. Jansen chooses to focus her attention on the European Union, especially on Estonia, where a population wide gene bank has been established; moreover, she also discusses what is happening in Macedonia, a relatively neglected country in Eastern Europe. And outside of Europe she considers the cases in Australia, India and Israel. This makes for a very wide view of the biobank situation in many countries across the world, and thus makes her study a valuable one.

Turning back from a multi-country, comparative study, Terry Kaan's chapter, *The Regulation of Genetic Testing and the Protection of Genetic and Medical Information in Singapore* is a very detailed analysis of what is happening in one country, Singapore. As Singapore is among the forefront of countries pushing forward its biotechnology capabilities as a means for economic development, Kaan's chapter presents a very interesting case study. Moreover, Singapore also has a well developed committee conducting studies and providing advice on bioethical issues. As a member of that committee and a law professor, Kaan is in a unique position to provide us with a view of what is happening there so that we could learn from the Singaporean experience. A particular focus of the chapter will be the tension between privacy concerns and the imperatives of access for biomedical research, given that biomedical research has been championed by the Singapore government as one of the future leading sectors of its economy. What is particularly interesting is Kaan's discussion of what is known as "genetic exceptionalism," the idea that genetic issues are special and deserve a number of special treatments. Kaan, however, thinks that the exception is not, or should not, be as strong as what its proponents take it to be. This weakening of genetic exceptionalism informs many decisions of the Singaporean Bioethics Advisory Committee.

The next chapter takes us from Singapore up north to Tajikistan. Firuza Nasyrova, in her *Legal Aspects of Bioethics in Tajikistan,* provides a detailed account of the available legal mechanisms on biotechnological issues in her country. She gives us a brief history of Tajikistan, telling us that one of world's greatest philosophers, known in Latin as Avicenna, had his hometown in Tajikistan. She also discusses biosafety regulations, especially the ones concerning the Cartagena Protocol, the topic also discussed by Kuanpoth, Rodriguez and Litaay in their respective countries and regions, as well as the governance structure in Tajikistan on these issues.

In the fifteenth chapter, Sergio Romeo-Malanda, Dianne Nicol and Margaret Otlowski, in *Genetic Testing and Protection of Genetic Privacy: A Comparative Legal Analysis in Europe and Australia* bring us back to the topics of genetic testing and privacy, this time focusing on the similarities and differences in the legal mechanisms and regulatory frameworks in both regions. This comparison is interesting because the rationale for examining genetic issues in the European context is that the framework for personal data protection was established much earlier there than in other jurisdictions, and has undergone much greater scrutiny at the policy level. As such, the European approach provides one possible model that might be adopted much more broadly, especially if we take into consideration that the key legislative development in the field of genetic data is the recognition of the human right to privacy. In addition, the Australian perspective is also interesting because of the considerable attention given to the protection of genetic information in this country in recent years. There has in fact been a major national inquiry into the protection of human genetic information jointly conducted by the Australian Law Reform Commission and the Australian Health Ethics Committee of the National Health and Medical Research Council. The ensuing report, Essentially Yours 2003, made numerous recommendations for reform, some of which have already been implemented with other changes in progress. These recent developments thus represent a useful counterpoint for analysis of the issues with regard to protecting genetic data.

The last chapter in the book, is Jürgen Simon's and Jürgen Robienski's *Property, Personality Rights and Data Protection with regard to Biobanks: A Layered System*, a discussion and analysis of the concept of property in the context of genetic research and biobanking. The authors are proposing a novel way of conceptualizing the concept of property, one which could unravel the difficulties that beset bioethical and legal debates on this issue. The issue concerns property rights to bodily material. Once a certain part of the body is taken out, to what extent could the person from whom the tissue has been separated claim property rights over the tissue? According to Simon and Robienski, if the tissue is completely anonymized, then the person loses her right to claim the tissue to be her property. In this case if the former carrier of the tissue wants to maintain some ownership right, there must be a way to link the tissue anatomically to the carrier herself. This could be done by way of pseudonymization. The trustee model where the tissue is safeguarded and pseudonymized by a trustee could act as the middle path between complete retention of personal and property rights of the alienated tissue on the one hand, and complete disregard of the property or ownership rights to the tissue on the other. By having a trustee, it would be possible to negotiate for a proper way of benefit sharing whereby the former carrier of the tissue could also get due shares of the benefits.

Further information about the Eighth Asian Bioethics Conference can be found on line at http://www.stc.arts.chula.ac.th/ABC2007/ More information about the workshop itself is available at http://www.stc.arts.chula.ac.th/ABC2007/AsiaEuropeWorkshop.html.

Soraj Hongladarom
Chulalongkorn University, Thailand

Acknowledgment

Many people have been very helpful in making this book possible. First of all I would like to acknowledge the generous support given by the Asia-Europe Foundation and the European Alliance for Asian Studies, which made possible the 2007 workshop that led eventually to this book. I also acknowledge the Commission on Higher Education of Thailand whose research grant allows me necessary resources to complete this project. Last of all, I would like to thank all the contributors to this book. Without their contribution and their commitment to academic excellence, this book would not have been possible. To them I owe a lot of gratitude.

Soraj Hongladarom
Chulalongkorn University, Thailand

Section 1
Theoretical Concerns

Chapter 1
Buddhism and Human Genetic Research

Somparn Promta
Chulalongkorn University, Thailand

ABSTRACT

What the author is trying to do in this chapter is to explore how Buddhism, especially Theravāda Buddhism as adopted in Thailand, responds to the advancements of human genetic research in the modern world. Buddhism has a certain number of doctrinal beliefs normally differing from those in the theistic tradition, making Buddhism respond to genetic research in a certain way. The way Buddhism responds to genetic research could be characterized as a kind of humanistic view. This kind of view is mainly based on human wisdom and rational investigation of the problem. Belief as normally understood in terms of religion plays a lesser role in Buddhist ethics. The following will show the positions of Buddhism on the problems raised by genetic research. As the concept of personhood plays the key role in the debates over human genetic research, the author will start with this point. As human genetic research raises so many issues that it is impossible to explore all of them, the chapter will then focus on some of these issues, namely human cloning and the use of embryonic stem cells in medical practice.

THE CONCEPT OF PERSONHOOD IN BUDDHISM

The concept of personhood plays a significant role in modern bioethical debate as a number of the biomedical problems are concerned with the question of what should be counted as a person. For example, the embryo explored by the scientist could be harmed in some cases. Normally such harm is meaningful only if it occurs to a person. The problem then arises that if the embryo is a person, then the work done by the scientist in these cases can be debated in terms of morality. Abortion seems to be an explicit case showing that the definition of personhood is the most basic task. To judge whether abortion is morally wrong or not, we must first decide whether the fetus is a person.

DOI: 10.4018/978-1-61692-883-4.ch001

The question concerning personhood is problematic in that it is closely involved with human biological developments in the womb. Certainly, at some stage of development we could argue that the fetus is person because he or she can express some basic human qualities, such as the response to external objects, the reaction implying the feeling of pain, and so on. But at some stage of development, the very beginning state in which the fetus has no biological properties indicating that it is different from a cluster of cells, the concept of personhood seems hardly attributable to the fetus. There have been attempts by philosophers and scientists to establish a clear-cut definition of personhood through empirical measures such as medical data. For example, they use the appearance of the nervous system as a clear indication that the fetus is person, whereas before that it is not. Even though this method greatly benefits us, that does not mean it is unproblematic. It could be said that such a definition of personhood is more practical than philosophical. Something practical does not necessarily need strong justifications. So those who adopt the definition of personhood as stated above can be questioned as to why something without a nervous system should not be counted as a person.

It seems that ultimately the views concerning the problem of personhood can be grouped into two sets. The first set looks at the issue in terms of convention. For the philosopher of this set, personhood is merely a convention of society. We stipulate conventions for the purpose of social utility. For example, to protect good people from harm by bad people, we stipulate that people have the right to their life and property, and we say that in such a case people are persons in the sense of those who can claim rights over their life and property when these things are violated. The murderer before committing murder is counted as a person also, but after that his personhood can be changed. In the case of the death sentence, it seems that we do not accept that the murderer is a person. If we accept him as a person we can never punish him that way. From the above, we find that one may be person at some time and not a person at another time. It is a convention of society to determine personhood through the process of law. Another set of philosophers do not agree with this theory of personhood. For them, the study of personhood should not be associated merely with legal convenience. On the contrary, legal reasoning must be based on metaphysical reasoning or something deeper than legality. It seems that for the philosophers of the second set ontological investigation must be inevitably applied to the study of personhood.

In general, Buddhism shares the idea of the second set. One of the major characteristics of Buddhist philosophy is its naturalistic feature. Being naturalistic in this context means that truths are out there in nature, not in human imagination. So, in exploring truth, Buddhism explores nature. In the case of personhood, what is explored by Buddhism is the nature of human beings. This leads to questions concerning the basic concepts of human life, such as: what is the meaning of personhood according to Buddhism; when does personhood occur; what should be counted as a violation of personhood. We will examine theses questions below.

The Meaning of Personhood

Normally, Buddhism is viewed as a religion that rejects the existence of the self. This sometimes leads to the belief that there is no concept of personhood in Buddhist teaching. As understood by Buddhism, there are two meanings of personhood. One is the substantial meaning, and another is the non-substantial one. The Hindu theory of personhood can be cited as an example of the first meaning. For Hinduism, the self *(ātman)* is the essence of human life. The definition of personhood in Hinduism is based on this self. The self as taught by Hinduism is rejected by Buddhism, as Buddhism states that human life is composed of the five aggregates–namely, materiality, feeling, perception, mental formation, and consciousness–and these aggregates are not substances. But the rejection of the self does not mean that there is no concept of personhood in Buddhist teaching. Personhood according to Buddhism is still possible even though there is no self in human life.

Buddhism defines personhood in terms of psychological facts. For example, somewhere in Buddhist texts the Buddha says that if someone tries to kill you and you feel that you dislike the action of that man, the same action done by you is also disliked by other people. Buddhism believes that all human beings share a set of psychological properties such as self-love, hatred of death, and desire for a good future. These psychological facts are something to be respected by other persons. Killing is wrong in Buddhist teaching because it violates self-love. Other moral tenets in Buddhism can be also understood in this light.

The concept of personhood in Buddhism can be better understood if it is related to the contents of morality taught by Buddhism. The Five Precepts constitute the basic moral code of Buddhism. They state that killing, stealing, sexual misconduct, lying and taking intoxicants is wrong. The first four precepts involve other persons, while the last one involves oneself. In the first four precepts, two things are mentioned: the person's life and the person's belongings. Killing is concerned with a person's life, and we see from the above that killing is wrong because it violates the psychological reality of self-love. Stealing, sexual misconduct with another's beloved and lying are wrong because they violate a person's belongings. It should be noted that when we say that killing is wrong, Buddhism does not think that it is wrong because it violates the self of another person. The transcendental self is something beyond our observation, but psychological facts are totally observable. So using these facts as the grounds of personhood is more reliable. The last statement of the Five Precepts is involved with oneself. Taking intoxicants is wrong because it violates self-love. The person who takes intoxicants does not love himself, Buddhism argues.

It should be noted that the concept of personhood in Buddhist teaching is in some sense closely connected with the concept of human life as the composition of the five aggregates. The connection between these two concepts can be illustrated as follows. First of all, the five aggregates function as the foundation of personhood. The dead man cannot be a person because there is only the body, which is just one component of the whole five parts. The man in a comatose state is regarded by Buddhism as a person because he possesses all five aggregates, even though he is not conscious. Buddhism believes that the five aggregates under some conditions may not function, but they exist. When we sleep and do not dream at all, it can be said that the mind and its components (mind and feeling, perception and mental formation) temporarily do not function. So killing a sleeping man is wrong because the man still has all five aggregates. This line of argument is applied to the case of person in a coma or in any deeply unconscious state. Euthanasia given to a person in such a state is viewed by Buddhism as no different from killing a conscious person.

It seems that such a definition of personhood in Buddhism may give rise to some epistemological problems. We know that a sleeping man has all five aggregates because he can awake from sleeping. With the patient in the comatose state, it may be very difficult to determine whether or not he or she will awake again. So the point is: we know that a person has all five aggregates after his or her return from sleeping or from a state of deep unconsciousness. If we have a comatose patient who finally dies in that state, the question is: in the view of Buddhism, does that person have all five aggregates, or does he not?

The answer to this question is partly based on a religious belief that cannot be justified by sense experience. Buddhism argues that the body of human beings cannot survive without the support of the mind. As long as the body of the patient still survives, we can assume that the mind still exists. As the five aggregates are equated to the body and the mind, so in such a case we can say that the person still possesses all five aggregates; and that makes him or her a person in Buddhist perspective.

When Does Personhood Occur?

Normally the theory of the soul claims that personhood occurs when the soul enters the body. In the Buddhist texts there are some passages indicating the same idea. The Buddha says that a person arises when three conditions appear: the mother and father have sexual intercourse, the mother possesses a good biological state, and the mind is present. This statement mentions two components of human life. The first is the biological (or material) process, and the second is the non-material one. What is called "mind" in Buddhism means something containing properties of energy rather than substance, like the soul. So the Buddhist image of "mind" could be likened to the image of electricity. According to Buddhism, mere biological fertilization is not enough to give rise to a new life. Modern Buddhist scholars seem to believe that when the egg and the sperm have united, if the mind does not enter as another condition the process of fertilization can never start. In the case of natural abortion, these scholars explain that it occurs because of the departure of the mind from the ongoing fertilization process.

The Buddha did not give an explicit statement about when personhood starts, but indirect sources seem to suggest that according to Buddhism personhood starts at the first moment of fertilization. It is recorded in the monastic rules that a monk once performed an abortion on a girl; the Buddha judged his action seriously wrong, which incurred him the highest offense in the monastic rule. A monk committing this kind of wrongful deed must be expelled from the monastic community. The Buddha considered the embryo to be a person like an adult, so the monk who killed the embryo through abortion was judged by Buddhist monastic rules as having committed a crime equal in gravity to killing an adult. In the commentary on the rule stated above, it is stated clearly that killing a human being means destroying human life from the first moment of fertilization to human life outside the womb. So, even though the Buddha himself did not give a clear-cut pronouncement about when personhood occurs, the Buddhist tradition, especially the Theravāda tradition, clearly states that personhood starts when the process of fertilization takes place.

The Conditions under which Personhood is Violated

Normally Buddhism views killing as a violation of personhood. The first precept in Buddhist morality prohibits killing on the grounds that it is a violation of personhood. It seems that killing in this context does not necessarily include suicide. In some religions suicide is prohibited as an evil. Buddhism regards suicide as something to be examined in detail before judging it in terms of morality. That is, Buddhism does not view all acts of suicide to be wrong. Taking one's own life for the benefit of other persons could be considered "suicide," but this kind of suicide is not wrong from the Buddhist perspective. In the Buddhist texts, there are a number of stories about the merit accumulation of the *Bodhisatta* (one with the intention to be a Buddha in the future). To be a Buddha in the future, the *Bodhisatta* must practice what are called "perfections" (*pāramī*). One of the major perfections is the donation (*dāna*). It should be noted that there are two kinds of donation in Buddhist thinking: donation of life and donation of property. Of these two, the former is superior. The stories relate that in some circumstances the *Bodhisatta* donates his life. This seems to imply that the taking of one's own life under reasonable circumstances is not a violation of personhood and counts as a good deed from the Buddhist perspective.

In modern genetic and medical research, sometimes questions concerning a possible violation of personhood arise. The use of stem cells from the embryo for medical purposes can be cited as an example. The major objection to the use of embryonic stem cells is that such use is no different from killing one person and using the body of that person to cure the body of another person. This objection is very strong and makes any attempts to support the use of stem cells difficult. Even though the concept of life donation as described above could be employed to provide a rationale for the use of embryonic stem cells, the embryo whose stem cells are used is in no position to judge whether or not he or she is willing to donate his or her life, so the use of stem cells can be construed as either (if the embryo is not willing) killing or (if the embryo is willing) donating life, and between these two possibilities we can never know which one is true.

According to Buddhist ethics, the killing of even a willing person is to be regarded as killing and therefore wrong. There is only one case in which the taking of a willing person's life is not considered as such. It is the taking of life done by the owner of that life for good reasons, such as to protect a great number of people or to save the life of someone more valuable than one's own. We find that this principle cannot be directly applied to the case of the embryo, as we cannot know how the embryo thinks. Actually, the embryo at the beginning stage, say within two weeks of fertilization, does not have any thoughts. So how should we deal with such a situation? Some people argue that this case is like the case of a person in a persistent vegetative state. A man in this state has no thoughts. So society must make a decision on behalf of such a person. Normally, when we have to judge on behalf of another person, we use ourselves as the frame of reference. Buddhism, Confucianism, and some other systems of belief share the ethical principle that the good thing is what we want other persons to do to us and what is bad is what we do not want other persons to do to us. In the case of the embryo, we could apply this principle thus: if the embryo were a member of society and shared our knowledge of the situation concerning the need for embryonic stem cells, how would he or she judge the matter? If the embryo in our imagination says that in such a case it would be unreasonable not to allow the use of the embryonic stem cells, what we can conclude is that the use of embryonic stem cells in such use is morally right.

Capital punishment by its very nature is a violation of personhood, but some of us think that society has to allow this practice on the grounds of social necessity. In terms of personal ethics, Buddhism views execution of a criminal who has committed a very serious crime as wrong, as it is a violation of personhood. But in terms of social ethics, Buddhism states that if the death sentence has been proved to prevent serious crimes, this practice can be allowed in a Buddhist community. So we can say that the violation of personhood is possible in some cases within social dimensions with regard to society's needs. The use of embryonic stem cells is somewhat like the case of capital punishment, abortion and euthanasia. These practices can be either socially moral or immoral depending on the reasons behind the actions.

This does not mean that Buddhist ethics is relativist or situational. Buddhism believes that things in nature have some essential properties and these properties will determine the results of what we have done. Actions performed by human beings are one kind of natural phenomena. Human actions in themselves contain certain moral properties. Killing regardless of conditions is a violation of personhood, so killing is a bad thing in itself to some extent. However, Buddhism teaches that killing when judged as a situation related to certain conditions can vary in terms of moral justification because of those conditions. It may be possible that in some cases the weight of necessity determined by conditions seems to outweigh the badness of killing; in such cases Buddhism teaches us to use wisdom. This principle is readily applicable to research in human genetics.

THE BUDDHIST APPROACH TO GENETIC RESEARCH

One of the most basic beliefs of Buddhism is that proper questions lead to proper answers. The ethics of Buddhism, looked at in light of this principle, could be viewed as an ethics of questioning, meaning that before answering any question of ethics we must first ask what is a proper approach to that subject. In this section we will consider the Buddhist way of viewing genetic research, with human cloning and embryonic stem cell research as two examples illustrating how difficult it is to judge research in human genetics in terms of ethics.

Human cloning and embryonic stem cell research are involved with the interpretation of human life and its values, such as personhood and human dignity. Normally, theistic religions seem to have more explicit religious grounds as bases of how human life should be respected by persons who are involved in research on human beings. God is the source of human dignity in the theistic religion. Buddhism, as a non-theistic religion, is based on another ground in moral reasoning. The Buddha says that what he teaches are natural phenomena. The *dhamma*, which refers to the teaching of the Buddha, is understood by Buddhists as natural things and natural laws. For non-Buddhists, the best way to understand Buddhist teaching is to view it as they view natural sciences such as physics, chemistry, and biology. Buddhism teaches that the universe is naturally given, and the Buddha himself clearly declares that he is not interested in exploring the origin and the end of the universe. What he wishes to explore is the universe as it appears. Great enlightenment brings him the insight that the universe is regulated by five kinds of natural law (*niyāma*): namely, the physical law (*utuniyāma*), the biological law (*bījaniyāma*), the law of action (*kammaniyāma*), the law of mind (*cittaniyāma*), and the law of *dhamma* (*dhammaniyāma*). Buddhist morality is based on belief in these natural laws. Goodness and badness in human actions are based on the laws of nature, especially the last three types.

Naturalness and Unnaturalness

The debate over human cloning and other human genetic research normally involves discussion about its unnaturalness. Some of the arguments against human cloning state that such a practice is unnatural in the sense that it is not provided by nature. Some people who believe in God might think that anything unnatural means it is not permitted by God and for that reason is dangerous. Sexual reproduction is natural in this sense and thus it is established by God. Human cloning is an attempt to produce a human being through unnatural means, and thus it is against the work of God. According to this line of argument, we will find that the concept of being moral is equated to the concept of being natural. By the same token, the concept of being immoral is equated to the concept of being unnatural.

This argument seems to be used by some people to argue against contemporary genetic research. Some scientists who support human cloning, for example, state that human cloning should not be viewed as unnatural because there is a kind of human cloning permitted by nature: the case of identical twins. According to these scientists, human cloning performed by scientists can be viewed as the making of identical twins. What is different is merely that natural identical twins are of the same age, while artificial identical twins are of different ages. Looked at from this point, human cloning is not immoral because it is natural in the sense that it follows the law of nature as found in the case of natural identical twins.

In Buddhism, morality can be separated from the concept of being natural because according to Buddhist teaching it seems impossible to say that such and such a phenomenon is unnatural.

Buddhism proposes that the moral goodness or badness attributable to any action depends solely on the moral properties. Actually, Buddhism does not think that there is anything unnatural. Buddhism believes in the Five Laws of Nature as we have observed previously, and thinks that there is nothing which is beyond these laws of nature. In Buddhist texts, for example, reproductive methods other than the sexual one we are acquainted with are mentioned. For those of us who never perceive such methods, they could be considered unnatural. But they are natural in the sense that they are permitted to appear in the universe through any of the five natural laws.

Man according to Buddhism is a natural thing. When man creates something, Buddhism regards that something as natural, too. So, natural things in the Buddhist perspective are of two kinds: those created by man and those not created by man. Between these two kinds of things, there is no difference in terms of ethics. That is, some natural things are good and some are bad. Likewise, some things created by man are good and some are bad. So, moral goodness or badness has nothing to do with being natural or unnatural.

Moreover, Buddhism considers man and nature to be a single system. That is, Buddhism accepts that nature has its own long history and that man in his present form has a much shorter history compared with nature. However, there is some potentiality in man which cannot be found in nature, namely, the potentiality of consciousness and intelligence (or wisdom, if that is the more preferred term). Through consciousness, man learns to solve certain problems which may take a very long time to solve by natural processes, or which may be impossible for nature to solve by itself. When we break our arm, surgery is done to heal the broken arm. This surgery is done by man, but it does not join the broken arm. Nature instead plays the role behind the process of joining the broken bone. So, it can be said that in the joining of the broken arm two things are equally needed – man and nature. Following this line of thought, Buddhism does not posit a separation between man and nature. The notion that we can trust only natural phenomena is viewed by Buddhism as extreme. Buddhism likewise views as extreme the idea that man can dominate nature or do anything unconditionally.

So Buddhist ethics does not consider the issue of human cloning through the concept of naturalness. In general, Buddhism admits that whatever happens in the world is natural. It does not matter whether or not something appears by virtue of human technology. Natural things in Buddhist perspective include both what is given by nature and what is created by human beings. The fact that Buddhist ethics does not utilize the concept of naturalness makes it harder for Buddhism to deal with modern bioethical problems. But this could be also considered a strong point as it provides more space for debate. That is, sometimes we might find that labeling something immoral because it is created by human beings is seemingly irrational. The world today has greatly benefited from these "unnatural" products of science and technology.

It seems that the basic difference between Buddhist ethics and ethics based on the theistic religion is that Buddhism holds a humanistic attitude while theistic religion does not. For the theistic religion, human beings are just like innocent babies whose knowledge of the universe is very limited, while God is the father who knows everything. The scientist's attempt to reveal the secret facts hidden behind natural things is considered no different from the action of an innocent baby who puts her fingers into the unknown holes in the walls of a room. Inside some holes there could be some dangerous things; we cannot know. For the sake of safety, we should not go beyond what has already been prepared by God as found in nature. Human cloning is forever questionable in terms of safety regardless of the data gained from scientific research because there could be danger within it, as it has not been prepared by God. Buddhism partly agrees with such a warning. However, the best way to decide whether such a thing contains danger or not is to undertake experiments. Buddhism supports attempts to gain new knowledge as long as such attempts are governed by wisdom. Wisdom in Buddhist teaching is a process of learning through doing, not imagination or speculation. The enlightenment of the Buddha is not a state of mind gained independently from a process of long-term learning. In short, wisdom in Buddhist teaching is a practical term.

The Harm Principle

The spirit of Buddhism is not-harming (*avihimsa*). Harm (*vihimsa*) is one of the major criteria used by Buddhism to determine the morality of an action. This principle says that any action which does harm is morally wrong; otherwise, it is not. Harm in Buddhist teaching can be divided into two main categories: harm to oneself, and harm to others. Harm to oneself means the action is intentionally performed by a person and that the action harms him in one of two ways. It harms him in terms of physicality, or it harms him in terms of dignity. Taking intoxicants is prohibited by the Fifth Precept in Buddhism on the grounds that taking such substances does physical harm to oneself. Selling bodily organs such as kidneys could be viewed as harm in terms of dignity. The person who does such a thing, for whatever reason, could be viewed as not respecting his own status as a human being. He treats his life as if it were a nonhuman product that can be sold. This interpretation makes it possible to state that the sale of human organs constitutes a harm with regard to one's dignity.

The second kind of harm, harm to others, can be of two meanings as well. Normally, in a free society, some personally harmful actions may be tolerated by the laws of that society. Drinking beer, for example, is harmful to one's own person; but this kind of action is tolerated by the law in Buddhist countries because it is accepted that only serious harmful actions should not be tolerated by the law. Using drugs is prohibited by law in Buddhist countries because it is believed that the harm resulting from drug use is much more serious than that which results from drinking beer. So it can be said that according to Buddhist morality personal freedom does not cover personally serious harmful actions, in terms either of physical damage or of damage to one's human dignity. Harm to others is more obviously seen as wrong by nature, whether it relates to physical damage or to damage to human dignity. However, as the intention behind an action plays a significant role in the Buddhist system of moral judgment, investigating harm to others cannot be separated from consideration of the intention of the doer.

Then the problem arises: is there some kind of harm allowable in the Buddhist community, or is any kind of harm strictly prohibited? In utilitarianism, it seems that some kinds of harm are permitted. That is, for the benefit of the greater number of people, a violation of the rights of the minority may be permissible. But the violation of rights involved in such a case is understood in terms of the right to property, not the right to life. Government policy in any country in the world is more or less utilitarian. The expressways in Bangkok are at the cost of people whose lands have been chosen by the state for this purpose. But for the benefit of the majority, this kind of harm can be accepted. Moreover, this kind of harm can be compensated by the state because it is an economic harm. By contrast, a harm to life seems to be immoral in every respect because it is a harm that we cannot compensate.

Applying the harm principle to the issue of human cloning, it seems that the first question is: can human cloning be interpreted in terms of harm? It is clear that the cloning of human beings in some cases could be questioned whether or not it is personal issue. For example, a man clones himself to use the embryonic stem cells. In such a case, can we say that it is really a personal matter, implying that the harm principle to be used for this case is the harm to oneself only? According to Buddhism, a clone is a person from the first moment of fertilization, so it is very difficult, if not impossible to locate human cloning within the area of personal activity. It seems obvious that the harm in the case of human cloning is the harm to others. However, this does not mean that any case of human cloning is viewed by Buddhism as a harm to others. Buddhism merely says that any harm caused by human cloning must be regarded as harm to others. Simply speaking, Buddhism does not accept that human cloning can be understood in terms of personal activity. Therapeutic human cloning and the use of embryonic stem cells could be considered in terms of harm to the life of the embryo. Can we accept such harm for the benefit of us or not? The same question may be rationally posted apropos other areas of human genetic research as well.

According to Buddhist teaching, we must distinguish between life and property. The right to the former is considered the primary right, while the right to the latter is the secondary one. The great difference between the primary and the secondary right is that the former can never be transferred, while the secondary right can be. In the Buddhist texts, it is recorded that voluntary euthanasia constitutes a violation of one of the Four Rules of Defeat (*pārājika*) for the monk who commits it. That is, in Buddhist monastic rules, a monk violates the Four Rules of Defeat if he engages in sexual intercourse, commits robbery, kills a human being, and denies the existence of a higher goodness. Killing a human being at his or her request is wrong on the grounds that the right to life cannot be transferred. Suicide is found in the Buddhist texts, and in some cases it could be argued that no guilt attaches to a monk who commits suicide. The difference between suicide and euthanasia according to Buddhist teaching is that, in committing suicide, a person is not violating the right to life because he is the owner of that right, while a person who commits euthanasia at the request of somebody else is violating that right. The request cannot justify euthanasia because the right to life can never be transferred. Only the owner of the right can forsake it.

Applying what we have considered above to the embryo, it could be the case that the embryo as a person must be accepted as the owner of a right to life. There are many sources in the Buddhist texts pointing out that killing an embryo is no different from killing an adult. So, the use of embryonic stem cells even for curing disease, according to Buddhism, is no different from sacrificing the life of one adult to save the life of another. If the use of an adult life for this purpose cannot be accepted, the question is: on what grounds can use of the embryo be justified?

As the Buddhist view on any subject is not absolute, in the sense that what the Buddha teaches is not dogma to be accepted unconditionally, Buddhism's view of the right to life could be discussed further. Actually, the sacrifice of one's life for the benefit of another can be found all over the world, including a Buddhist country like Thailand. We have the soldier acting as the guardian of the country. The death of a soldier for his country suggests that in some cases the sacrifice of one's life for the benefit of one's country or the majority of people in one's country may be necessary. In Buddhist literature, a life donation is sometimes found. As we have said previously a *Bodhisatta* sometimes donates his life for the benefit of another and such doing is deemed good. This seems to imply that the right to life in some cases could be transferred.

By distinguishing between life donation and euthanasia, it may be possible to perceive how Buddhism regards the issue. What is the difference between these two issues? In a donation of life, the donor is fully aware and understands what merits will accrue as a result of his action. In euthanasia, a man who requests death is understood by Buddhism as acting so from an unwholesome impulse, and a man who commits euthanasia is understood as doing so without moral authorization. That is, no one can take another's life without violating the other's right to life and personhood, regardless of conditions. But the case will change if the owner of life donates it. Therapeutic human cloning and embryonic stem cell research, if they can be justified, seem to obtain such justification within the category of life donation.

Taking one's life for the benefit of another is not necessarily evil in Buddhist perspective. At least, Buddhist doctrine permits the taking of life under certain circumstances. However, donation is a concept in personal ethics. Donation must come from consent and wisdom. We do not know whether embryo is willing or not. This is the most difficult problem to overcome. Maybe the idea of forced donation could be a way out of this difficulty. Forced donation is self-contradictory in Buddhist personal ethics, but it could be possible in the social ethics of Buddhism. In Buddhist Thailand, a girl who gets pregnant as a result of rape has the right to abort the child. The child in this case can be understood as being the "forced donor" of his life for the benefit of the mother. Why do we think the mother deserves such protection? The answer is: because not giving her the right to abortion is socially immoral. If we can prove that in some cases not giving a person the right to benefit from therapeutic human cloning or the use of embryonic stem cells is socially immoral, it means that we have found a way to justify these practices.

The Principle of Analysis

Sometimes the Buddha identifies Buddhism as "a religion that teaches analytical morality." The term "analytical morality" is a rough translation of the Pāli word "*vibhajja.*" This term, as understood by Buddhist scholars, denotes a system of thought that does not look at the world through a prism of black and white. Actually, Buddhist logic, or the Buddhist epistemological outlook, has much criticized the black and white logic found in the work of great thinkers such as Aristotle. Applying the principle of analysis to the case of human cloning, the advice from Buddhism is that, first, to assume that all kinds of human cloning and other genetic research are solely right or wrong is not valid. Buddhism considers all that happens in the world as matters of varying complexity; some events may be less complicated, while some are much more complicated. Analysis will reveal the proper way to deal with specific events. The cloning of human beings or any kind of human genetic research has originated in the human mind, and the human mind must always have its reasons for thinking in a particular way. From this viewpoint, it could be that in some cases the human mind behind these activities has a good intention, while in some cases everything is directed by a bad intention. So, what we must do is to analyze the given case and find the details within.

As noted above, Buddhist ethics is analytical ethics. After the process of analysis, it may be found that in some cases the cloning of human beings or other kinds of human genetic research does not harm anyone, that all persons involved are happy, and thus that such research is tolerated by Buddhist ethics. The problem is: when we talk about the concept of harm from the Buddhist perspective, does such harm involve only the person, or can it be extended to society? This question is important because some arguments against cloning and other human genetic research suggest that though we possibly cannot find any obvious victim of harm in terms of individuals, it can be said that society is harmed by allowing such activities. Legal moralism, as presented by legal philosophers such as Patrick Devlin, is of the view that one of major structures that support the existence of society is the moral structure. Devlin argues that even though drinking intoxicants can be viewed as personal freedom, we should remark that if most members of society are those who persistently exercise such a habit, our society must be weak. In this case, intoxicants cannot be viewed in terms of freedom only. It can be related to the moral structure of society as well. Human cloning and other human genetic research considered in this light could be viewed as being harmful to society, even in a case where we think that everyone involved is happy and no one is harmed at all.

One of the Five Laws of Nature taught in Buddhism, the Law of *Dhamma* (*dhammaniyāma*), seems to shares the above view in part. For Buddhism, the moral tendencies found in society have an effect on the well-being (or otherwise) of people in that society. In short, Buddhism agrees that human society is not just a place where people gather and do only what benefits themselves; on the contrary, society has a spirit, and this is nothing but the common ideal to meet certain moral standards. We are not just living, but we are living a good life as noble humans. However, the moral structure that supports the existence of society in Buddhist perspective must be identifiable, not merely an abstract vision of our imagination. One process that helps us to reduce the degree of abstract imagination is to relate the moral structure of society to what individuals do. Cloning and other forms of human genetic research can be viewed as something that points out the level of morality in the minds of people. So, we can say that allowing such research for any purpose other than that which promotes human welfare affects the moral structure of society. As the actions of individuals in society are related to law in the sense that the law must determine what actions are permissible and what are not, so the law practiced in society can be viewed in part as an indicator of the moral structure of that society as well.

Social necessity is a notion Buddhist ethics tolerates in some cases. Buddhism teaches that killing is an evil; but Buddhism never teaches against having an army. Reasonable capital punishment is sometimes interpreted by Buddhist thinkers as a social necessity, implying that it should be tolerated or deemed legal in a Buddhist community. If in some cases we can rationally prove that cloning or other human genetic research is a social necessity, Buddhist ethics would appear to deem it acceptable as shown in other cases mentioned above. Analysis of the context and surrounding data will help us to classify the various categories of human cloning, and other human genetic research, some of which categories may meet the conditions tolerated by Buddhist ethics. At this point, we find that human cloning and related genetic research are an open-ended subject in the Buddhist community, meaning that some doors are open for the further exploration of these activities in Buddhist society. What is required are merely the reasons and explanations for why such should be allowed?

Individual and Social Dimensions of Ethical Problems

As noted previously, in the Buddhist community the ethics of Buddhism is considered in two dimensions: individual and social. The ethics taught by the Buddha is in the first place intended for personal use. Buddhist ethics in this respect considers human lives as individual units; each of them facing some common problems, and every individual bears responsibility for solving these problems by him- or herself. So, what is good and what is bad within this dimension of Buddhist ethics are personal matters in the sense that if something is considered good, its goodness is explained with reference solely to its effect on the individual. In the Buddhist community, when people request the Five Precepts, the request form states that these precepts are to be adopted by each person individually (*visuṃ visuṃ rakkhanatthāya*). So, it is understood among Buddhists that goodness or badness in one's life is a personal matter. Each person must monitor his or her own life.

Consider this example: It is very clear that abortion is wrong according to Buddhist ethics. But to say that abortion is an impermissible sin in Buddhist perspective could be misleading, as sometimes people understand this to mean that abortion must be illegal in Buddhist society. To say that abortion is a sin is to say it is so within a personal code of morality. That is, abortion is equated to killing a human being, so committing an abortion violates the first clause of the Five Precepts. The Buddha says that to attain the state of Liberation, or *nibbāna*, one should avoid unwholesome actions. Abortion is an unwholesome action, meaning that committing it will lead to a path away from *nibbāna*. Following the advice of the Buddha, a Buddhist who is confronted with a dilemma concerning abortion should consider by herself how to handle the problem. The Buddha never claims that a woman confronting such a dilemma must not commit abortion. He just says that a woman should consider by herself what is the best thing within response to such a condition. Suppose finally she finds that the best response is to commit abortion, then Buddhist ethics will have nothing to say. It is her choice and her own responsibility. However, Buddhist ethics still continues to claim that as abortion is the killing of a human being, the woman who decides to choose abortion must be responsible for the consequences of her choice. She must know that doing this is a bad *kamma*, and its result is already determined by the law of *kamma*.

The above remarks represent a consideration of abortion in terms of personal morality. As there is a fetus who is to be killed, abortion then cannot be considered solely with regard to the personal morality of the mother. The Buddhist Harm Principle should be brought in to deal with the issue. It is evident that abortion is very harmful to the fetus, even in his/her earliest stages of development. So the state, as the authoritative power whose most basic function is to provide justice for the people involved in a conflict of interest, has the right to prohibit abortion if it is deemed that between a mother and a fetus the latter's right to life is more deserving of protection.

Human cloning or any other kind of human genetic research can in one sense be considered in this light. That is, if the morality of these activities is raised in the Buddhist community, one possible approach to this problem is: let it be personally judged by each member of the community. Certainly, different persons can have different views of the same matter. But Buddhism believes that ultimately enlightened Buddhists will arrive at the same conclusion of the same ethical dilemma. Looked at from this perspective, the way to deal with ethical problems raised by human cloning and the like does not necessarily require the establishment of certain rules to be followed by all members of the community. On the contrary, these problems can be resolved through the moral education of each member. For Buddhism, solving ethical problems by changing people's minds is evidently more effective than passing laws to regulate external behavior.

However, if we accept that a community is composed of different members, some of whom are bad and some are good, then an understanding of Buddhist ethics in terms of the personal dimension alone will leave problems behind. How to judge human cloning and other kinds of human genetic research may not be a problem for the enlightened members of a community, but it may be very different for the unenlightened ones. Without rules or laws, the unenlightened members of a community may sometimes harm others, intentionally or unintentionally.

It is clear that Buddhist personal ethics is based on the law of *kamma* taught by the Buddha. To judge whether the response to a given moral question is wrong or not according to personal ethics is not difficult. Human cloning according to personal ethics is not immoral insofar as it is undertaken for reproductive purposes. Buddhism adheres to a moral principle that what is conducive to the harm and suffering of oneself and others is unwholesome. By the same token, that which is conducive to the benefit and happiness of oneself and others is wholesome. Destroying life or prohibiting birth can be considered as harmful, while prolonging life or giving birth is beneficial. Reproductive cloning could be judged as not immoral in this sense. By contrast, stem cell research could be interpreted as harmful, since the embryo is destroyed. This is not to suggest that according to Buddhist personal ethics reproductive human cloning is totally right and stem cell research totally wrong. It just means that we can interpret the issues in both directions.

According to Buddhist social ethics, on the other hand, stem cell research could be viewed differently from what we have seen in the Buddhist personal ethics. In reliable reproductive cloning no one is harmed, so it is against neither the personal nor the social morality of Buddhism. It is only therapeutic cloning, the cloning for medical use in which the clone (including the clone generated solely from a woman's egg) is destroyed, that could be problematic.

Modern ethical dilemmas are usually concerned with the conflict of interest between two persons or two groups of persons. In the issue of abortion, the two parties involved are a mother and a child. The mother's interest is protected if an abortion is permitted, while the child's interest is protected if an abortion is prohibited. Likewise, in therapeutic human cloning and embryonic stem cell research, there are two persons or two groups of persons involved. In terms of rights, the patient's right to health is protected if therapeutic cloning and stem cell research are allowed. But in carrying out such practices, the clone's or embryo's right to life is violated. The hard task to be undertaken by any ethical school or ethical theory, including Buddhist ethics, is to decide, between the two sides in the conflict of right or interest, whose right or interest should be protected and on what grounds.

At this point, we find that the ethics at the heart of the issue is social ethics, and socially ethical dilemmas are more difficult to solve compared with personally ethical dilemmas because in personal ethics only a single person is involved. It is much easier to find a solution to a conflict affecting solely one's own life. When a man is deciding whether or not he should clone himself to have a clone for purposes of medical healing, the principles of wholesome and unwholesome deeds given by Buddhism seem sufficient to provide him with a solution. Religious ethics normally endorses the altruistic way in moral decisions. So, the devout Buddhists are those who prefer not to clone themselves, for the reason that death is not dreadful compared with the sin committed in cloning an embryo for medical use. But when society tries to judge the claim of some of its members that they have the ultimate right over their own bodies, and thus the right to clone themselves for medical use, finding a solution is not easy. Whose rights should be protected between the patient and the clone? Between the benefit of the greater number of people and the violation of the embryo's rights, which should be chosen? How Buddhist social ethics should deal with such a dilemma is not easy to answer even for those who are well-versed in Buddhist doctrines.

Principle of Freedom

One of the basic features of Buddhism is that freedom is highly valued. The principle of freedom in Buddhism is closely related to the humanistic tendencies generally found in Buddhist texts. There are two meanings of freedom: positive and negative. Positive freedom means freedom to do something. Negative freedom is freedom from that which does not allow us to do something. Ultimately these two meanings are undividedly related to each other. Buddhism does not think that all forms of freedom are right. It is merely some kinds of freedom that are valuable. Freedom in Buddhist perspective can be both a means and an end. An enlightened person in Buddhist view is one who is free. He is free in two senses. First, he is free in the sense that he is not under the influence of anything, especially the desires which Buddhism considers to be the blind forces that push sentient beings into the struggle for desirable things. Secondly, he is free in the sense that his actions are totally pure. This kind of person can never harm anything. We will see that freedom as the highest quality of life is the end. In Buddhist perspective, the end and the means must share some basic nature. Freedom as the means will support freedom as the end. This is why in Buddhism religious dogmas are less influential. The Buddha gives his disciples the freedom even to argue against what he has stated. It could be said that the first kind of freedom (methodological freedom) is required to attain the second kind of freedom (ultimate freedom).

In the Buddhist community, the personal freedom of believers is accepted through social tolerance of some kinds of evil. For example, even though the Fifth Precept says that taking intoxicants is wrong, intoxicants can still be sold in the Buddhist community. This does not mean that Buddhist ethics accepts that Buddhists are free to take these substances. It just means that the freedom to learn about moral lessons in one's life is needed to be a free person in the future. Taking intoxicants is always wrong, but it might be more wrong if society did not give its members the freedom to learn this lesson by themselves.

Applying the principle of freedom to the cloning of human beings or any kind of human genetic research, two things should be considered. Cloning and other genetic research can be viewed both as an activity and as an object. Cloning as an activity means that it reflects an attempt by scientists to search for something relevant to the advancement of scientific research. Cloning as an object means that it produces something and introduces it to society. The debate over human cloning appears to stress the second meaning of the term. We look at the product resulting from the process (the clone) and ask: should we tolerate this kind of thing? It may be that the most important meaning of the term is the first one. The serious question then arises: should we tolerate an attempt to search for something valuable in terms of scientific advancement?

The history of science is filled with results that we had feared at the beginning but that was proved not wrong over time. The technique of fertilization called IVF at first was expected to produce a monster without a human soul. Nowadays, such a fear has been proved not true. The process of learning something does not necessarily yield pleasurable results. But if we are not free to learn, how shall we know what is right and what is wrong?

Buddhism believes in human wisdom and considers the history of humankind in terms of a learning process. Wisdom includes knowing to stop at the point when the inner moral whisper advises us to stop. However, the inner moral whisper about something never occurs without serious study of that subject. The serious study of any subject can never occur without freedom to study. Today we have many conceptions of the activities conducted by scientists, some of them negative and some positive. So long as actual study has not yet started, conceptions will forever remain conceptions. The problem is: should some conceptions be the dominant idea, under which the real study of the subject must be aborted? There can be different answers to this question. But the answer from Buddhism appears to be: to abort the study of something on the grounds of negative conceptions is unreasonable and unjustifiable.[1]

REFERENCES

Keown, D. (2001a). *Buddhism and Bioethics*. Palgrave Macmillan.

Keown, D. (2001b). *The Nature of Buddhist Ethics*. Palgrave Macmillan.

ENDNOTE

1 This paper is mainly based on the interpretation of the Theravāda Buddhist Pāli Canon used in Thailand and other Theravāda Buddhist countries such as Burma and Sri Lanka. For English translation, see *The English Tipitaka* by the Pali Text Society. There are a very few scholarly books on Buddhism and genetic research. For further reading, the works of Damien Keown (2001a; 2001b) are recommended.

Chapter 2
Genomics and Population Health: A Social Epidemiology Perspective

Chan Chee Khoon
Universiti Sains Malaysia, Malaysia

ABSTRACT

Imagine being able to find out how a drug will affect you before you take it... receiving a medication that is specifically tailored to treat your disease, while minimizing your risk of developing adverse effects. Although a person's environment, diet, and general state of health can all influence how he or she responds to medicines, another important factor is genes. Pharmacogenetics is the study of how your genes affect the way your body responds to a medicine. Pharmacogenetics helps to determine what the right medicine is for you, based on your own genes.[1] The Pharmacogenetics and Pharmacogenomics Knowledge Base http://www.pharmgkb.org/resources/education/phar-genetics.jsp.

SARS (Severe Acute Respiratory Syndrome) can be contained despite the absence of robust diagnostic tests, a vaccine, or any specific treatment. When awareness, commitment, and determination are high, even such traditional control tools as isolation, contact tracing, and quarantine can be sufficiently powerful to break the chain of transmission. - Gro Harlem Brundtland Director-General, World Health Organisation. (WHO website, accessed on July 5, 2003)

DOI: 10.4018/978-1-61692-883-4.ch002

SOME PERTINENT QUESTIONS FROM THE SARS EPIDEMIC

In the SARS epidemic of 2002-2003, the microbial agent involved (SARS coronavirus) was swiftly identified and sequenced in a remarkable collaboration between otherwise highly competitive laboratories in Asia, Europe, and North America (World Health Organization, 2003).

Notwithstanding the rapid success in isolating (Peiris, Lai, and Poon, 2003) and sequencing (Ruan, et al., 2003) the SARS coronavirus, the epidemic quickly subsided in the absence of reliable diagnostics, vaccines, or efficacious therapies.

WHO gave much credit to institutional responses such as isolation, contact tracing, ring fencing, and quarantines (i.e. centuries-old techniques), with lesser mention of personal risk avoidance and the possible contributions of seasonality effects or cross-reacting immunity from related endemic micro-organisms (Ng, Turinici, and A Danchin, 2003).

Most importantly, the economic and financial stakes involved (Chan, 2003) ensured that SARS would not be a "neglected disease" [2].

A number of pertinent questions arise from these observations, which could be asked more generally of emerging biomedical technologies:

- How important are biomedical advances (incl. genomics) to population health and to patient care? What is the relative significance of genetics in the etiology (and social ecology) of health and disease?
- What are realistic expectations of the advances that genomics can contribute to disease control, diagnostic aids, and treatment? In what ways can pathogen genomics be most useful in epidemic control strategies?
- What are the likely trajectories of genomics R&D in the foreseeable future, given the current modalities for funding of biomedical research, the associated regimes of patents, intellectual property rights, and market-driven product development, and the chronically unresolved problems of neglected diseases of the poor?
- What would be an enabling environment for the realization of the useful potential of genomics, for an equitable harvest of benefits and a humane deployment of genomic technologies that can avoid the emergence of a marginalized genetic underclass and the imposition of arbitrary, constructed norms?

A partial response to the above questions was offered by Holtzman and Marteau (2000) who argued from a clinical perspective that *"the new genetics will not revolutionize the way in which common diseases are identified or prevented... only a small proportion of the population has Mendelian disorders, and this will limit the ultimate impact of the Human Genome Project. Our doubts stem from the incomplete penetrance of genotypes for common diseases, the limited ability to tailor treatment to genotypes, and the low magnitude of risks conferred by various genotypes for the population at large. Consequently, most people will have little interest in learning their genotypes."*

The completion of the sequencing of the human genome in 2000 provided the occasion for extravagant claims for genomics as an all-round panacea for the major health (and social) problems of humanity in the 21st century. Notwithstanding this *genohype*,[3] there has been limited success thus far with gene-based therapies, and few promising candidates on the horizon.[4,5,6,7] Commercial interest is thus likely to shift towards genetic testing for 'disease susceptibility' alleles in line with a 'paradigm shift' towards 'predictive medicine' (genetic profiling of individuals for assessing risk of future illnesses). This has the added attraction that mass markets are involved, since the genetic testing for 'disease susceptibility' may be applied in a routine manner as part of well-person (or well-child) care and screening. Accompanying this almost certainly will be corporate research and development (R&D) aimed at producing 'pills for the healthy ill' (the worried well)[8] to carve out sizeable new markets not just for screening tests but also for 'prophylactics' for those deemed to be 'at risk' and consequently anxious for the availability of some (commodifiable) risk reduction options. Conversely, corporate R&D will continue to ignore and bypass the "neglected diseases" of the poor, a scandalous situation which has been well documented by *Médecins Sans Frontières* (MSF).[9]

Notwithstanding these reservations, some remain optimistic that genomics could add to the options and capabilities for coping with challenges to individual and population health, even as it engenders vigorous debates over the ethical safeguards needed for a humane, responsible and equitable deployment of genomic technologies.

LESSONS FROM POPULATION HEALTH

A population health approach differs from traditional medical and health care thinking in two main ways. Population health strategies address the entire range of factors that determine health. Traditional health care focuses on risks and clinical factors related to particular diseases. Population health strategies are designed to affect the entire population. [Traditional] health care deals with individuals one at a time, usually individuals who already have a health problem or are at significant risk of developing one. – Strategies for Population Health: Investing in the Health of Canadians. Federal, Provincial & Territorial Advisory Committee on Population Health, 1994 www.phac-aspc.gc.ca/ph-sp/phdd/pdf/e_strateg.pdf

In the 1960s and 1970s, the hitherto unchallenged presumption that improvements in human health by and large flowed from advances in (bio)medical knowledge, feeding through to professional practice and individual care, came under scrutiny.

The debate was launched by Thomas McKeown who began publishing his findings in the 1960s on the historical decline of tuberculosis mortality in England and Wales.[10] McKeown had noted that tuberculosis mortality in England and Wales over the period 1838-1960 had declined by more than 85 per cent by 1945. Since this occurred well before the discovery and isolation of streptomycin by Waksman and Schatz in 1947 (one of the early antibiotics effective against the tubercle bacillus), and well before the widespread availability of BCG vaccination from the 1950s onwards, McKeown reasoned that factors other than medical intervention had been paramount in the historical decline of tuberculosis mortality in the two countries.

McKeown's writings on the main drivers of this declining mortality spawned vigorous debates (Colgrove, 2002). In addition to economic growth and improvements in food intake and nutritional status which McKeown himself highlighted, others argued also for the population health impacts of birth spacing and family size, housing and sanitary reforms and clean water supplies (most importantly, the social movements and interventions that inspired and sustained these campaigns), and safe milk supplies (pasteurization and eradication of bovine TB from livestock herds). The early conflation of TB mortality with pneumonia and bronchitis (affecting the timing of TB's decline), the contribution of isolation and quarantine to the control and reduction of infectious disease, and the decline of child labor and its associated early life effects on adult health were highlighted in other contexts.

Mortality from typhus fever, another major killer in the 19th century which was tabulated separately as a cause of death in UK mortality statistics from 1869, showed continuous decline over the ensuing decades such that by 1906, three years before Charles Nicolle's discovery that the body louse transmitted typhus, London County Council reported no more deaths from that disease (Rosen, 1971). Typhus fever, closely associated with poverty, destitution, poor housing, overcrowding, and poor personal hygiene was much less common among the middle and upper classes in 19th century England. Its decline was arguably linked to the increased availability of public baths, wash-houses, and widening use of cotton clothing particularly underwear which allowed for improved personal cleanliness.

In the US, the Sonja and John McKinlay similarly concluded from their historical analyses that medical intervention had only a minor impact (about 3%) on the overall decline in infectious mortality in the US between 1900-1973, which in turn accounted for 69% of the overall decline in US mortality during that period (McKinlay and McKinlay, 1977)[11].

Richard Lewontin, in generalizing the argument to the major causes of infectious mortality, gave lesser weight to potable water supplies and sanitation, at least among the early industrializing countries. He may however have downplayed the synergistic interactions of diarrhoeal disease and malnutrition on childhood mortality, [12] especially in poorer countries:

The history of tuberculosis is the history of nearly all the major killers of the nineteenth century. Whooping cough, scarlet fever, and measles, all with death rates in excess of 1,000 per million children, and bronchitis, all declined steadily with no observable effect of the discovery of causative agents, of immunization or of chemotherapy. The sole exception was diphtheria which began its precipitous decline in 1900 with the introduction of anti-toxin and which was wiped out in five years after the [US] national immunization campaign. The most revealing case is that of measles which killed about 1,200 in every million children in the nineteenth century. By 1960, despite the complete absence of any known medical treatment, it had disappeared as a cause of death in Britain and the US while in much of Africa it remains the chief cause of death of children. The causes of the tremendous decline of mortality from infectious diseases in the last 100 years are not certain. All that is certain is that 'scientific medicine' played no significant part. Water supply and sanitation are not involved, since water-borne diseases have not been the major killers. The suggestion that a reduction in crowding may have reduced the rate of transmission of respiratory diseases is not altogether convincing, since measles remains pandemic although it kills virtually no one in advanced countries. The most likely explanation, both for the historical trend and for the differences between regions of the world today, is in nutrition, although hard evidence is not easy to come by.[13]

On the limited contribution of medical interventions to population health, Simon Szreter, who had played a prominent role in the critical re-appraisal of McKeown's work, summed up the consensus thus: *"The medical profession's scientific leaders have, since McKeown's time, had to change their tack and concentrate on the future, rather than the past, as the field in which they can stake the claim that they can save humanity from all its ailments with science."*[14]

This conclusion however is not so easily generalizable to the late industrializers and poorer countries in post-World War II circumstances, at a time when their burdens of infectious diseases could be addressed with modern biomedical technologies such as vaccines, antimicrobials, vector control technologies, diagnostics, etc. which progressively came on the scene.

The ongoing campaigns for access to antiretrovirals for instance testify to the potential impact of treatment for the public health control of the HIV/AIDS pandemic. Access to lifesaving treatment for infected individuals is emphatically a moral and ethical imperative. Beyond that, the availability of effective therapy may also encourage those at high risk to come forward for voluntary testing, and hence reduce the pool of infected-but-unaware individuals who constitute one of the drivers of the pandemic.

It is nonetheless noteworthy that the SARS outbreak of 2002-2003 subsided largely in the absence of reliable diagnostics, vaccines, or efficacious therapies. Likewise, the Nipah outbreak in Malaysia (1998-1999) was rapidly brought under control without vaccines or efficacious therapies, once the modes of transmission were established. The knowledge that *Nipah encephalitis* was linked to a newly recognized *paramyxovirus* which could be transmitted through close proximity to live, infected pigs but not via insect vectors, fomites or suspended airborne particulates, or contact with raw or prepared meats (ascertained from virological studies, field epidemiology, medical ecology, and clinical medicine), allowed for its rapid control in humans, even as this control decimated the pig farming industry in parts of Southeast Asia.

The lesson perhaps is that in appraising the contribution of modern biomedical science to disease control and to population health, it is important to distinguish between *knowledge-based practices and coping responses*, as opposed to an undue focus on *commodifiable consumables*. This distinction is especially pertinent when we consider the strategic priorities of market-driven research and product development, in contrast to publicly funded needs-driven research in the biomedical sciences.

RE-EMERGENCE OF SOCIAL EPIDEMIOLOGY

In 1980, the UK Dept of Health published the Black report which documented health inequalities between occupational classes in the United Kingdom. Appointed in 1977 by the Labor Secretary of State David Ennals, the Black committee (chaired by Sir Douglas Black, chief scientist at the Department of Health and Social Security) submitted its report in 1980 to a newly installed Conservative government led by Margaret Thatcher. Viewed as an ideological inconvenience by the new government, the Black report was sidelined with only 260 duplicated copies of the typescript being made available for distribution, rather than a proper print run by Her Majesty's Stationery Office bearing the imprimatur of the Department of Health and Social Security. Peter Townsend, a member of the Black committee recalled that no press release or press conference accompanied its release on a Friday just before the August Bank Holiday, which virtually guaranteed the lowest possible level of publicity. Notwithstanding this, the report was picked up by the mass media, and a follow-up and expanded report (Townsend and Davidson, 1988) was eventually "outsourced" to Penguin publishers in 1988 in a belated and perhaps unintended privatization.

In the event, the salient findings of the report (below) were well circulated among health professionals and researchers, and eventually received considerable publicity in the print media as well:

[the] data show marked differences in mortality rates between the occupational classes, for both sexes and at all ages. In the first month of life, twice as many babies of "unskilled manual" parents (class V) die as do babies of professional class parents (class I) and in the next eleven months 4 times as many girls and 5 times as many boys. In later childhood the ratio of deaths in class V to deaths in class I falls to 1.5-2.0, but increases again in early adult life, before falling again in middle and old age. A class gradient can be observed for most causes of death, being particularly steep in the case of diseases of the respiratory system.

[As regards health services,] underutilization by the working classes is a complex resultant of under-provision in working class areas, and of costs (financial and psychological) of attendance which are not, in this case, outweighed by disruption of normal activities by sickness. The evidence suggests that working class people make more use of GP services for themselves (though not for their children) than do middle class people, but that they may receive less good care. It is possible that this extra usage does not fully reflect the true differences in need for care, as shown by mortality and morbidity figures. Similar increases in the use of hospital services, both in-patient and out-patient, with declining occupational class are found, though data are scanty, and possible explanations complex.

The Black Report thus raised important questions, among them:

- Were there greater exposures and risk factors faced by the lower classes?
- Was there differential resistance to disease?
- Was there differential availability and access to health services?
- Was there differential use of curative, preventive and promotive health services?
- Was there differential quality of service?

Over the next decade, some answers began to emerge, most notably, from the Whitehall epidemiological studies of British civil servants led by Michael Marmot. Begun in 1967, the first phase involved a twenty-five year follow-up of the health experience of 18000 civil servants who were classified into four employment grades: (1) administrative (2) professional and executive (3) clerical, and (4) other. These were in other words, all non-industrial, office-based jobs with high job security.

Remarkably, a social gradient in health was also evident among these civil servants. Among the age group 40-64, there was a four-fold difference in mortality rates between the top and bottom employment grades, with a consistent social gradient in between (Marmot, 2004). The first phase of the Whitehall study enrolled only men as subjects. In the second phase, one-third of the subjects were women, and a similar social gradient was again evident.

These results, consistent with the earlier findings of the Black report were particularly noteworthy considering that this was a cohort of London-based white collar employees who had, in principle, the same access to the National Health Service (NHS). Indeed, the study showed that among those who had evidence of heart disease, the lower employment grades were just as likely to be investigated and treated as the upper grades, although there was limited information on the quality of service received.

The Whitehall study also established that obesity, smoking, poorer diet, and reduced leisure time and physical activity were more common among the lower classes. Nonetheless, the risk gradient for coronary heart disease mortality persisted even after controlling for smoking, blood pressure, plasma cholesterol, blood sugar, and height, which in total accounted for less than one-third of the difference in mortality rates that existed between the highest and lowest employment grades.

After considering and dismissing a number of alternative hypotheses (good health and robust constitution lead to high social position; favorable genotypes confer both good health and high social position), Marmot and his colleagues concluded that the psychic benefits of "being in control" of one's life (autonomy in working and living circumstances), often correlated with high social status, occupational rank, and income level, was a significant factor contributing to the health status differentials.

This interpretation which invokes a psychosocial health effect (Marmot and Wilkinson, 2001), mediated via neuro-endocrine pathways and the body's physiological responses to stress, is controversial. Extended to a de-contextualized "social capital" and its etiological claims in social epidemiology, it has provoked vigorous responses from "neo-materialists" who lay greater emphasis on the interpenetrating unity of the social, psychological, and material dimensions of income inequality and deprived environments (Lynch, et al., 2000; Muntaner, 2004).

The debate over social gradients in health and its underlying etiology and pathogenetic mechanisms will continue, but regardless of how it is resolved, it has reinforced the shift from individual care to macro-social trends in epidemiological assessments of their contributions to population health.

To the extent that genomics individualizes care in its search for customized commodifiable products, it is at odds with this emerging current within public health. It may add to the armamentarium for individual patient care and treatment, a perfectly legitimate concern of a responsible clinician, but its contribution to population health along with its claims for priority in resource allocation for health will be scrutinized and contested.

REFERENCES

Chan, C. K. (2003). Risk perception and coping responses in a SARS infectious outbreak. *Third World Resurgence*, (July/August): 28–30.

Cohen, R. (2002, June). Neglected diseases and the health burden in poor countries. *Multinational Monitor.*

Colgrove, J. (2002). The McKeown thesis: a historical controversy and its enduring influence. *Am J Public Health*, 92, 725–729.

Davey Smith, G., & Lynch, J. (2004). Social capital, social epidemiology, and disease aetiology. *International Journal of Epidemiology*, *33*, 691–700. doi:10.1093/ije/dyh261

Holtzman, N. A. (1999). Are genetic tests adequately regulated? *Science*, *286*(5439), 409. doi:10.1126/science.286.5439.409

Holtzman, N. A., & Marteau, T. M. (2000). Will genetics revolutionize medicine? *The New England Journal of Medicine*, *343*(2), 141. doi:10.1056/NEJM200007133430213

Kremer, Michael &Glannerster, Rachel. (2001). Creating a market for vaccines. *New York Times*. June 1.

Lynch, J., DaveySmith, G., Kaplan, G., & House, J. (2000). Income inequality and mortality: importance to health of individual income, psychosocial environment, or material conditions. *BMJ (Clinical Research Ed.)*, *320*, 1200–1204. doi:10.1136/bmj.320.7243.1200

Mackenbach, J. P. (2007). Sanitation: pragmatism works. *BMJ (Clinical Research Ed.)*, *334*, s17. doi:10.1136/bmj.39044.508646.94

Marmot, M. (2004). *Status Syndrome: How Your Social Standing Directly Affects your Health and Life Expectancy*. London: Bloomsbury.

Marmot, M., & Wilkinson, R. G. (2001). Psychosocial and material pathways in the relation between income and health: a response to Lynch *et al. BMJ (Clinical Research Ed.)*, *322*, 1233–1236. doi:10.1136/bmj.322.7296.1233

McKeown, T. (1971). A historical appraisal of the medical task. In McLachlan, G., & McKeown, T. (Eds.), *Medical History and Medical Care*. London: Oxford University Press.

McKeown, T., Brown, R. G., & Record, R. G. (1972). An interpretation of the modern rise of population in Europe. *Population Studies*, *26*, 345–382. doi:10.2307/2173815

McKinlay, J. B., & McKinlay, S. M. (1977). The questionable contribution of medical measures to the decline of mortality in the United States in the Twentieth Century. *The Milbank Quarterly*, *55*, 405–429. Retrieved from http://www.manhattan-institute.org/html/medical_progress.htm accessed July 19, 2006.

Muntaner, C. (2004). Social capital, social class, and the slow progress of psychosocial epidemiology. *International Journal of Epidemiology*, *33*, 674–680. doi:10.1093/ije/dyh200

Ng, T. W., Turinici, G., & Danchin, A. (2003). A double epidemic model for the SARS propagation. *BMC Infectious Diseases*, *3*, 19. doi:10.1186/1471-2334-3-19

Peiris, J. S. M., Lai, S. T., & Poon, L. L. M. (2003). Coronavirus as a possible cause of Severe Acute Respiratory Syndrome. *Lancet*, *361*, 1319–1325. doi:10.1016/S0140-6736(03)13077-2

Rosen, G. (1971). Historical trends and future prospects in public health. In McLachlan, G., & McKeown, T. (Eds.), *Medical History and Medical Care*. London: Oxford University Press.

Ruan, Y. J., Lin, W. C., & Ling, A. E. (2003). Comparative full-length genome sequence analysis of 14 SARS coronavirus isolates and common mutations associated with putative origins of infection. *Lancet, 361*, 1779–1785. doi:10.1016/S0140-6736(03)13414-9

Szreter, S. (2002). Rethinking McKeown: the relationship between public health and social change. *American Journal of Public Health, 92*, 722–725. doi:10.2105/AJPH.92.5.722

Szreter, S. (2004). Debating mortality trends in 19th century Britain. *International Journal of Epidemiology, 33*, 705–709. doi:10.1093/ije/dyh143

Townsend, P., & Davidson, N. (Eds.). (1988). *Inequalities in Health: The Black Report and the Health Divide*. Harmondsworth: Penguin.

Trouiller, P., Olliaro, P., Torreele, E., Orbinski, J., Laing, R., & Ford, N. (2002). Drug development for neglected diseases: a deficient market and a public health policy failure. *Lancet, 359*, 2188–2194. doi:10.1016/S0140-6736(02)09096-7

World Health Organization. (2003). A multicentre collaboration to investigate the cause of severe acute respiratory syndrome. *Lancet, 361*, 1730–1733. doi:10.1016/S0140-6736(03)13376-4

ENDNOTES

1 Interestingly, this may lead to fragmented and diminished markets for customized (but more expensive) therapies targeted at responsive individuals, whilst creating or expanding the market for pharmacogenomic profiling of individuals.

2 Neglected diseases, as highlighted by *Médecins Sans Frontières (MSF)*: of the 1,393 new drugs approved between 1975 and 1999, only 16 (or just over 1 percent) were specifically developed for tropical diseases (such as malaria, sleeping sickness, Chagas' disease, kala azar) and tuberculosis, diseases that account for 11.4 percent of the global disease burden. For 13 out of those 16 drugs, two were modifications of existing medicines, two were produced for the US military, and five came from veterinary research. Only 4 were developed by commercial pharmaceutical companies specifically for tropical diseases in humans. These tropical diseases mainly affect poorer communities in countries of the South, which do not constitute a valuable enough market to stimulate adequate R&D by the multinational pharmaceutical companies. (See Trouiller, et al., 2002; Cohen, 2002; Kremer and Glannerster, 2001.

3 "*Genohype"*, expressive term introduced by N. A. Holtzman to denote the overblown expectations of the benefits that genomics can confer on patient care and population health. N. A. Holtzman (1999) "Are Genetic Tests Adequately Regulated?" [editorial], *Science* 286(5439):409.

4 N. A. Holtzman & T. M. Marteau (2000) "Will Genetics Revolutionize Medicine?", *New England Journal of Medicine* 343(2):141-144

5 R. Hubbard & R. C. Lewontin (1996) "Pitfalls of Genetic Testing", *New England Journal of Medicine,* 334(18):1192-1194 (Correspondence, *New England Journal of Medicine* 335(16):1235-1237 – H. J. Stern, A. Maddalena, J. D. Schulman, W. D. Foulkes, H. F. Bunn, T. P. Stossel, B. G. Forget, G. Stamatoyannopoulos, D. J. Weatherall, R. Hubbard, R. C. Lewontin).

6 "Patients have yet to benefit from genome research", *Miami Herald,* 12 October 2004.

http://www.miami.com/mld/miamiherald/news/nation/9895562.htm. Gilbert Omenn, a cancer specialist and president-elect of the American Association for the Advancement of Science was quoted as saying that despite an *"avalanche of genomic information... cancers remain a largely unsolved set of medical problems [for which] we continue to rely on highly toxic drugs"* (accessed on 23 November 2004).

7 Nicholas Wade (2010) "A Decade Later, Genetic Map Yields Few New Cures", *New York Times,* 12 June 2010.

8 H. Wallace (2002) *"Genetics and 'Predictive Medicine': Selling Pills, Ignoring Causes"*, GeneWatch UK, Briefing Paper Number 18, dated May 2002; World Health Organization (2002) *Genetic Technologies: Implications for Preventive Health Care* (A Report for WHO prepared by GeneWatch, UK), Geneva: World Health Organisation, Human Genetics Program.

9 P. Trouiller, P. Olliaro, E. Torreele, et al. (2002). "Drug Development for Neglected Diseases: A Deficient Market and a Public Health Policy Failure", *Lancet* 359:2188-2194.

10 Thomas McKeown (1971) "A Historical Appraisal of the Medical Task" in G. McLachlan & T. McKeown (ed) *Medical History and Medical Care,* London: Oxford University Press, pp. 29-50; T. McKeown, R. G. Brown & R. G. Record (1972) "An Interpretation of the Modern Rise of Population in Europe", *Population Studies* 26:345–382.

11 See http://www.manhattan-institute.org/html/medical_progress.htm (accessed July 19, 2006).

12 N. S. Scrimshaw, C. E. Taylor & J. E. Gordon (1968) *Interactions of Nutrition and Infection*, Geneva: World Health Organization; N. S. *Scrimshaw* (2003) "Historical Concepts of Interactions, Synergism and Antagonism between Nutrition and Infection", *Journal of Nutrition 133*:316S-321S

13 R. C. Lewontin (1979) "Death of TB", *New York Review of Books* 25(21 & 22), 25 January 1979. http://www.nybooks.com/articles/archives/1979/jan/25/death-of-tb/ (accessed on 26 July 2010)

14 See Szreter, 2002. Indeed, Johan Mackenbach has remarked that "effective intervention does not always need accurate knowledge of disease causation (the development of sanitary measures largely preceded the germ theory)... at the time, popular explanations [of disease transmission] included the "miasma" theory that fevers were caused by foul damps arising from decaying organic material... Chadwick [a champion of the "sanitary revolution" in Britain nonetheless] succeeded despite his defective theory of disease causation..." (Mackenbach, 2007).

Chapter 3
International Organizations as Fora for International Bioethical Debate:
Towards a Just International Bioethical Law?

Chamundeeswari Kuppuswamy
University of Sheffield, UK

ABSTRACT

The role of international bodies as the place where international debates on bioethical issues take place is investigated. In theory these international venues are supposed to be neutral in that they do not favor any particular traditions or belief systems. As a neutral venue for debates, these international arenas should recede to the background, so to speak, and let all the voices from among all parties involved in the debate and discussion be heard. However, in practice such a scenario scarcely happens, as these fora and venues are often criticized as being dominated by the West and their claim of universal values. In order for the international venues to be a really just place where all the voices are given due prominence, the roles and arrangements of these international organizations need to be scrutinized.

INTRODUCTION

Technological developments in regenerative medicine[1] and reproductive medicine[2] are causing numerous ethical and social controversies. Embryonic stem cell research and cloned embryonic stem cell research for therapeutic purposes, cloning for reproduction, altering the genetic code by mixing other genes in or eliminating faulty genes from the genetic code, altering genes such that the alteration is carried forward into the next generation are all controversial. Advances in these areas means that we can alter and enhance human capacities, both physical and cognitive, extend life spans by replacement organs and radically become different as a species from what we currently are. Such transformation and change is a subject of debate, and such debate, if not taking place globally is both inefficient and an opportunity missed to enlist the wisdom and knowledge of the missing sections of humanity and human thought.

DOI: 10.4018/978-1-61692-883-4.ch003

How and where do different world cultures meet? How is our common future being shaped? Who are the participants and how are representations made? What framework is being appropriated in order to make sense and promote arguments towards building a certain reality?

This paper proposes to discuss, analyze, and briefly evaluate the role of international organizations as facilitators and fora for international bioethical debate, where the use and application of new technologies such as biotechnology is being scrutinized by the international community with a view to developing governance and legal frameworks. The paper was inspired by the call made by the editor of this volume who wrote: 'The voluminous literature on the ethical, social and legal aspects of life sciences and biotechnology show that there is indeed a very large variety of problems related to the social and cultural contexts of science and technology. What is most interesting is how one could understand these social and ethical ramifications in the context of the world's cultures and historical traditions.[3] While in no way seeking to detract from the contributions already made to soft law on biotechnology by international organizations, this paper maintains that the work of international organizations should be examined to assess their role and function in facilitating the development of robust and just international bioethical law. By bioethical law, reference is being made to the regulation of issues that are widely discussed within the field of bioethics, mainly in the West in the last couple of decades.

States seek to achieve cooperation through membership in international organizations (Akande, 2006, p. 277). The membership of organizations is strong. The United Nations has a membership of over 190 States. Similar numbers make up the membership of organizations such as the United Nations Educational Scientific and Cultural Organization (UNESCO), the World Health Organization (WHO), the International Civil Aviation Organization (ICAO), the World Bank Group, the World Trade Organization (WTO) and the World Intellectual Property Organization (WIPO). There are autonomous institutions created under treaties such as the Law of the Sea Convention and the Climate Change Convention, which have also attracted substantial membership from States. Regional organizations exist to promote cooperation on regional matters. The European Union, the Gulf Cooperation Council, the African Union, the South Asian Association for Regional Cooperation (SAARC) are all examples of regional organizations. The subject matter for cooperation is wide ranging, from fisheries and telecommunications to trade, disarmament and human rights (Sands and Klein, 2001). Such cooperation is beneficial and useful, as evidenced from the output of these organizations and their continued existence, supported by states. Cooperation through intergovernmental organizations could be particularly useful in the context of novel technologies that have the potential to usher in tremendous change. These technologies could change human nature, our moral framework and our idea of perfection. The problem with these technologies is the problem is incoherent (Saletan, 2007) the contours of issues are not easily identified. Biotechnology and other technologies that involve biological material or technologies that have an impact on our biology are intimately tied to humans, and by doing research using such material we are venturing into unexplored territory. Given what is known theoretically, the results could bring about massive change. To engage with the ethical, social and other issues that arise in the context of the potential massive change, a number of mechanisms are employed. Some states expressly record in law by promulgating moratoriums on research that they deem sensitive or controversial. A number of moratoriums on research on stem cells research and human reproductive cloning are in place. Russia and Israel are two countries that recently extended their law prohibiting human reproductive cloning for another 5 and 7 years respectively (Israel extends its human reproductive cloning moratorium, 2009;

Russia extends human cloning ban, 2009). The moratoriums exist in order to assess technological progress, for technological progress to provide technical answers and for wider engagement with their implications. Others have in place laws and regulatory bodies that constantly assess the progress being made in the field on a case by case basis, as research is carried out.[4]

Through cooperation between states in international organizations, social, ethical and cultural engagement can be enriched. Through cooperation, various possibilities can be considered, international and multicultural perspectives taken into account and issues shaped. Ideas are generated and opinions are reshaped. In fact, it can be argued that such ideas are foundational to development of robust and just rules. Ideas change positions and positions are in turn built on ideas. In particular the cross linkages between issues happen. When human cloning was considered by the United Nations General Assembly, human dignity was discussed in the context of resource allocation and women's welfare. Many states questioned the wisdom of spending resources on such research when numerous global health crises such as HIV/AIDS, tuberculosis and malaria affected a large number of peoples in the developing world. States also argued that the potential for exploitation of women from poor countries should be anticipated given that embryo research involves the use of a very large number of embryos, and poor women could become the source of embryos in order to maintain a constant supply of embryos as such research becomes widespread. The resolution eventually featured both of these issues[5], although the initial draft did not contain such references. It highlighted an important point about bioethics, and its relative emphasis on "developed country issues" (Turner, 2004) which was what essentially reflected in the draft agenda item. When it went through debate in the General Assembly, different facets of ethical and social issues surrounding embryo research emerged. By bringing women and resources issues directly to the heart of the debate on dignity and human cloning, the international forum has facilitated and "made a statement" that such issues are relevant not only within poverty alleviation and public health. The General Assembly process provided a new orientation for human dignity to chart a course through the highly compartmentalized bioethics discourse within which it currently thrives. Not only did experts and state representatives bring their expertise and positions to the table in the course of developing rules; new ideas were forged and shaped in the process (Weiss & Carayannis, 2001). In yet other forums, such distinct issues as human rights and intellectual property agendas have come together to affect changes in both fields. Large intergovernmental organizations such as the UN and its specialized agencies facilitate bringing into focus a panoramic view of the big picture, so to speak.

International organizations provide a forum for identifying and deliberating on matters of common interest, act as vehicles for taking action on international or transnational problems, provide a forum for developing rules on matters of common interest. They also, at a later stage, provide mechanisms for promoting, monitoring, supervising State compliance (Akande, 2006, p. 278). It is the first of the roles of international organizations that has transpired most in the context of genetic and reproductive technology. Often an effort to develop binding rules has resulted not as much in binding rules but soft law and deliberation. The 2005 cloning resolution is a case in example. A working group of the sixth (Legal) committee of the UN General Assembly were working on an international convention against the reproductive cloning of human beings since 2002. Eventually, the plan for a convention was replaced by a GA resolution, and in this case, even its status as soft law is diminished because of the fractured support it obtained in the forum (Kuppuswamy, 2006). For soft law on cloning, it is best to look at an earlier GA resolution from 1997. The debates leading up to the 2005 Declaration also demonstrate the limited effectiveness of such fora when there are deep seated differences in views held by States. Such differences can be addressed by focusing more on the first of the functions as above mentioned, i.e., identifying and deliberating on matters of common interest. For Lister, international organizations are useful to the extent that a sense of social solidarity broad enough to encompass human beings of all races and conditions has developed, i.e., outside of the fora (Lister, 1990). But it is arguable that such organizations can also be instrumental in achieving or at least working towards social solidarity through encouraging and providing a space for exchange of information and views, and by promoting healthy debate. Solidarity is achieved by working both ways. The UN Commission on Human Rights (currently Human Rights Council since 2006) is a UN organ established for the promotion and protection of human rights, and has addressed bioethical issues, focusing on the need to identify and address problems for human rights protection from new and emerging technologies. WHO has promoted dialogue on the impact of trade and intellectual property rights on the protection of public health through development of new treatments and interventions. UNESCO has succeeded in fostering the broadest scope of debates, with a view to developing a convention to regulate genomics and biotechnology. Its output has been widely recognized as the most comprehensive treatment of the subject, resulting in three declarations on the human genome, human genetic data and bioethics, between 1997 and 2005 (Universal Declaration on the Human Genome and Human Rights, 1997, p. 41; International Declaration on Human Genetic Data, 2003, p. 39; Universal Declaration on Bioethics and Human Rights, 2005, p. 74). The UN General Assembly has, for many decades, addressed human development, particularly in the vast number of developing countries that make up its membership. It has considered the role of science and technology for development and more recently became involved in the regulation of human reproductive cloning. In sum, various international organizations have had or continued to have bioethics and new human technologies on their agenda, thus providing numerous multinational fora for deliberation on matters of common interest from various perspectives.

The nature of the debate in these organizations depends on the composition of the members participating and voicing their views in the forum. The leadership provided in such fora are also important (Weiss & Carayannis, 2001). The permanent bioethics committee set up at UNESCO in 1998 was to have a membership that was culturally diverse and geographical representative of various regions in the world (Statute of the International Bioethics Committee, 1998). The recruitment policy of the bioethics committee since its inception in 1993 has been based on obtaining a culturally and geographically diverse set of scholars and experts. This was later institutionalized when the committee became a permanent committee of UNESCO through the adoption of its statute in 1998. The policy of diversity extends to municipal deliberations as well. The National Bioethics Advisory Commission in the United States included world religious perspectives in their reports on stem cell research and cloning, alongside scientific and ethical reports. Diverse representation and worldviews contribute to an all-round debate offering the widest possible coverage of human thinking. For a debate to be meaningful, especially in international organizations, the participation must be diverse and representative of its constituency and stake holders, which includes a vast majority of States and cultures. To provide for a just and robust bioethical law, international organizations providing a platform for such deliberation should actively ensure that they become forums for diverse engagement on issues.

How inclusive are international organizations? How do international organizations facilitate wider participation in their deliberations? The role of non members has increased since the inception of these organizations. Autonomous organizations, banks, governance authorities of disputed territories, civil society organizations, regional state groupings, judicial bodies and other entities have been included in deliberations in the General Assembly and other UN organs. Vatican is a permanent Observer at the UN. Not just states, who although are the most influential in intergovernmental organizations, but others take part in the proceedings too, also in the decisions taken in these bodies. There is an increasingly diverse range of actors in international organizations, thus promoting inclusive debate. But inclusiveness also depends on the nature of the issues too. In the case of biotechnology, embryo research and related areas, it is arguable that we need more actors and more views in order to be truly inclusive. An indication of this already comes from the active participation of the Vatican in the proceedings leading up to the 2005 Declaration on Human Cloning (A/RES/59/280). The Holy See or the Vatican is a permanent observer at the UN, although without a vote. It played a sustained role through the negotiations in the General Assembly, forwarding proposals and submitting written statements.[6] The Vatican professes a clear ruling that all embryonic research gravely offend the dignity of the human race and the dignity of human life (Views of the Holy See, 2003). It seeks "a complete and explicit prohibition on all techniques of creating new individual human embryos by cloning, including somatic cell nuclear transfer, embryo splitting, and other similar techniques that may develop in the future" (View of the Holy See, 2003).

In the issue of human cloning, these statements played an active role in the decision. This is only fitting because ethical opinions and moral views are influenced by religious belief. Religion plays a prominent role in bioethics. It should play a role, states James Childress, a pioneering bioethicist (Childress, 2001). In Europe, States with predominantly catholic populations have a pattern of laws that prohibit embryo research, while others States have a diversity of laws generally regulating such research. Objections to genetic engineering and cloning are often formulated on the basis that it is wrong to play God. The President's bioethics advisory commission in the US also sought religious perspectives. It is arguable that the Hole See's strong views and active mobilization of support for its proposals threw a spanner into the works. An otherwise straightforward convention to prohibit human reproductive cloning turned into a frustrating and unrewarding regulatory effort, which currently leaves the field open to a wide interpretation, bringing uncertainty to the international position on human cloning. However, despite the result, which will take years to come to fruit, with hindsight we may be able to conclude that it is the right result. But the question here is one of fairness. When one of the world's major religions is able to participate in the forum, the others' views should be elicited too. A cross section of what could be on offer from the other religious perspectives will also show that these perspectives are very different indeed, and it will also show that taking into account one religion could not possibly account for others views (in case the view is that all major religions hold common values and even identical positions). There is a rich trove of arguments to be explored by deliberating on these various religious positions. This was highlighted in the US bioethics advisory commission report, which stated that "religious positions on human cloning are pluralistic in their premises, modes of argument, and conclusions" (Cloning Human Beings, 1997).

The Jewish tradition, for example, sets up a different set of issues for discussion in the context of cloning. The Jewish religious commandment of procreation and the grave doubts over using gametes of unknown filiation play decisive roles in determining whether human reproductive cloning (HRC) should be permitted. Currently, the overriding objection against human reproductive cloning in Israel is based on the experimental status of the technology, and the predicted dangers posed to health of any offspring produced by use of this technology. But were the obstacles in early reprogramming of the DNA of a cloned embryo to be cleared and were the scientific community able to declare it to be safe to proceed to undertake further development of the cloned embryo, the above mentioned statements would then form a basis of discussion on whether HRC should be permitted or banned. Jewish religious authorities would advise against the use of anonymous sperm or oocytes obtained from a third person if it were possible to use material from either of the infertile couple, who should want a child (Revel, 2003). It is not very clear as to whether the religious objection is towards merely anonymous donation, or if it applies more strictly to any third person donation. One of the reasons against anonymous donation arises from the possibility of embryonically related individuals marrying incestuously. Coupled with the commandment to procreate, the preference for proximate genetic material for genetic procreation would lead to a declaration that human reproductive cloning is a desirable procedure for infertile couples in Israel. Indeed, anywhere else, where such preferences are welcomed.

This is a radically different way of looking at HRC, and it does not sit comfortably alongside an argument that HRC is against human dignity. But this is not all captured through Israel's actions and statements in the General Assembly in the course of the passage of the 2005 resolution on human cloning.

Another entrenched idea in bioethics and human rights is the notion of privacy. Yet another would be autonomy. A variance in the understanding of privacy and autonomy is seen in Buddhism where the notion of the individual self is inherently different from our ordinary conception of the self. The Buddhist view on privacy is fundamentally different from the traditional understanding of the concept. Privacy in Buddhism is more of a "conventional" nature, than of an "inalienable" or "inherent" nature. It is not as foundational to the person as we currently envisage it, but only important because of a certain set of objectives we as a society have set for ourselves. The self in Buddhism does not exist as concretely as it does within the system where privacy attaches to the person and personality of the person (Hongladarom, 2009). Autonomy, understood as individual independence in decision making, is built on the presumption of a continuous self. For example, the self is the same all through one's health and sickness. But in Buddhism, "the search for real preferences, central to the identity of the person, is a pointless one" (Hugues & Keown, 1995). The argument is that the differences between an individual self separated in time is also the difference between oneself and another. Therefore an individual self's "real" preference does not exist, so the exercise of autonomy is unreal. Similarly the distinction created between living beings and humankind is a construct not fully accepted in many eastern religions. Marking the distance between societies in terrain of ethics, the Japanese international lawyer Ida observes that "ethical appeals to human welfare or individual happiness to justify science and technology may have intuitive force in the West, but may seem alien to a non-Western audience" (Ida, 2009, p. 70).

With such pluralistic ideas to consider and discuss, it would not be unwarranted to convene a standing committee on religion and bioethics to gain insight into these traditions and ideas. Science is still not developed to a level where all the complex questions of life can be answered. While we await such information for organizing evidence-based analysis, an engagement through world religions perspective would be in the interest of States. The international bioethics committee is the only international forum involved in "in-depth bioethical reflection." It is a more conventional forum where science meets philosophy and policy. Not only would a religion and bioethics committee provide a specialist deliberating space and possibly diverse ideas to enable decision making on new and emerging technologies, it would also build and strengthen religious understanding and tolerance.

The question remains open as to whether Human Rights, sometimes referred to as the secular religion of the West, is indeed the framework that is capable of providing a framework of do's and don'ts for these manipulative technologies. Perhaps we have to go back to the drawing board to sketch out a life sciences ethic (Commission on Human Rights Resolution, 1993, 91) at the national and international level. The consensus is that life sciences should develop in a manner as to ensure respect for human rights. While Article 10 of the 1997 Human Genome Declaration states that no research shall prevail over respect for human rights and freedoms, there is no stipulation for how a determination of disrespect can be made. Therefore unless the stipulations are clearly laid out, it is left to interpretation as to whether a particular activity or research is in violation of human rights. Human reproductive cloning or human enhancements per se are not violations of human rights. To object to these procedures in principle, because of their negative effects on human rights is based on the reaction that such cloned beings may face. The dangers of discrimination and stigmatization may arise, but are not present in the act of cloning itself. The occurrence of violations may then require intervention by the state. However, as Marks succinctly points out: "The issue then becomes whether the potential humiliation from being a clone is sufficient to justify characterizing cloning as a human rights violation by law or treaty" (Marks, 2002). This is echoed by Savulescu, who writes that any problem would stem from how we treat these cloned beings. He goes on to say, "We should not be clonists, just as we should not be racists or sexists" (Savulescu, 2005). While there are implications for human rights from the advancement of biotechnology, it does not provide adequate tools to address the challenges and risks posed by such technologies. The danger of narrowing our gene pool by selecting out genes can pose a threat to the human species, if carried out in a large scale. The role that freedom of scientific research plays in creating and perpetuating inequalities is another cause of concern. These and other similar issues are not covered within the scope of human rights law. It may be that we extend the scope of "human rights" law to include related entities, i.e., embryo and the genome. The focus of UNESCO's soft law instruments have been on human rights and bioethics, with an emphasis on human rights. This has attracted criticism from bioethicists and human rights scholars. The former contend that the discussions have been hijacked by human rights language and formulation of issues. The latter contend that the language is too vague and imprecise. A current member of UNESCO's bioethics committee writing about the lack of regional and ideological diversity and an overrepresentation of human rights lawyers in the drafting committee for the 2005 declaration on bioethics and human rights comments about the lack of philosophical rigor in the document (Snead, 2009).

If we are intending a revision of human rights law by extending the scope of protection afforded to embryos and the genome, then we would do well to consider the discourse on cultural relativism and the related but distinct notion of moral fundamentalism. Often the international human rights framework is criticized on the basis that claims about universality mask ignorance of non-Western beliefs and traditions, and are more partisan than the claim to universality seeks to reveal. The universality claim is that human rights are rights that humans everywhere hold as a result of being born human. Therefore a formulation of human rights as found in international human rights instruments should be universally accepted and respected. The objection on grounds of cultural relativism is "primarily concerned with the validity of the source of a moral rule or, in a weaker form, with interpreting human rights norms with sensitivity to the cultural specificities of a particular state" (McGoldrick, 2005, p. 33) If human rights are based on moral positions (Shestack, 1998), then the nature of the moral rule and its source brings about the human right and also a corresponding understanding of the right. As an illustration, if the right to life is envisaged in the context of a moral and philosophical position that life is only born from other life, or in other words, rebirth occurs, then it will be different to a moral position that a human being is born and dies only once. The objections to human rights law from the historical cultural relativism angle echoes concerns that I have raised above in respect of unequal participation of religious and cultural perspectives in bioethics debates in international organizations. The argument for supporting cultural relativism here is not made in order that this justifies actions of states like Burma which disregard their human rights obligations (Sweeney, 2005, p. 461). It is also not made in the context of upholding internal exercise of sovereign rights (Sweeney, 2005, p. 461). However it must be relevant that the cultural underpinning of a rights-system is determinative of its contents, at the least a major influence on it. It is still possible to us the human rights framework as a system, but on the condition that 'essentially infinite, open-ended, and highly experimental in nature' (Mutua, 1996, p. 593).

On the one hand these adverse observations can be applied to international human rights and bioethics instruments. On the other, it distracts from the potential for pioneering human rights contribution made possible by the debate on technology centered on human dignity, human genome and human embryo. Constructive engagement with diverse cultural, religious and philosophical positions can only contribute to the open-ended human rights project that states have embarked on since the end of the Second World War.

In conclusion, a number of international organizations have considered and continue to consider bioethical issues related to reproductive and regenerative medicine. While they provide platforms for discussions and debate the influence of some views is greater than others. The human rights framework is neither adequate nor sufficient to address the rapid changes predicted through technological progress. Organizations could open the possibilities for exploring supplementary forms of governance by re-orienting the debate in different directions. The obvious and ubiquitous religious and cultural perspectives need to be brought more center stage. The avenue is already open via the presence of the Holy See in the General Assembly; a broader participation of faiths would contribute towards inclusive debate. Technologies such as biotechnology, information technology, nanotechnology and neuroscience are set to further join forces by converging together to have an impact on the human. It seems that there is only one choice - that of transcending political and historical divides to come together in solidarity, to debate and develop a truly 'international bioethical law' based on debate and intercultural understanding.

REFERENCES

Akande, D. (2006). International organizations. In Malcolm, D. (Ed.), *Evans, International Law* (2nd ed.). Oxford, UK: Oxford University Press.

Childress: Religion has important role in bioethics debate. (2001). Speaking at a conference at Princeton University in 2001. Retrieved from the news section of the Princeton University Website http://www.princeton.edu/main/news/archive/A97/89/83Q20/index.xml?section on 10 November 2009.

Cloning Human Beings. (1997). Report and Recommendations of the National Bioethics Advisory Commission, 1997. Retrieved from http://bioethics.gov/reports/past_commissions/nbac_cloning.pdf on 15 August 2009.

Commission on Human Rights Resolution. (1993).

Hongladarom, S. (2009). Privacy, the individual and genetic information: A Buddhist perspective. *Bioethics*, *23*(7), 403–412. doi:10.1111/j.1467-8519.2009.01716.x

Hughes, J. J., & Keown, D. (1995). Buddhism and medical ethics: a bibliographic introduction. *Journal of Buddhist Ethics*, 2 (Online Journal retrieved from http://www.buddhistethics.org/2/dkhughes.html).

Ida, R. (2009). Should we improve human nature? an interrogation from an Asian perspective. In Savulescu, J., & Bostrom, N. (Eds.), *Human Enhancement*. Oxford University Press.

International Declaration on Human Genetic Data. (2003). Records of the General Conference 32nd Session, 29 September - 17 October 2003, adopted 16 October 2003.

Israel extends its human reproductive cloning moratorium for another 7 years till 2016. (2009). Retrieved from *Jerusalem Post.* Retrieved from http://www.jpost.com/servlet/Satellite?cid=1255450644033&pagename=JPArticle%2FShowFull on 1 November 2009.

Kuppuswamy, C. (2006). The role of international institutions in the formation of international bioethical law: UNESCO and the United Nations General Assembly attempt to govern human cloning. *Journal International de Bioethique*, *18*(1-2), 139–170.

Lister, F. (1990). The role of international organizations in the 1990s and beyond. *International Relations*, *10*, 101–116. doi:10.1177/004711789001000201

Marks, Stephen P. (2002). Human rights assumptions of restrictive and permissive approaches to human reproductive cloning. *Health and Human Rights*, 6(1), 80-100, p. 87.

McGoldrick, D. (2005). Multiculturalism and its discontents. *H.R.L. Re*, *5*(1), 27–56.

Mutua, Makau wa. (1996). The ideology of human rights. *Virginia Journal of International Law*, 36. 593. Retrieved from SSRN: http://ssrn.com/abstract=1525598, p.

Revel, M. (2003). Human reproductive cloning, embryo stem cells and germline gene intervention: an Israeli perspective. *Medicine and Law*, *22*, 701–732.

Russia extends human cloning ban. (2009). Retrieved from *Ria Novosti,*http://en.rian.ru/russia/20091002/156329425.html on 28 October 2009.

Saletan, W. (2007). Tinkering with humans. *New York Times.* Retrieved from http://www.nytimes.com/2007/07/08/books/review/Saletan.html?pagewanted=1&_r=1 on 2 December 2009.

Sands, P., & Klein, P. (2001). *Bowett's Law of International Institutions*. London: Sweet and Maxwell.

Savulescu, J. (2005). How will history judge cloning? Retrieved from the *Times Higher Education*, 6 May 2005 at http://www.timeshighereducation.co.uk/story.asp?storycode=195874.

Shestack, J. J. (1998). The philosophic foundations of human rights. *Human Rights Quarterly*, *20*(2), 201–234. doi:10.1353/hrq.1998.0020

Snead, O. C. (2009). Bioethics and self-governance: the lessons of the Universal Declaration on Bioethics and Human Rights. *The Journal of Medicine and Philosophy*, *34*, 204–222. doi:10.1093/jmp/jhp024

Statute of the International Bioethics Committee. (1998). Retrieved from the UNESCO IBC portal on the 10 July 2009 http://portal.unesco.org/shs/en/ev.php-URL_ID=2026&URL_DO=DO_TOPIC&URL_SECTION=201.html.

Sweeney, James A. (2005). Margins of appreciation: cultural relativity and the European Court of Human Rights in the post-Cold War era. *I.C.L.Q.*, 54(2), 459-474.

Turner, L. (2004). Bioethics needs to rethink its agenda. *BMJ (Clinical Research Ed.)*, *328*(7432), 175. doi:10.1136/bmj.328.7432.175

Universal Declaration on Bioethics and Human Rights. (2005). Records of the General Conference, 33rd session Paris, 3-21 October 2005, adopted 19 October 2005.

Universal Declaration on the Human Genome and Human Rights. (1997). Records of the General Conference, Twenty-ninth Session Paris, 21 October to 12, adopted 11 November 1997.

Views of the Holy See on human embryonic cloning, July 17, 2003. (2003). Retrieved from http://www.holyseemission.org/cloning2003eng.html, retrieved 7 December 2009.

Weiss, T. G., & Carayannis, T. (2001). Whither United Nations economic and social ideas?: a research agenda. *Global Social Policy*, *1*(1), 25–47. doi:10.1177/146801810100100103

ENDNOTES

1 Regenerative medicine aims to repair, replace or re-grow tissues and organs, and restore function, thus contributing to quality of life, as well as facilitating life extension.

2 An emerging field in reproductive medicine is reprogenetics, which is the combination of genetic technology with reproductive technology. Using our knowledge about the function of genes, we are able to 'screen' embryos before implantation into the womb to look for any 'undesirable' or disease associated genes, and avoid implanting these embryos.

3 Public call for proposals retrieved from http://www.stc.arts.chula.ac.th/Genomics/, first accessed 5 June 2009.

4 The Human Fertilisation and Embryology Authority (HFEA) in the UK is an example of tight regulation in biotech research.

5 A/RES/59/280, 23 March 2005, paragraphs (d) anf (f).

6 See Proposal submitted by the Holy See (A/C.6/57/WG.1/CRP.4) in A/C.6/57/L.4); See Annex I in A/C.6/58/L.9, Written amendments and proposals submitted by delegations, Paper submitted by the Holy See (A/C.6/58/WG.I/CRP.1).

Chapter 4
The Value of Life of the Embryo Observed from Two Different Lenses:
From its Own Potential to Develop, and from the Context in which it is Embedded

Elena Ignovska
Faculty of Law "Iustianus Primus," Macedonia

ABSTRACT

The chapter deals with moral deliberation over the status of the embryo, observed from two perspectives, namely the inner context of physical and biological composition including the argument of potentiality as a driving force of development, and the external context within lived and experienced practices in which an embryo is inevitably embedded. Both components are integral parts of what constitutes the life of the embryo, and therefore any separated observation is biased and does not fulfill the demands of the universal truth. Hence, the usual argument that focuses exclusively on the embryo itself, whether the embryo deserves moral right as a result of its potentiality for autonomy, is misguided.

INTRODUCTION

Ethical and legal discrepancies over the status of the embryo has always been full with different interpretations and applied practices based on the never ending story about the beginning of the human life. This eternally mystical question has been tackled by philosophers, ethicists and lawyers, but what urges for a precise resolution is the scientifically immense progress towards introduction of new ways of life creation.

DOI: 10.4018/978-1-61692-883-4.ch004

The moral status of the human embryo, and therefore the definition of personhood represent a core issue for the ethical debate around the techno-sciences. By fixing a broader or narrower concept of the human person, the group of human beings that hold an ontological status can be enlarged or diminished.

Under these circumstances, respect for the embryonic life will vary according to the model we follow, in terms of belonging to the group of human persons worth respecting. An extreme position could imply full moral recognition of embryos as already being human persons, and consequently, minor or nonexistent competence on the progenitors to decide over their parenthood regarding already created embryo, whereas on the other side, we may experience an excessive lack of respect towards embryonic human life, and primary respect for the progenitors and their autonomy in the process of decision making over the life of that very same embryo. Therefore, the determination of the status of the embryo does not affect only the future life of the embryo, but the future life of the progenitors, as being parents or not.

The concept of personhood is ambiguous for several reasons. As a key element that launches all the other consequential misinterpretations is the context in which a person developed. By investigating cultural attitudes, Vasil Prodanov distinguishes between individualists and collectivists, as related to Catholic, liberal and Western tradition on the one hand, and Orthodox, communitarian Eastern European tradition on the other (Prodanov, 1996). According to him, while "the former sees the person as a bundle of rights," the latter perception of a person is identified with a "bundle of his social identities, affiliations and roles." From this quotation, it is understandable why the actualization of personhood in Western countries begins with the moment when one is entitled to acquire her own rights, and in Eastern countries only when one coexists in the group, class or community and consequently becomes a recognized subject, not as a holder of rights, but primarily as a holder of obligations. Therefore, if for the individualists, the human essence originates from within, as inherited and self-sufficient element for individual development, for the communitarians, the same human essence is a substrate of social identities, affiliations and roles. Bearing in mind this historical background of development of the thought and science, it is easy to understand why some authors hold firmly on ontological properties, claiming individuality as a supreme value and promote observation over isolated oneself, while others always interrelate the individual with society, as a mutual mirror in which both the individual and the society look upon and, therefore shape themselves.

This paper will try to settle both levels of perception by setting forward two perspectives: the rights oriented one that primarily concentrates on the debate on the progenitors as already established human persons, and the moral one that originates from the intrinsic capacity of the embryo itself to develop, from the argument on potentiality. Elaboration of that kind will attempt to explain that both perceptions are important and interconnected for creating overall picture, and even more, that any other restrictive observation will lead to biased conclusions.

ETHICAL INTRODUCTION: AN ARGUMENT OF POTENTIALITY

In a circumstance where the universe consists of you, the others, and myself, the task of sharing the same lenses of perception towards an object under observation is a difficult one.[1] Therefore, the process of articulating and balancing different moral values has to take into consideration all the parties, starting from their intrinsic capacities, and extending towards the contexts in which they appear to act.

To clarify what the concept of moral status means, I will use the definition that Mark T. Brown offers: "To ascribe moral status to a class of entities is to rank them on an internal scale of moral values that determines how the obligations owed to these entities can be reconciled with those owed to entities located at other points on the scale" (Brown, 2007). According to this definition, the paper will try to locate the moral position of the embryo in constellation with its peers group, gametes as antecedents, newborns and adult human persons as its descendants from their potential to develop from one to another[2].

In the ethical deliberation, the paper will focus on observing and analyzing two points of view:

- Isolated interpretation of embryo's potentiality *per se*, in which I will trigger 1) The ontological position of embryo and the application of the term *"potentiality"* in the discussion, and 2) relatedness and discrepancies among in vivo and in vitro embryos, and
- Broader interpretation of the embryo's potentiality: the context of realization of the potentiality.

After clarifying the substantial elements in the terrain for discussion over the embryo's status, Brown's internal scale of moral values, in relation with the other protagonists on the developmental scale and outside of it, will be applied. Comparing different entities on their developmental path for becoming one from another, and even more, comparing embryo with its progenitors, will be instrumental to examine the question whether the life of a human embryo, as supported with its own potential to develop, can outweigh the free, private and autonomous decision of an actual person to become/not to become a parent.[3]

Terminology and Clarifications for Understanding the Attitudes in the Discussion

The term 'embryo' refers to the stage of prenatal development between the time of the implantation of the fertilized ovum in the woman's uterus, until approximately the end of the seventh or eighth week, and from then on, it is instead called fetus. In the historical academic background, the application of this term went through many variations, as used in various medical and legal contexts, and therefore its meaning was interpreted in a very inconsistent manner. In accordance with the afore-mentioned definition, the American Fertility Society (AFS) made a distinctive definition of 'embryo' and 'pre-embryo.' They relate the second term to the pre-embryonic stage that lasts until 14 after fertilization (Malo, 2000, pp. 301-312), but that may vary for several days. Nevertheless, the term 'embryo' is used very often with a broader meaning, depicting the developing human fertilized ovum from the conception until approximately the end of the second month, and therefore as synonymous with multiple other terms such as 'pre-embryo' and 'pre-zygote.' According to Arthur C. Guyton, there should not be differentiation among the above-mentioned terms, since all necessary chromosomes for developing an embryo in the human body are present from the moment of fertilization (Guyton and Hall, 1999, p. 944; Hollowell, 1999, 319). By this account, Guyton refers to the comprehension that the pre-embryo\zygote already has the same characteristics as the necessary chromosomes do not alter trough normal cell division. For the sake of consistency and facilitated discussion, and due to the acceptance of the potentiality argument as a tool for arguing in the following debate, this paper will accept Guyton's attitude, and the terms 'pre-embryo' or 'zygote' and 'embryo' will be identified and encompassed into the term 'embryo' only, referring to the state of the embryo from the moment of fertilization until approximately the end of the seventh or eighth week.

In an attempt to investigate the "investment of genetic material" of the progenitors, and consequently their share in the "joint creation," Sara Chan and Muireann Quigley distinguished two conceptual components with which the progenitors participate in the newly formed embryo, namely *1) physical:* made out of tissue samples, cells, DNA and the atoms that make the nucleic acid molecules that hold genetic code as unique genes, and *2) informational*: the code itself, the particular arrangement of atoms that make the DNA sequence that forms individual genome (Chan and Quigley, 2007, pp. 439-448). According to them, physical components might be a subject of property rights and from here on, they support the "property-based approach" as an answer to the question "what rights over embryos arise from property in genetic material?" In same line of comprehending goes Steiner, concluding: "Our bodies are factories: They produce things like blood, skin, hair.. Self ownership gives us the title to these, and protects our liberty to dispose of them" (Steiner, 1994, p. 233). Nevertheless, according to them, informational components are controversial.

The property-based approach, as a simple mathematical formula of genetic investment of progenitors in a mutual project is blind towards the project itself: the new complex alive mechanism. Therefore, the following text will focus on the life during initial stages of prenatal development of the embryo as existing *per se*.

- *Prenatal development* is the entire process of growth, maturation, differentiation and development that occurs between conception and birth.
- *In vivo fertilization* is a natural biological fertilization within women's body. Through a complex interactive process, the embryo or blastocyst becomes embedded in the uterine by the end of the first week of the prenatal development. Since pregnancy refers to the condition of having a developing embryo in the body, after the union of egg and sperm cells, a woman is considered pregnant after approximately one week of embryo's prenatal development.
- *In vitro fertilization* refers to the meaning 'outside the body;' it attempts to stimulate in vivo fertilization in a laboratory setting. This process stimulates attempts to substitute the in vivo process to the extent modern science allows.[4]

Ethos

The entire social and political structure in which we live in stands on some prerogatives that are considered to be inherent and inalienable to all human persons equally. These are the cornerstones of the liberal Western civilization, that is, each human person is entitled to live and not to be harmed by others, and further on, to be free from third party oppression: and in these terms, to build a family through consensual procreation.

Ethos Regarding the Moral Status of the Embryo

This query can be faced from three diverse ethical positions, among which no sharp boundary exists. Discrepancies among them mostly depend on the favorable argumentation. Therefore, three focal argumentations are appointing the direction: those based on the biological premises, stressing the genetic unique make up, personhood arguments, that above all, value the autonomy of the rational human beings, and potentiality arguments, emphasizing the growing potential.

At one extreme the embryo is considered as a human subject soon after fertilization because from that very moment it presents all the genetic information that will be present through the entire life. This position entails an obligation to provide an opportunity for implantation to occur and tends to ban any action before transfer that might harm the embryo. Within this perspective, the fertilized egg and the embryo have equal inviolable value as all other human beings, and therefore a right to life (The Protection of the Human Embryo In Vitro, 2003).

At the opposite extreme, we find the view that the embryo has no different status from any other human tissue, based on the fact that it is not an individual until is implanted in the female's womb. Then, with the consent of those who have decision-making authority over it, justified with the exclusivity of their own interests, no limits should be imposed on actions taken with it. This approach claims that embryos do not have a right to life, therefore, they do not need any protection (The Protection of the Human Embryo In Vitro, 2003).

The third view holds that the embryo deserves respect greater than that accorded to human tissue because of its potential to become a person and because of its symbolic meaning for many people, yet it should not be treated as a person, because it has not yet developed the features of personhood, it is not yet established as a developed individual, and may never realize its biologic potential (Hellegers, 1978; The American Fertility Society Ethics Committee, 1988, p. 34). This concept may be supported with the *gradualist* that treats both sperm and egg as living entities, and the fertilized egg as gradually developing human being. Because of that, they affirm significance of the right to life and the right to develop to embryos, though not as rights that originate from an absolute value. Due to this relativity, the gradualist position entitles variable degrees of protection afforded to embryos, progressively increasing through development, until the birth, when they became absolute. Nevertheless, through the continuous process of development, embryo's rights are challenged from the stronger rights or interests of the mother (The Protection of the Human Embryo In Vitro, 2003).

Ethos Regarding Parenthood

Parenthood is one of the major role transitions in adult life for both men and women (Demyttenaere, 2000, pp. 371-379). This means that the experience of sub fertility can be viewed as a non-event transition. A transition is defined as "an event or non event that alters the individual's perception of self and of the world, that demands a change in assumptions or behavior, and that may lead either to growth or to deterioration" (Koropatrick et al., 1993).

In Harry Frankfurt's scheme, the wish for a child belongs to the category of a constrained volitional need, because it is generated by a very strong and urgent desire that remains present until it is not fulfilled. In addition, Frankfurt stresses that the non-fulfillment of a constrained volitional need causes distress, and in case of involuntary childlessness, one suffers harm (Frankfurt, 2004, p. 104).

The strong preference by prospective parents to have their own genetic children is often expressed in terms of reproductive freedom and human rights (Chan and Quigley, 2007). The liberty to procreate can be expressed positively, as a right to reproduce, or negatively, as a right not to reproduce. According to Yvonne Denier, if someone has a right to reproduce, then society and medical experts in assisted reproductive technology would have a rights-based duty to provide that to which one has right.[5] Furthermore, she imposes the question "Can one legitimately speak of a positive right to a child? Does society indeed has a rights-based duty to prevent involuntary childlessness?" On these bases, the procreation is seen solely from the lenses of the progenitors, those that strive to fulfill their own wish for a child, and as such the perception and the role of the child is merely instrumental towards satisfying someone else's wishes and desires.

Nonetheless, this should not imply that such desire has to be fulfilled at all costs, in particular when it might harm others. Indeed, parenthood, and thus procreation, is a private matter. We will agree that nobody can interfere in the private sphere of an individual without his or her consent. Then, the decision to procreate and form a family is a basic and intimate human prerogative that requires consensus between the parties involved. However, it also happens to be fundamental to the very existence and survival of the human species.

In Western culture, parenthood is widely accepted as a matter of bilateral consensus, thus in the case one of the parties refuses to become a parent, then the negative decision should prevail over the other's desire, because nobody can impose to others a personal wish on their detriment. Hence, we must respect the free choices of other individuals. The problem appears when the decisions of others affect the growing life of the embryo. Whether this intrusion is morally justified or not will depend upon the answer on the question whether the embryo's life should be treated equally as the life of other human persons. In order to investigate whether the growing life is relevant for treating embryo as it was another human person, the following text will deal with the argument of potentiality of the embryo on its path of becoming one.

Relevance of the Biological Construction of the Embryo for its Potential as a Criterion that Stresses Respect for Growing Life

An ontological concept is required in investigating what characteristics are being *qua* possessed by embryos, and how they correlate with the different potentialities that drive them to develop into the next stage.

Logical explanation of the potentiality to develop is offered by Michael Lockwood. His thesis about transitivity of potentiality states: "if x has an active potentiality for giving rise to Y, and Y has an active potentiality for giving rise to Z, then it must follow that X itself has an active potentiality for giving rise to Z" (Lockwood, 1988; Warnock, 1985, p. 197) According to this formula, gametes, embryos, newborns and adult human beings are on the same path to develop from one to another. An *argumentum ad absurdum* then follows that gametes should be treated as persons already just because they have the possibility to develop into a human embryo.

An opposition to this notoriously false argument is offered by Gomez Lobo who affirms: "Respect for embryos does not logically entail respect for gametes" (Gomez-Lobo, 2004). In defending his standpoint, he elaborates the biological essence of both entities. Notorious fact is that biological configuration of the gametes is represented with 23 chromosomes of the sperm and ovum separately, while embryo is completed with the structure of 46 chromosomes, same as possessed by newborn, infant or an adult human person. From this point, Gomez concludes that biological code infiltrated in embryos and later in humans is what makes them the same, and grants them the same state of beings: Embryos deserve respect because they already possess potentially the features that in adults are fully actualized (Gomez-Lobo, 2004). According to Gomez, and according to all Human Being theorists[6] that grant full moral status to the embryo, the biological code of 46 chromosomes is a feature that is inherited from conception, the gradualists afford variable degrees of protection preciously kept and preserved till death, a characteristic that distinguishes one person from any other and that will never repeat again. In this sense, the Human Being theorists are right, but the question whether this biological code is the criterion for being a morally recognized human person still remains open. But if this is the case, then one can argue that every cell of our body should be respected as a human person, even when separated from the body from which originated. In these terms, Stephen Hanson disagrees with Gomez's argument, stating instead, that same (genetic) potential is also possessed by every cell in the body, and this fact does not lead us to the conclusion that every drop of our blood has the same moral status as adult humans, comparing it with the totally absurd position to consider blood transfusion as immoral (Hanson, 2006).

According to the established relation between gametes and embryos on the one hand, and between embryos and each cell of one's body on the other hand, the conclusion is as follows: Gametes cannot posses any remote potentiality to drive changes from within and to develop into an embryo, due to the lack of half of the chromosomes, while embryos and human cells have the complete chromosomal structure, and therefore, possess the same remote potentiality (Gomez-Lobo, 2005). Nevertheless, their way on the path toward human persons is incomparably different. Obviously, something is wrong with the established connection between the biological essence and the exclusivity of the remote potentiality to develop, since ascribing moral recognition and respect to every human cell as being a human is absurd.

From this perspective, one cannot oppose Lockwood's transitivity of potentiality just by ascribing different potentials for developing from gamete to embryo, and from embryo to human person due to the "basic natural capacity" as a necessary condition for possessing the remote potentiality. Nevertheless, one cannot avoid the validity of the *argumentum ad absurdum* of Lockwood's conclusion that x has active potentiality for giving rise to z, nor can one dismiss the conclusion that gametes have active potentiality to give rise to human beings as totally incongruous. Therefore, we should examine instead the meaning and application of the term potentiality as such.

Potential properties are dynamic properties always pointed towards some future events that determine their development into what they are designed to be. According to Mark T. Brown, "potential has consequential moral value, not because of what it is, but because of what it can become" (Brown, 2007). Western philosophical tradition has always associated personhood with qualities already captured in the observed persons, and therefore an existing organism cannot become a person, because it lacks those qualities. Ascribing potentiality to an embryo is just a pathway to get to the aim. If the aim is becoming a human person, then the argument on potentiality is just a means to get there, even though is obvious that the aim is in the future and not in the present moment.

H. T. Engelhardt argues: "If X is potential Y, it follows that X is not Y. Consequently, it follows that X does not have the actual rights of Y, but only potentially rights of the Y (Engelhardt, 1989, p. 171). If an embryo is only a potential person, and not yet a person, then it follows that the embryo does not have the rights of a person, but only the potentiality to gain the rights of a person, once it becomes a person. Therefore, if X has the potential to become Y, it does not follow that we can treat X as if it was already Y. In these terms, a doctoral student that "has potentiality" to become a professor, does not hold the rights and duties of the professor he or she has the "potential" to be. While "having potentiality" describes the possession of features already captured in the entity, being a reason that grants possibility for growth and change, "being a potential" is a static feature of the current moment that very precisely describes what one is not. If possession of potentiality is pointed towards becoming something that now is not, it is very clear that the dynamic force of potentiality is just a way of becoming something that one strives to become, as guided by already determined agenda. We may decide to treat X as it was Y, supported with a good reason for doing so, but still the decision will fail to be logically supported, since it will remain only a political act of consensus.

This elaboration clearly depicts that the future moment of development is desired, but not actualized yet.

From this argument, the conclusion follows that even if the embryo possesses intrinsic biological properties as a driving force to become what at the moment it is not, as prescribed merit to its potentiality, still the argument of advanced qualification in the personhood world is forced and its argumentation weak. Biological configuration of 46 chromosomes of the embryo is a necessary condition and initiator for its growing potential to be realized, but not sufficient one to treat the embryo as person already.

From here on, the problem of our ethical deliberation appears at another level. Once we have the essence of what is considered to be a growing life, but not a person already, the ethical question requires us to investigate when this growing life appears to matter morally. According to John Harris, "life itself does not begin at fertilization, since egg and sperm are alive also (Harris, 1985). Life continues, and so what we need is not an account of when life begins, but of when life begins to matter morally." Therefore, the question is not whether life has begun at the stage of being an embryo, but rather whether that life matters morally, as do the lives of all other human persons. That is the reason why the remaining part of the paper will try to see how the life of the embryo matters, and which are the components relevant to make it matter. For the sake of completing the argument on potentiality and its application on embryos, the text that follows will compare its manifestation of the very same biological entity, but in two different settings, namely in vitro and in vivo. Then the paper will pose some questions regarding the relevance of the context.

Relatedness and Discrepancies Among In Vivo and In Vitro Embryos

The value and meaning of the process of becoming a human person is being initially triggered in the discussions over embryo's protection in abortion cases. With assisted creation of the embryos the "final product" is the same, in terms of holding the same biological construction, though the process is facilitated from in vitro technology, and therefore the creation is just "assisted" and not "artificial." This fact does not allow us to draw a conclusion that in vitro embryos are being "less natural" than those created by coition, but the relevance of the question that confronts us is inevitable: Is the potentiality of those embryos during the facilitating process still on equal moral footing with those already created and implanted in the woman's womb with the very same assistance? Is the process of assistance part of the internal driving potentiality of embryos, or is it the final outcome of the implantation that grants them with such a capacity?

Even if one starts from the position of Gomez attributing biological properties to embryos, is the "necessary condition" for being a human person at the same time a sufficient one to prescribe same potentiality to embryos in different settings, namely in vivo and in vitro (Gomez-Lobo, 2004)? Gomez conflates the meanings of active and remote potentiality; however, from Aristotelian point of view, their similarities are only in their intrinsic nature, while the other element that grants them with significant differences, and is crucial when observing the potentiality of the embryo, is cast away. According to Aristotle, active potentiality refers to the independence from the external causes and as such, changes are driven from within and are sufficient for further development without any other external facilitators (Aristotle, 1936). Could one state that in vitro frozen embryos can trace their path towards development into human person without the assistance of complex in vitro procedure that includes their artificial preservation in frozen condition, disruption of their frozen state and reversal process to their normal condition, nutrition and implementation in the women's womb as steps prior to being in vivo embryos? What is important for an in vivo embryo to develop is the current situation to remain the same; therefore, none of the external factors has to be modified from the position where they are. On the other hand, in vitro embryos have to change their environment if they are to develop.

Stephen Hanson argues that embryos in frozen state have no meaning in preserving their current condition since at that point they are neither capable for growth, nor for development over time (Hanson, 2006). According to him, their dynamics is being paralyzed by the fact that they are frozen and their developmental process has been "switched off." Therefore, the need for external "activator" is of crucial importance for activating their potentiality in the first place.

Monika Bobbert's attempt to stress the argument on potentiality as powerful enough to establish recognized personhood of embryos draws a very clear distinction between in vitro and in vivo embryos, but the same time counter strikes her own argument. Her observation that "embryos are self purposeful in a reproductive context, because they have the status of potential human being who could have developed an existence," depicts very clearly what the difference is between embryos in different contexts (Bobbert, 2006). This paper is in line with the statement that "the context in which an embryo develops is crucial for its potentiality" (Bobbert, 2006). Nevertheless, the problem appears again at the level of who determines the context. The meaning of the "reproductive context" is variable in vivo and in vitro circumstances, due to the tendency for the context to be lost during in vitro fertilization once the consent of its progenitors is withdrawn or re-directed, the purpose accomplished and spare embryos still remained, or even in the case of extreme approach when embryos are being created completely outside any "reproductive context."[7]

One might argue that even in vivo embryos are not self sufficient but dependent on the biological environment of the mother. What makes the dependency different is precisely the unity of the embryo and the mother established with the biological tie so that physical separation at this level cannot be claimed. Therefore, if a situation like this continues to exist, the potentiality for development will be executed, and no additional external changes are needed.

With regard to the dependency of the nutrition provided in the natural in vivo environment, Mary Mahowald states that the biological tie is not equivalent with the genetic tie and the "actual connection between a woman and an embryo is as crucial to the latter's biological potentiality to become a person as is the fusion of sperm and ovum" (Mahowald, 2004). According to her, only if successful human gestation is accomplished *ex corpore,* or within a non-human's body, can the genetic code, all together with the biological configuration, be sufficient for possessing active potency.

This paper does not reject the fact that biological construction in both settings is the same. Nevertheless, one cannot rely on biological construction alone, as independent from the context in which exists. The environment is not an additional element to the biological configuration of embryos, but is an essential component of their becoming.

Therefore, the problem is shifted once again to another level. It is not the embryo as such but the environment in which it exists and flourishes that is the field where the discrepancies arise in first place, and that is where the further argumentation should be focused.

BROADER INTERPRETATION OF THE EMBRYO'S POTENTIALITY: THE CONTEXT OF REALIZATION OF THE POTENTIALITY

Observing the life of the embryo in an isolated *per se* framework appeared to be almost impossible due to the dependency of the potentiality to develop from its context. In these terms, The Report of the Council of Europe on Protection of the Embryo in Vitro emphasized the necessity of the wider social consideration as a background to medically assisted procreation, and therefore warned that protection of embryos in vitro should not be seen in isolation (The Protection of the Human Embryo In Vitro, 2003).

The external factors are not limited by environmental elements, essential from biological point of view, but enriched with the context in which the purpose to execute different human practices is embedded, as well as with the complex network of parties that participate with their interests and cause direct clash of moral rights between confronted fronts of those that claim to enjoy such moral rights. In same line, Devolder and Harris affirm that "solely dependence on an entity's inherent dynamic to become a human person ignores the immense importance of diverse external factors that play a role in the actualization of this potential, as well as the substantial differences in potential at various stages of development" (Devolder and Harris, 2005).

Different Human Practices in the Context of Realization of the Potentiality of the Embryo

Ontological philosophical approaches deal exclusively with the intrinsic structure of the subject of perception. This kind of observation appears to be too narrow for ethical deliberation over the status of the embryo, especially if the argument of potentiality is the key concept. Potentiality as such strives to develop into something that it is not at the moment. Martin Heidegger embedded the embryo in the "totalities of relevance" (Heidegger, 1996). For him, the ontological and ethical status of the embryo is determined by the way human activities shape the embryo's place in a horizon. Therefore, Fredrik Svenaevs (2007) interprets phenomenological tradition of Martin Heidegger, urging for phenomenological back-up for potentiality argument in the determination of the embryo's destiny. According to him, the context is always predetermined by the human practice and the goal of the embryo will depend upon the goal of the human practice as such. This approach insinuates that the purpose of the human practice tends to appear as an arrow that determines the final goal of becoming. In a situation like this, the embryo's moral significance is predetermined by the attitudes established towards it, in the course of specific human activities. He even goes further by interpreting Aristotle's active and passive potentialities from the perspective of their predetermined goals, ascribing active potential to things that when left alone can develop in predictable way, and ascribing passive potential to embryos, as entities that can develop in a number of different ways depending on what will happen to them. In these terms, he identifies in vitro embryos with stem cell lines, due to the reasons that both must be combined with concrete activities, in which humans play part and predict their goal of development. Although this point of view might be considered as instrumental, Svenaevs justifies his way of reasoning under the cover of "phenomenological life-world tradition." He argues within Martin Heidegger's concept of "totality of relevance," that being a tool means more than being an instrument. In this sense, the observed entity acquires meaning in the totality of relevance, but in relation with the practices in which is engaged. He transferred the internal ontological reality of the embryo into the phenomenological world of practices and culture as being totally interdependent and influential to one another. This way of reasoning, according to him, is on the same line with Aristotelian *phronesis* practical wisdom that articulates the goals of a practice central to the making of the ethical choices.

Even the above elaborated Bobbert's argumentation over independence of embryos within reproductive context gives rise to an effect opposite to the one that was meant to be achieved (Bobbert, 2006). It clearly indicates that the concept can be reproductive but also can be other, and that the development is dependent precisely on the existing context as determined by the human practices.

Through accepting the existence and importance of different human practices in relation with embryo's determination, one leans towards phenomenological outer world of embryo, as opposite to embryo's ontological predisposition as an isolated island, self-sufficient for survival, development and elaboration, and therefore respectful for life. Accepting the middle ground in between these two extremes is facilitated by the notion of transferring the harsh instrumental role into the role that attributes a meaning, and therefore the context in which the role is played determines the position of the particular embryo under investigation.

The Potentiality of the Embryo in the Context of the Other Concerned Protagonists

The developmental energy of growing from one form to another in an interrelated world is itself also interrelated. One thing does not develop into another in discrete moments, but rather in a process. During this process, the "thing" may participate in the modification under its own intrinsic energy, or it may be modified by others; however, it is almost impossible in an interrelated world, such as the one in which we live, to find its independence. Whenever life exists, it is embedded within an interconnected world in which we confront interests and values of others.

Joel Feinberg proposes the "interest principle" as a suggestion for interpretation of the question "What kind of being can have moral rights?" (Feinberg, 1980). According to his "interest view," we owe moral duties to only and all those beings who are capable of having interests, and no other being can be harmed in a morally relevant way. In the discussion about the embryo's moral perplexity, it cannot be avoided that interests also possessed by its progenitors. In such a structure, avoidance of the existence of the interests of all other protagonists here is equivalent to avoiding the moral discussion over the embryo's place in the horizon, in the first place.

Deckers proposes a double moral standard, in dependence of the justifiable reasons and beneficiaries that might be gained as measured with against that might be lost (Feinberg, 1980). He stresses that "there is a morally relevant difference between sacrificing an embryo for the sake of those who may benefit from the embryo's research, and sacrificing an embryo to save the mother."

It might seem that having a double moral standard for embryos in different contexts does not offer neither unique and equal, nor profound moral deliberation, but on the other hand, not having one will mean total ignorance of the progenitors and the rest of the living world in which the embryo strives to live in. It follows that the moral deliberation of the embryo is interconnected and as such cannot be observed in isolation from all the other participants. In a situation like this, favorable moral recognition of the embryo will infringe upon moral rights of the human persons that are not external actors but intrinsically part of the same mosaic.

Potentiality of the very same embryo, the one with firm biological configuration of 46 chromosomes, is once again challenged by the context of its existence. This external virulence may infringe its potentiality in many different ways, as well as can promote it. The possibility of such disruption in the context of the other parties concerned cannot be ignored, and the moral acceptability will have to deal with promotion of best interests of the overall situation in the context in which appears. Once again, human practices in the totality of relevance appear to be determinant for appointing the direction of development.

Bridging Both Observations: The Isolated Embryo and the Context of the Other Protagonists

Mark T. Brown's internal scale for ascribing moral prerogatives may be very versatile and therefore a useful formula in ethical deliberation over the embryo. According to his concept of moral status (Brown, 2007), the following text will navigate the moral position of the embryo in constellation with, on the one hand, its peers group—zygotes as antecedents, and newborns and adult human persons as its descendants from their potential to develop, and on the other hand, with the other protagonists—potential mother and father, as immediately concerned with the outcome.

Brown's Internal Scale of Moral Values with the Peers Group of Zygotes, Embryos, Newborns and Adult Humans from their Potentiality to Develop from One to Another

Potentiality as such urges for a comparison among the different stages of the development, which in the case of embryo stretches from what once was a less developed biological form to what now may be a much more developed one. Therefore, comparison within its peers group in the process of inevitable interaction might purify our moral sight. Furthermore, the proposed scale ranks the moral capacities of the protagonists that are already human persons and of those that manifest life, by the power of development from one to another. The argument of potentiality facilitates their connectedness and belonging on the same internal scale, bridging two different worlds, i.e., the world of those that are "still developing" in a dependent way, and the world of those that are already developed and independent human agents, separated from the body of their "facilitator."

The embryo is located around the middle part of the developmental scale, between the position of the zygote (which lacks even the biological essence), and the newborn. The adult person at this stage might only be an imaginary goal of the development.

The boundary between the zygotes, which are subjects towards which humanity does not owe the same obligations as to human persons on the one hand, and the newborns and adult humans as subjects deserving the moral respect on the other, is delineated by the following two components: 1) the biological construction of 23/46 chromosomes described in the first chapter, and as a trigger component for launching the argument on potentiality in the discussion, and 2) the set of interactions with the other already recognized moral subjects in which the zygotes cannot participate, and humans are automatically participating solely by virtue of their being. While enjoying respect and protection by virtue of their being, human persons have the obligation to restrict themselves from infringing upon other morally recognized subjects that hold the same moral rights, in order not to offend the same moral status of the other human persons.

The embryo, as a transitional form from being a zygote to human person, has fulfilled the first necessary element of the biological essence, but still lacks the capacity to execute moral obligations. In these terms, Alicia Przyluska–Fiszer states: "biological essence is considered a necessary condition but not a sufficient one, while rationality is usually treated as a sufficient but not necessary a condition for moral standing" (Przyluska-Fiszer, 1996). Consequently, the question that is imposed in the discussion is as follows: On which grounds may we be justified in maintaining that moral obligation is owed to the embryo? One might argue that a newborn does not have the same capacity to execute moral obligations as well, but the only evident difference with the capacity of embryo, under the same circumstances, is that the newborn is already separated from the body of the woman that carried it, and as such, moral obligations are not owed to the woman with competing interests in relation to her offspring any more. Therefore, obligations owed to the newborn are being already differentiated to someone that is a unique individual, while on the other hand, obligation owed to the embryo is still under postponed condition–its development and further individualization. This condition is less physical in terms of separation of the woman's body, and more metaphysical in terms of referring to the connotation of identifying embryo's and mother's interests into mother's only, and separating them into embryo's and mother's, respectively.

When applying Brown's internal scale of moral values, one might conclude that the privilege to owe obligations to zygotes is not justified, while one is obligated to respect the newborns. The embryo, lying on the middle path of development, should be respected as a growing life when its own biological possessions are supported with additional elements, namely the reproductive context to execute the potentiality within the reproductive purpose, and only when the overall situation will grant rights to them that outweigh those of their progenitors. That would be the case when, within an in vivo environment, the embryo claims greater rights over the rights of the woman that carries it, or within an in vitro environment, when the best interest of the overall situation will balance the beneficence of such a growing life with the autonomy of the progenitors. However, the potentiality argument is not sufficient to ascribe personhood prematurely to the embryo, due to its agenda of only becoming one in the future dimension of existence.

Can the Life of a Human Embryo (Potential Person) Outweigh the Free, Private and Autonomous Decision of an Actual Person to Become/not to Become a Parent?

For Søren Holm what determines the moral importance of an entity is not only the properties of the entity, but also its relationship with the other entities (Holm, 1996). On the same line are Simon Beteman and Tania Salem, who state that the number of protagonists in the conception of the embryo is already a network, even more in the case of assisted reproduction where the number of actors conceiving the embryo increases. They even distinguish the number of parties involved in assisted fertilization from the dual number in a sexual intercourse. While the natural conception is accomplished by the couple only, reproductive technologies include "unprecedented protagonists into the network, couple, physicians, and biologists, sometimes even donors contributing genetic material or physical processes." They furthermore state that "each protagonist puts forward different criteria to justify and establish its relationship with the embryo, the infertile couple–their genetic link, and the woman–her bodily implication in the embryo's coming to life" (Novaes and Salem, 2003). Therefore, the actualization of the potentiality of the embryo is at least dependent at two levels: 1) parental level compounded of genetic parents, man and a woman, and 2) woman's only level, additionally supported with her bodily integrity.

When applying Joel Feinberg's interest criterion for ascribing moral relevance to a subject, what strikes attention is that all of the participants in the story of the embryo already have well established interests that strive to protect and even more, are entitled to do so (Feinberg, 1980). On the woman's only level, their interest does not manifests in a form "my right against yours" or "the right of a human person against a right of another human person," but are intrinsically interrelated in a way one cannot divide them as being separate. The following example should illustrate the point: Abortion decision of the woman made upon estimating her poor chances of survival, as prevailing to her wish to bring the child into the world. The pregnant woman is already a moral agent that represents her own interests and the interests of the embryo at the same time, as intrinsically connected with her own. Under these circumstances, the interest of the embryo is interconnected, dependent and represented by the mother. The mother not only represents hers and the interests of the embryo, but both interests coincide into one.

Furthermore, Joel Feinberg observes that the concept of potentiality is too "promiscuous" (Feinberg, 1974). to be used as a universal point making outcome, backed up with Aristotelian premises that any matter is potentially anything (*Metaphysics* IX.7, 1048b35-1049b1). Because of that, he distinguishes between physical possibility, as in accordance with the law of nature, and logical possibility, as additionally needed to actualize the existing possibility. In these terms, whenever human rational agents are involved, the process of actualization depends upon exercising discretion in the decision making. Therefore, on the parental level, whenever embryos are created in an artificial setting, their reproductive or non reproductive context is predetermined by the human practices in which it is engaged by the parents. Whenever the legitimacy of their right to procreate, as positively or negatively expressed, is respected, the actualized potentiality will depend on their decision. In these terms, the parents are the main agents who determine the context, and therefore, the cradle of totality of relevance of the embryo, not as an omnipotent being who decides about life destruction, but as parents who decide about life creation of their offspring.

In a situation like that, it seems as if the obligations that are being owed at the parental level to the potential mother and father as potential parents, and at the woman's only level, to the woman that carries the embryo inside her own body, are prevailing over the obligations owed to the embryo. Obligations owed to them in the discussion over the embryo cannot be waived, but instead have to be reconsidered together with the rights that the embryo strives to claim for. The embryo's location on the scale will be dependent upon the context in which is embedded by the main protagonists: progenitors. Therefore, obligation is owed to the embryo only when within the reproductive context its rights are not confronted by the greater rights of its progenitors.

CONCLUSION

In ancient Greece, the very broad and ambiguous term of life was specified and narrowed down into two categories as described by two words: *zoe*, with meaning of physical or biological life, and *bios*, with meaning of life as 'lived, and made upon actions, decisions, motives and events that compose what we now call a biography' (Dworkin, 1994, referring to distinction made by Rachels and Ruddick, 1989 and explored in Rachels, 1986, pp. 24-27). In order to discuss the value of life of the embryo, this paper analyzed both of the terms as used in ancient Greece: the biological configuration of the embryo, and its contextual constellation, accompanied by the potentiality to develop. The eternal energy to develop and grow, as ascribed to potentiality, is part of the never ending *perpetuum mobile* process of life as energy that exists and modifies constantly. Therefore, life beginning and life ceasing should be considered as processes of this transformation, not a simple moment. What should morally matter in the deliberation over the embryo's importance to be respected and protected, should not be reduced to the moment of life beginning, or in its biological configuration, supported with the potentiality to develop, but rather should be contextualized in the world in which exists, where its own *tabula rasa* biography is about to be written, as dependent on the overall context in which it exists.

The problem was lifted up on two levels: the ontological one, where the biological structure of the embryo claims potentiality for development *per se*, and the phenomenological one, where the embryo manifests its potentiality in a relation with its own world. Conclusions are made in the direction that the potentiality of the embryo is not a sufficient argument to claim that it should be treated as if it were already a human person, even more when such obligations are interrelated with its progenitors as autonomous human persons. On the contrary, any separation of the intrinsic and extrinsic components in the embryo's deliberation is narrow in scope and furthermore myopic for the consequences that might invoke. Nevertheless, embryonic life should be respected when the best interest of the overall situation will balance the beneficence of such a growing life with the autonomy of the progenitors within the reproductive context.

REFERENCES

Aristotle (1936). *Metaphysics. (Tredennick Transl.)*. Cambridge: University Press.

Aristotle (1941). *Metaphysics.* (W. D. Ross Transl.). In McKeon, R. (Ed.), *The Basic Works of Aristotle*. New York: Random House.

Aristotle (1999). *Nicomachean Ethics. (Terence Irwin Transl.)*. Indianapolis, IN: Hackett.

Bobbert, M. (2006). Ethical questions concerning research on human embryos: embryonic stem cells and chimeras. *Biotechnology Journal, 1*, 352–1369. doi:10.1002/biot.200600179

Brown, M. T. (2007). The potential of the human embryo. *The Journal of Medicine and Philosophy, 32*, 585–618. doi:10.1080/03605310701680973

Chan, S., & Quigley, M. (2007). Frozen embryos, genetic information and reproductive rights. *Bioethics, 21*(8), 439–448. doi:10.1111/j.1467-8519.2007.00581.x

Council of Europe. (2003). *The Protection of the Human Embryo In Vitro. Report by the Working Party on the Protection of the Human Embryo and Foetus.* CDBI –CO- GT3, 13. Strasbourg.

Demyttenaere, K. (2000). Anxiety and depression in subfertility. In Meir Steiner, Kimberley A. Yonkers and Elias Eriksson (Eds.), *Mood Disorders in Women* (371-379). London: Martin Dunitz.

Denier, Y. (2006). Need or desire? a conception and moral phenomenology of the child wish. *The International Journal of Applied Philosophy, 20*(1), 81–95.

Devolder, K., & Harris, J. (2005). Compromise and moral complicity in the embryonic stem cell debate. In N. Athanassoulis (Ed.), *Philosophical Reflections on Medical Ethics* (88-108). Palgrave Macmillan.

Dworkin, R. (1994). *Life's Dominion: An Argument about Abortion and Euthanasia.* Vintage.

Engelhardt, H. T. (1989). The context of health care: persons, possessions, and states. In Beauchamp, T., & Walters, L. (Eds.), *Contemporary Issues in Bioethics* (3rd ed.). Belmont, CA: Wadsworth.

Feinberg, J. (1974). The rights of animals and future generations (appendix: the paradoxes of potentiality). In W. T. Blackstone (Ed.), *Philosophy and Environmental Crisis* (67-68). Athens, GA: University of Georgia Press.

Feinberg, J. (1980). Abortion. In T. Regan (Ed.), *Matters of Life and Death* (183-216). Second Edition. New York: Random House.

Frankfurt, H. (2004). *The Importance of What We Care About.* Cambridge University Press.

Gomez-Lobo, A. (2004). Does respect for embryos entail respect for gametes? *Theoretical Medicine, 25*, 199–208. doi:10.1023/B:META.0000040038.52317.08

Gomez-Lobo, A. (2005). On Potentiality and Respect for Embryos, A Reply to Mary Mahowald. *Theoretical Medicine and Bioethics, 26*(2), 105–110. doi:10.1007/s11017-005-1235-9

Guyton, A. C., & Hall, J. E. (1999). *Textbook of Medical Physiology.* 10th Ed.

Hanson, S. S. (2006). More on Respect for Embryos and Potentiality: Does respect for Embryos Entail Respect for In Vitro Embryos? *Theoretical Medicine and Bioethics, 27*, 215–226. doi:10.1007/s11017-006-9001-1

Hanson, S. S. (2006). More on respect for embryos and potentiality: does respect for embryos entail respect for in vitro embryos? *Theoretical Medicine and Bioethics, 27*(3), 215–226. doi:10.1007/s11017-006-9001-1

Harris, J. (1985). *The Value of Life.* London: Routledge.

Heidegger, M. (1996). *Being and Time. (J. Stambaugh Transl).* Albany, NY: State University of New York Press.

Hellegers, A. (1978). Fetal development. In T. L. Beauchamp (Ed.), *Contemporary Issues in Bioethics* (194-199). Encino, Calif.: Dickenson.

Hollowell, K. (1999). *Cloning: Redefining when Life Begins. Exposing Flaws in the Pre embryo-Embryo Distinction.*

Holm, S. (1996). The moral status of the pre-personal human being: the argument from potential reconsidered. In *Conceiving the Embryo, Ethics, Law and Practice in Human Embryology* (193-220). Martinus Nijhoff.

Interview with Victoria M. Sopelak, Ph.D. (2002). University of Mississippi Medical Centre (UMC), June, 11, 2002.

Koropatrick, S. (Ed.). (1993). Infertility: a non-event transition. *Fertility and Sterility, 59*, 163–171.

Langley, L. S., & Blackston, J. W. (2006). Sperm, egg and a petri dish: unveiling the underlying property issues surrounding cryopreserved embryos. *Journal of Legal Medicine*, 167–206. doi:10.1080/01947640600716408

Lockwood, M. (1988). Warnock vs Powell (and Harradine): when does potentiality count? *Bioethics, 2*, 188–213. doi:10.1111/j.1467-8519.1988.tb00048.x

Mahowald, M. B. (2004). Respect for embryos and the potentiality argument. *Theoretical Medicine and Bioethics, 25*(3), 209–214. doi:10.1023/B:META.0000040065.84498.4c

Malo, P. E. (2000). Deciding custody of frozen embryos: many eggs are frozen but who is chosen? In Care Law, H. (Ed.), *J. DePaul* (pp. 307–312).

Novaes, S. B., & Salem, T. (2003). Embedding the embryo. In J. Harris and S. Holm (Eds.), *The Future of Human Reproduction, Ethics, Choice, and Regulation* (101-126). Oxford.

Prodanov, V. (1996). Cultural pro-attitudes, reproductive ethics and embryo protection. In Evans (Ed.), *Conceiving the Embryo, Ethics, Law and Practice in Human Embryology*. The Hague/London/Boston: Martinus Nijhoff Publishers.

Przyluska-Fiszer, A. (1996). Human embryology and the criterion of moral standing. In *Conceiving the Embryo; Ethics, Law and Practice in Human Embryology.* Martinus Nijhoff: 165-172.

Rachels, J. (1986). *The End of Life*. Oxford: The Oxford University Press.

Rachels, J., & Ruddick, W. (1989). Lives and liberty. In Christman, J. (Ed.), *The Inner Citadel: Essays on Individual Autonomy*. New York: Oxford University Press.

Steiner, H. (1994). *An Essay on Rights*. Oxford: Blackwell.

Svenaeus, F. (2007). A Heideggerian defence of therapeutic cloning. *Theoretical Medicine and Bioethics, 28*, 31–62. doi:10.1007/s11017-007-9025-1

The American Fertility Society Ethics Committee. (1988). Ethical considerations of the new reproductive technologies: biomedical research and respect for the pre-embryo. In *'Fertility and Sterility: 'Official Journal of the American Fertility Society,* 49(2) suppl. 1, The Protection of the Human Embryo In Vitro. (2003). Report by the Working Party on the Protection of the Human Embryo and Foetus. CDBI–CO- GT3, 13, Council of Europe, Strasbourg, 19 June 2003.

Warnock, M. (1985). *A Question of Life: The Warnock Report on Human Fertilisation and Embryology*. Oxford: Blackwell.

ENDNOTES

1. As written by Aristotle: "decision rests with perception." See Aristotle (1999), II,9: 1109b 24; IV,5: 1126b 4-5.
2. According to Brown (2007), "the intrinsic properties of the entity itself justify the ascription to moral status." Further on, Brown affirms that "a hierarchy of moral status merely reflects a commitment to a background hierarchy of intrinsic moral values." Therefore, Brown refers to moral status as a general claim about "how moral agents ought to conduct themselves toward entities that have certain kinds of intrinsic properties," that implies their moral recogni-

tion and attitude from/to already established moral agents, and therefore examines their moral position.

3 Meaning newborn from foetus, foetus from embryo, embryo from zygote.

4 The egg is being extracted from the women and laboratory fused on a Petri dish with the sperm. The created embryo divides in approximately 8 cells. These cells can be implanted in a precise day of the women menstruation cycle and afterwards the process takes the same path as In Vivo process. Even more, they can be cryopreserved, frozen in liquid nitrogen, safely preserved in a suspended biological state, kept in containers theoretically forever, but it is considered that their quality diminishes gradually after 5 years. Interview with Victoria M. Sopelak, Ph.D., Associate Professor of Obstetrics and Gynaecology at University of Mississippi, Medical Centre (UMC), June, 11, 2002. See Langley and Blackston (2006, p. 167)

5 Denier (2006) states that "[a] positive right is generally defined as 'a right to.' To this right corresponds a duty to provide that to which one has a right. Opposed to this, there is a negative right: "a right not to be harmed." To this right corresponds the duty to protect this legal area in order to legally obstruct illegitimate interference."

6 Human being theorists hold firmly on the intrinsic biological essence of all members of the group of H*omo Sapiens* a threshold condition for ascribing moral status to an entity. Therefore, for them a newborn or an infant is a human being deserving respect and protection just because of its unique DNA as a hallmark of the human kind, and in spite of the cognitive capacities. On the other side, Person theorists, represented by Utilitarian and Kantian traditions, glorify rationality, as a cognitive capacity standard for full moral status as a person. For Kant, personhood is absolute and possessed equally by all rational beings and only by them. According to this view, newborns do not fulfil the condition of being rational, and therefore, fail to classify as being worthy for moral respect as the rest of the other human persons.

7 If embryos are self sufficient only in a reproductive context, than it would mean that they are not self sufficient in a different context. Does that imply that one is authorized to avoid the reproductive context and generate embryos for other purposes?

Chapter 5
Direct-to-Consumer Genetic Testing

Richard A. Stein
Princeton University, USA

ABSTRACT

The chapter discusses the benefits and concerns associated with the availability of genetic tests that are commercialized directly to consumers, in a practice that became known as direct-to-consumer genetic testing. While it is every individual's right to learn about genetic predispositions that could shape their future, a number of concerns emerge, and include the clinical relevance of specific gene-disease associations, the sensitivity and specificity of genetic tests, the involvement of health professionals, the availability of counseling, and the danger of genetic determinism. The implications of incidentaloma, where a genetic test conducted for one purpose turns out to yield additional information, originally not sought and perhaps not even intended to be known, are also discussed. The author underscores the importance of educational campaigns, targeting professionals and the general public, in providing a better understanding of the relationship between genetics and disease susceptibilities.

INTRODUCTION

The 1953 discovery of the DNA double-helical structure by James Watson, Francis Crick, Maurice Wilkins, and Rosalind Franklin, represented one of the most significant advances in the biomedical world (Watson and Crick 1953; Maddox 2003). Almost half a century after this landmark event, in February 2001, the initial draft sequences of the human genome were published (Lander et al., 2001; Venter et al., 2001) and, in April 2003, the International Human Genome Sequencing Consortium reported the completion of the Human Genome Project, a massive international collaborative endeavor that started in 1990 and is thought to represent the most ambitious undertaking in the history of biology (Collins et al., 2003; Thanga-

DOI: 10.4018/978-1-61692-883-4.ch005

durai, 2004; National Human Genome Research Institute). The Human Genome Project provided a plethora of genetic and genomic information that significantly changed our perspectives on biomedical and social sciences. The sequencing of the first human genome was a 13-year, 2.7-billion-dollar effort that relied on the automated Sanger (dideoxy or chain termination) method, which was developed in 1977, around the same time as the Maxam-Gilbert (chemical) sequencing, and subsequently became the most frequently used approach for several decades (Sanger et al., 1975; Maxam & Gilbert, 1977; Sanger et al., 1977). The new generations of DNA sequencing technologies, known as next-generation (second generation) and next-next-generation (third generation) sequencing, which started to be commercialized in 2005, enabled the cost-effective sequencing of large chromosomal regions during progressively shorter time frames, and opened the possibility for new applications, such as the sequencing of single-cell genomes (Service, 2006; Blow, 2008; Morozova and Marra, 2008; Metzker, 2010).

While $10 allowed the sequencing of a single base in 1985, approximately 10,000 bases were generated, at the same cost, two decades later (Shendure et al. 2004; Pettersson et al. 2009). Next-generation sequencing methodologies, in addition to their lower costs, enabled large chromosomal regions to be sequenced in shorter time frames – for example, over 100 million base pairs can now be read, in 200-400 base-pair fragments, in as little as 4 hours (Imelfort et al., 2009).

As a result of these advances, it took only two months in 2008, and less than 1 million dollars, to sequence a diploid human genome, that of James Watson (Wheeler et al., 2008). All these developments promise to bring the $1,000 genome goal, an important milestone of personalized medicine, closer to reality.

The Human Genome Project revealed that 20,000-25,000 genes are encoded in the genome, less than the approximately 50,000-100,000 that were predicted a few years earlier (Fields et al., 1994; International Human Genome Sequencing Consortium, 2004). As a result of gene-disease and gene-phenotype connections that are continually unveiled, genetic testing has experienced an unprecedented development.

INTER-INDIVIDUAL GENETIC VARIATIONS

The Human Genome Project unveiled similarities and differences at the DNA level within and between populations. Most importantly, it revealed that two unrelated individuals are approximately 99.9% identical at the DNA level (Shastry, 2002; Feuk et al., 2006). A very interesting part of the genome revolves around the remaining 0.1%, which constitutes inter-individual differences. Several million such differences, which occur approximately once every 300-1000 base pairs, exist between two random genomes, and they became known as **S**ingle **N**ucleotide **P**olymorphisms (SNPs) (Brookes, 1999; Manolio et al., 2008). SNPs were linked to many medical conditions including diabetes, cancer, and asthma, and some of them increase, while others decrease the risk to develop a specific disease (Kaklamani et al., 2008; Ionita-Laza et al., 2009; Kilpivaara et al., 2009). To characterize genomic variations and provide a public database of common genome-wide sequence variants, the International HapMap Project was initiated in October 2002 (The International HapMap Consortium, 2003). Its second-generation map, published in 2007, reported over 3.1 million SNPs, and provided a valuable resource to study gene-disease interactions (The International HapMap Consortium, 2007).

While SNPs were previously thought to represent the only source of inter-individual genetic variation, it was recently revealed that various genes or chromosomal regions exhibit inter-individual variations in their copy numbers, and this became known as **C**opy **N**umber **V**ariation (CNV). For example, a cluster of several antimicrobial β-defensin genes was shown to be present in 2 to 12 copies per diploid genome in the chromosome of different individuals, and this variation is thought to shape the inter-individual differences in susceptibility to infectious diseases (Hollox et al., 2003). CNVs range in size from 1 kb to 3 Mb (Feuk, 2006; Freeman et al., 2006; Kehrer-Sawatzki, 2007) and are thought to account for more genomic variation, as far as the number of nucleotides is concerned, than all SNPs combined. Recent estimates predicted that two human genomes might differ from each other by as much as 20 million bases due to CNVs alone, which would thus account for approximately 5 times more variation than all SNPs together (Kehrer-Sawatzki, 2007). CNVs were described in almost 3000 genes (Kehrer-Sawatzki, 2007), and while some of them do not appear to be associated with disease, others were linked to conditions that include cancer, autism, schizophrenia, epilepsy, and amyotrophic lateral sclerosis (Gonzalez et al., 2005; Freeman et al., 2006; Blauw et al. 2008; The International Schizophrenia Consortium, 2008; Ionita-Laza et al., 2009). Copy number variations in the gene encoding complement component 4 were shown to modulate the risk to develop systemic lupus erythematosus, and several studies linked copy number variations in *CCL3L1,* encoding a chemokine, to HIV susceptibility. Lower *CCL3L1* copy numbers were associated with increased susceptibility for acquiring HIV-1 and with more rapid progression to AIDS, while higher copy numbers were reported to reduce the mother-to-infant *in utero* transmission of the virus (Kuhn et al., 2007; Yang et al., 2007; Nakajima et al., 2008). Lower *CCL3L1* copy numbers compared to the population median were also shown to predispose to hepatitis C infection (Grünhage et al., 2010).

In addition to shaping various medical conditions, SNPs and CNVs are also thought to explain why certain individuals are more sensitive to environmental chemicals, or develop more frequent or more severe adverse effects to medications. A recent study examined copy number variations in *CYP2D6*, encoding a hepatic enzyme involved in the metabolism of several drugs, and reported duplications of this gene in almost 10% of patients who do not respond to antidepressant treatments, but in only 0.8-1% of the general population (Kawanishi et al., 2004). Inter-individual genomic variation emerges as a promising tool for guiding treatments. High copy numbers of the *HER2* gene enhance sensitivity to gefitinib therapy in a subset of patients with non small-cell lung carcinoma; *UGT1A1* polymorphisms were able to predict medication-induced liver toxicity during the phase III clinical trial of an investigational cardio-protective compound; and a polymorphism in the *CCR2* chemokine receptor gene was linked to the risk of renal allograft rejection (Danoff et al., 2004; International Human Genome Sequencing Consortium, 2004; Cappuzzo et al., 2005). Moreover, inter-individual genomic variation has also become an important tool in exploring the molecular mechanisms of chromosome rearrangement during evolution (Murphy et al., 2005; Shastry, 2007).

As genome-wide association studies constantly generate large amounts of data, and new associations are reported on a monthly basis, the challenge is shifting towards the need to better understand the significance of the various findings and to reliably identify the ones that are biologically meaningful and clinically promising (McCarthy et al., 2008; Ioannidis, 2009; Zeggini and Ioannidis, 2009). At the same time, it is becoming increasingly clear that most common diseases are being shaped by multiple genetic factors, each contributing to a limited extent, and acting in combination with environmental influences (Manolio et al., 2008). The vast majority of genetic risk factors shaping most common conditions are still unknown and, as Zeggini and Ioannidis (2009) so insightfully remarked, it is difficult to know whether, with existing experimental approaches, we are visualizing "the tip of the iceberg" or "the bottom of the barrel".

GENETIC TESTING

Historically, genetic diseases were classified into monogenic, chromosomal, and multifactorial but, increasingly, this division appears to represent an oversimplification that mostly serves didactic purposes (Scriver & Waters, 1999; Badano & Katsanis, 2002). In reality, complex interplays between multiple genetic and environmental factors appear to shape most medical conditions.

The Human Genome Project catalyzed the development of a new era, in which increasing numbers of mutations linked to human disease and to sensitivity to therapeutic compounds and environmental toxins were unveiled, and genetic testing has witnessed an unprecedented expansion. Over 1200 different genetic tests are currently offered in clinical settings and more are continually becoming available (Hogarth et al., 2008). Genetic tests are grouped into several categories that include newborn screening, diagnostic tests, carrier, pre-implantation, and paternity testing. A defect in phenylalanine hydroxylase, the hepatic enzyme responsible for converting phenylalanine to tyrosine, causes phenylketonuria (PKU), an inherited metabolic condition in which the inability to metabolize dietary phenylalanine and its subsequent accumulation at toxic levels in the blood causes severe behavioral and physical developmental delay in the affected children (Ding et al., 2004; Paul, 2008). Proper management subsequent to genetic screening revolutionized the prospects for the affected children. Early diagnosis and the timely implementation of dietary changes prevent neurological damage, and newborn screening has virtually eliminated the developmental delay caused by this condition, although subtle neurological developmental delay was reported even with early-onset treatment (Levy & Albers, 2000; Clague & Thomas, 2002; Ding et al., 2004; Williams et al., 2008). PKU represents the classical example of a genetic condition that is easily detectable at birth, and for which the timely initiation of effective therapy provides a valuable therapeutic option. Annually, newborn screening programs in the United States detect approximately 3,000 new cases of metabolic disorders, hematologic disorders such as sickle-cell disease, and endocrinopathies, and result in improved diagnosis, treatment, and outcomes for these infants (Centers for Disease Control and Prevention, 2001).

A number of genetic tests, such as the one for Huntington' disease, can detect a mutation in asymptomatic and pre-symptomatic individuals with a relevant family history, decades before the onset of clinical disease. For this autosomal dominant condition, a single copy of the mutation, inherited from either parent, is sufficient to cause clinical disease (Jameson, 1998). A child with one affected parent has, therefore, a 50% chance to inherit the mutation, which invariably causes disease later in life. The benefits of genetic testing for this condition are widely debated, particularly because, even if a mutation is found, it is impossible to predict the timeframe for clinical disease onset, and no cure or treatment are currently available for this progressive and invariably fatal neurological condition (van Dellen & Hannan, 2004; Pfister et al., 2009). In another group of disorders, known as autosomal recessive, such as cystic fibrosis, individuals harboring a mutation are healthy carriers, and do not develop the disease, but can transmit the mutation to their children (Jameson, 1998). Children may be healthy carriers, if they inherit one copy of the mutation from either parent, or they can develop clinical disease when they inherit two copies of the mutation, one originating from each asymptomatic healthy carrier parent.

Other types of genetic tests can be performed in the embryo prior to implantation, in the developing fetus, to establish paternity, or to conduct forensic analyses (Pena and Chakraborty, 1994; Sobrino et al., 2005; Sobrino and Carracedo, 2005; Basille et al., 2009; Benn and Chapman, 2010). Moreover, genetic associations were recently unveiled for conditions and behaviors that are still poorly understood. For example, infidelity and the number of sexual partners were recently reported to be under a moderate genetic influence, and a genetic component was revealed for compulsive hoarding, defined as the tendency to accumulate large numbers of physical possessions without discarding them, and also for trichotillomania, or repetitive hair-pulling, and pathological gambling (Cherkas et al., 2004; Hemmings et al., 2006; Iervolino et al., 2009; Lobo and Kennedy, 2009). The geometric system that humans use for reorientation in space was also shown to be under the influence of genetic factors (Lakusta et al., 2010).

The advent of genetic testing opened a completely new area, and catalyzed the emergence of new concepts and topics in many professional disciplines. Each category of genetic test is accompanied by very specific medical, social, ethical, and legal challenges, and no single set of general guidelines can be developed for all tests. For example, the discovery of *BRCA1* and *BRCA2* opened the possibility to conduct carrier testing to identify individuals at risk for breast and ovarian cancer. However, the test cannot predict whether a person will develop cancer, neither can it provide any insight into when, if ever, disease would develop. Importantly, some women with *BRCA1/2* mutations never develop breast or ovarian cancer (Nicoletto et al., 2001; Brose et al., 2002). At-risk women harboring a mutation can use this knowledge to conduct more frequent surveillance, adopt lifestyle changes, or undergo prophylactic surgery. At the same time, a negative genetic test could create a false sense of security and lead to failure to undergo regular screenings, despite the fact that most breast cancers appear not to be inherited. *BRCA1* and *BRCA2* mutations are present only in 5-15% of ovarian cancers, and only 7% of breast cancers and 10% of ovarian cancers were linked to susceptibility genes (Claus et al., 1996; Ramus and Gayther, 2009). Besides the implications for the individual being tested, the presence of *BRCA* mutations is also relevant for first-degree relatives, and genetic testing can open significant psychological and emotional challenges for all those involved.

DIRECT-TO-CONSUMER GENETIC TESTING

During recent years, in what became known as "direct-to-consumer" (DTC) genetic testing, a number of companies started advertising and/or offering genetic tests directly to consumers. While certain companies advertise the availability of a genetic test, which consumers subsequently have the option to order through a healthcare provider, other companies, in addition, also allow consumers to directly purchase the test from them, and to subsequently receive the results directly (Hogarth et al., 2008).

Wasson (2008) describes three broad categories of DTC genetic tests. Some tests examine mutations that have established, robust connections to medical conditions, and examples are cystic fibrosis, Huntington's disease, breast cancer, and ovarian cancer. Other tests examine mutations reported by some studies to be implicated in certain medical conditions or adverse effects to medication, but the clinical relevance of these associations is not always certain, and the associations are not always reproducible. Finally, a third category includes genetic tests that claim to provide more general health related advice, such as bone health, skin health, or nutritional counseling performed based on an individual's genetic makeup.

A separate type of DTC marketing, which promotes pharmaceutical compounds, has been present for several decades in many countries, and the practice is fervently debated (Wolfe, 2002; Almasi et al., 2006; Adeoye & Bozic, 2007; Frosch et al., 2010). DTC campaigns for pharmaceuticals have been criticized for several aspects, which include creating tension in the physician-patient relationship, encouraging medication overuse, using vague terms, and often lacking balance –

which, realistically, is difficult to achieve in the brief commercial spot or limited space available for the advertisement. At the same time, these campaigns could be valuable in educating the public, encouraging patients to talk to their health care providers, or reminding them to take their medications (Gollust et al., 2002). However, compared to the marketing of therapeutic compounds directly to consumers, the use of the same practice for genetic testing represents an entirely new phenomenon, one that opens a completely new set of controversies and concerns. DTC genetic testing campaigns do not involve a therapeutic compound but, instead, commercialize an instrument that is part of the medical decision-making process. The information provided by genetic tests is intended to be offered and interpreted by health professionals in a much more complex setting, while taking into consideration additional personal medical information and relevant family history. While medication-disease links are better understood, and their validity has in many cases been established for decades, the same cannot be said about most DTC genetic tests. Furthermore, consumer education with respect to genetics and genomics lags far behind the overall health-related knowledge within the general population. In the absence of a very thorough educational campaign to target consumers and health professionals, and to clearly explain benefits and limitations, genetic testing not only cannot be used in a meaningful way to enable individuals to make informed decisions, but it could result in serious adverse effects and harm, which is often irreversible.

DTC genetic tests differ substantially, in several respects, from genetic tests performed in traditional health care settings. Instead of being requested by professionals, who would also analyze and interpret the results, DTC genetic tests are initiated by and offered directly to consumers (Gurwitz & Bregman-Eschet, 2009). Many DTC genetic tests do not examine a specific disease or gene, but often provide large volumes of information, and consumers may face difficulties understanding and interpreting their potential medical and clinical relevance. Wasson (2008) emphasizes that many links between genes and specific medical conditions are not firmly established, and this makes many genetic tests, particularly the ones offered for health related condition, have unclear validity, reliability, and clinical utility. Furthermore, the rate of false positive and false negative test results is often not addressed for genetic tests offered directly to consumers (Wasson et al., 2006). Altman (2009) points out that companies are generally more lightly regulated on privacy issues than health care establishments, which have very tight privacy guidelines, and while the quality of tests is not always regulated when companies administer the test, hospitals have strict quality control mechanisms. In addition, while DTC genetic tests are often provided for medical conditions that are also tested in healthcare settings, there are many other DTC genetic tests that have not yet been accepted for use in clinical medicine (Hogarth et al., 2008). Another consideration is that genetic tests commercialized by companies directly to consumers may be performed for many reasons, including what recently became known as "recreational genomics" - and, as Lee and Crawley (2009) point out, social networking platforms that allow consumers to share and discuss their test results have already emerged. In comparison, genetic tests in health care establishments are generally performed only for health-related purposes. Altman (2009) points out that the process of conducting a genetic test, which is relatively inexpensive as a result of the decreasing costs of DNA sequencing, and the process of interpreting the test result, which is still complex and insufficiently understood, are currently "bundled together", and proposes that the two be separated from each other, as an important step to enable oversight and regulatory efforts.

Benefits of Direct-to-Consumer Genetic Tests

DTC genetic tests increasingly started to emerge, over the past decade, in several countries worldwide (Schickedanz & Herdman, 2009). Several arguments exist in favor of genetic tests being available directly to consumers. One is every individual's right to learn about his or her own genetic predispositions, which shape many aspects relevant to personal and professional life. Most medical conditions are complex, and determined not only by genetic factors, but also by lifestyle and environmental influences. Even though genetic predispositions cannot be altered, learning about genetic risk could help people modulate the influence of environmental factors by implementing lifestyle changes, such as dietary restriction or supplementation, smoking cessation, more frequent screenings, refraining from medications with potential adverse effects, or career-related decisions such as avoiding chemicals that enhance specific risks. Ultimately, these active decisions can improve health and the quality of life. Another argument is that DTC genetic testing could help avoid genetic discrimination, and ease people's fears about this issue, because consumers would obtain information about their genetic predispositions directly, without their test results becoming part of their medical record. By ordering genetic tests directly from companies that commercialize them, hospitals and health care professionals are not involved and, thus, consumers have better control over who has access to their test results (Wasson, 2008; Altman, 2009). In addition, DTC genetic testing appears to raise awareness and encourage consumers to learn more about their disease predispositions (Hudson et al., 2007). Mouchawar and colleagues (2005) reveal that during a July 2002-March 2003 DTC campaign conducted by Myriad Genetics, Inc. in Denver, Colorado, referrals for BRCA testing increased 244% as compared to the same time period during the previous year, indicating that the campaign encouraged women with a family history of breast cancer to talk to their health care providers about the possibility of genetic testing. Another study, that examined surveys completed by 1,635 women and 1,054 healthcare providers, reported that awareness among consumers and health professionals increased as a result of a DTC genetic testing campaign. Consumers asked their providers more questions about testing, requested more referrals for counseling, and more tests were ordered and performed (Centers for Disease Control and Prevention, 2004). An additional benefit reported for DTC genetic testing is the enhanced convenience, since no appointments are required to provide the sample or to receive the test results (Berg and Fryer-Edwards, 2008).

Concerns about Direct-to-Consumer Genetic Tests

1. Clinical Relevance

As research studies increasingly uncover links between mutations and disease predispositions, it is still unclear how many of those associations are clinically and medically relevant. It was reported that the number of articles describing gene-disease associations doubled between 2001 and 2008, and was estimated to recently exceed 34,000, but many of those genes are part of very complex gene-gene and gene-environment pathways (Little et al., 2009). Several types of DTC genetic tests exist. Some of them, such as genetic testing for cystic fibrosis, Huntington's disease, breast and ovarian cancer, or *HER2* testing to guide breast cancer therapy, are also offered in hospital settings by health care providers, while other tests, such as the ones that examine the risk for bipolar disorder, athletic performance, periodontal disease, bone health, susceptibility to addiction, and restless leg syndrome, are often provided solely in the DTC setting (Caulfield et al., 2010). In addition, it is expected that genetic tests for major depression and schizophrenia will soon become available (Anonymous, 2008).

Marietta and McGuire (2009) organize the genetic tests provided directly to consumers into four categories. There are tests that examine susceptibility to disease; a second category includes tests conducted to assess nutritional and metabolic characteristics, such as skin health or caffeine metabolism; a third group includes individual characteristics, such as the type of ear wax or athletic performance; and a fourth type of testing explores information related to ancestry. While a few companies perform genetic testing that is informative about one condition, others perform genome scans that reveal several categories of information (Marietta and McGuire, 2009). For consumers, it is crucial to distinguish between genetic tests that are relying on peer-reviewed, validated scientific studies, and tests with an uncertain significance or validity. Bolnick and collaborators (2007) make a very relevant point for tests that explore genetic ancestry, and the same consideration should guide consumers and professionals in their evaluation of all genetic tests: "it is important to determine what these tests can and cannot determine" (Bolnick et al., 2007).

Janssens and colleagues (2008) identified seven companies offering DTC services for genomic profiling and conducted a meta-analysis of the respective gene-disease associations that were examined in 260 studies published between 2000 and 2007. The authors revealed that 24 (43%) of the 56 genes that the companies tested were not reviewed in the literature. For the remaining 32 genes, only 38% of the associations with diseases were statistically significant and, even among those, the significance was not strong (Janssens et al., 2008; Mitchell et al., 2010). Moreover, some mutations, such as the *MTHFR C677T* polymorphism, were significantly associated in the peer-reviewed literature with up to seven different conditions, and represented a risk factor for six conditions and a protective factor for one.

A 2007 report of the UK Human Genetic Commission examined 26 companies that were advertising genetic testing directly to the public, and noted that several of the tests were for nicotine addiction, response to nicotine replacement products, caffeine metabolism, athletic performance, skin DNA profile, gum disease, and bone health (UK Human Genetic Commission, 2007). Geransar and Einsiedel (2008) described a company providing genetic testing for *APOE* to assess susceptibility to Alzheimer's disease, even though several professional organizations recommended against conducting this test, at the time, due to the uncertain validity of the association. Moreover, it was reported that many companies provide genetic tests to assess the risk for cardiovascular conditions, even though health professionals generally agree that genetic factors are less important, and lifestyle factors are more relevant in shaping these diseases (Berg and Fryer-Edwards, 2008). Most genetic variants identified to date were estimated to confer a small relative risk, approximately 1.5 or lower (Schickedanz and Herdman, 2009). For example, a sequence variant on the short arm of chromosome 9 was associated with myocardial infarction, and the estimated risk in carriers is approximately up to 1.64 times higher than in non-carriers. While this finding is statistically significant, it does not necessarily mean that it is clinically relevant as well (Helgafottir et al., 2007; Cho, 2008). In addition, at least 2 companies were recently reported to offer DTC genetic testing for *CYP450* to guide the choice of antidepressants, even though a meta-analysis of 37 articles concluded that no evidence supports the use of the test for this purpose (Thakur et al., 2007; Hogarth et al., 2008; Kastanis et al., 2008).

As more research studies are published, it is becoming increasingly clear that gene-disease associations are very complex. Sometimes, individual mutations in several different genes may shape the same medical condition. For example, mutations in 11 genes are in some way linked to type 2 diabetes mellitus, at least 13 genes that shape hypertrophic cardiomyopathy were identified to date, over 15 genes are involved in autism spectrum disorders, and approximately 35 robust genetic associations were established for systemic lupus erythematosus (Risch et al., 1999; Kaput, 2008; Tsoutsman et al., 2008; Moser et al., 2009). There are other instances, when the same genetic variant might not necessarily be associated with disease in all populations, and taking into consideration population-specific differences could become a crucial part of interpreting genetic test results. Polymorphisms in tyrosine kinase 2 (*TYK2*), a type I interferon (IFN) signaling pathway gene, were found to be a risk factor for systemic lupus erythematosus (SLE) in Caucasian populations, but were not found to represent a risk factor in Japanese people (Sigurdsson et al., 2005; Graham et al., 2007; Kyogoku et al., 2009). Lee and collaborators (2009) reported that *PTPN22* polymorphisms and SNPs in several other genes, which are strong genetic risk factors for rheumatoid arthritis in Caucasians, are not associated with this condition in Korean populations, and a meta-analysis that examined the involvement of *SLC22A4* in rheumatoid arthritis confirmed its significant association with disease in Japanese but not in Caucasian individuals (Okada et al., 2008).

2. Involvement of Health Professionals and Counseling

Genetic counseling involves much more than simply communicating the outcome of genetic testing to the individual being tested: it needs to address complex medical, emotional, psychological, social, cultural, and economic aspects relevant to the specific medical condition being tested for, and unique to the person undergoing testing (Cummings, 2000). The psycho-emotional impact of genetic testing is often difficult to predict in advance. For example, Dorval and collaborators (2000) found that *BRCA1* carriers felt more distressed, after testing, than they had previously anticipated. Schickedanz and Herdman (2009) emphasize that genetic tests should be tailored to everyone's individual needs, and the "one-size-fits-all" model, offered by most DTC campaigns, is problematic. Genetic testing should be performed within health care facilities, under the direct supervision of health care professionals who interpret the results by taking into consideration additional information, including the patient's comprehensive medical and family history. Counseling should be provided before, during, and after the administration of the test, and it sometimes needs to involve not only the individual being tested, but family members as well. Many DTC genetic test providers do not involve health professionals and do not offer pre- and post-test counseling (Gollust et al., 2003; Wasson et al., 2006; Ries and Castel, 2008; Goddard et al., 2009).

Gollust et al. (2003) examined 105 sites providing DTC genetic services and found that from 14 sites that offered health-related genetic tests ranging from standard tests, such as the one for cystic fibrosis, to more unconventional tests related to nutrition, behavior, and aging, only five described the risks associated with testing, six described the availability of counseling or consultation with a physician, and one specified that they would not be responsible in any way for providing advice or for analyzing the test results. One company offering tests for DNA damage also marketed a "nutraceutical" compound that can decrease DNA damage, increase DNA repair, and improve immune function, even though such claims are not supported by the peer-reviewed literature (Gollust et al. 2003). Goddard et al. (2009) found, as of August 2007, 84 instances of health-related genetic tests that were sold on 27 different web sites without the involvement of a health care professional. This represented a total of 53 tests, because some tests were commercialized by more than one site. The authors noted that approximately two thirds of the tests, including the ones used to predict athletic performance, smoking cessation, or guide the preparation of skin care products, were not offered in traditional health care settings. Geransar and Einsiedel (2008) reported that less than half of the 24 companies they examined, involved in direct-to-consumer genetic advertising, required the involvement of a physician, and the ones commercializing enhancement tests, such as nutrigenetic tests or tests evaluating fitness, were the least likely to have this requirement. The authors noted that while most companies offered background information about the medical condition that was being tested, the information provided was not always accurate or complete and, even when disclaimers were provided about how to use the test results, they were often contradicted by other statements of an emotional nature found on the same site. In addition, over one third of the companies did not provide references to support their statements, and approximately one quarter of the companies were referencing unpublished materials or internal company findings (Geransar and Einsiedel, 2008). In a study that analyzed 13 companies offering DTC genetic tests, Berg and Fryer-Edwards (2008) reported that only four companies mentioned anything about risks associated with genetic testing, and only three offered counseling, via e-mail or phone. Liu and Pearson (2008) examined 46 DTC web sites and reported that only 15.2% described the potential harms associated with genetic testing.

Goddard and colleagues (2009) examined five Internet sites that commercialized genetic tests predicting the risk for thrombosis, and found that three distributors did not mention any risks or benefits associated with testing, and even for the companies that informed about these issues, the information overlapped with professional recommendations for only one company. As the study pointed out, the distributors used less than 9 of the 13 practices that were recommended by several professional societies. Disclosures about the risks and limitations of genetic testing and about the clinical validity of the test represented the practices least likely to be used (Goddard et al., 2009).

The American College of Medical Genetics (2004) recently pointed out that genetic testing should only be provided by qualified health care professionals, who should be responsible for several aspects pertaining to the testing process, including ordering the test, pre- and post-test counseling, interpreting the results, and follow-up. At the same time, the Human Genetics Commission in the United Kingdom (2003) recommended that most genetic tests have some form of involvement by a physician.

Counseling represents one of the essential facets of genetic testing, and should occur prior to performing the test, during the testing process, and subsequent to receiving the results. Since genetic tests, by definition, provide information not only about the individual, but also about other family members, counseling might need to involve relatives who could also be impacted by the results. While several studies did not find significant psychological and emotional consequences subsequent to genetic testing, other authors reported that individuals may experience anxiety and become worried after learning about a positive test, and a risk for depression was found even among individuals who did not undergo genetic testing but have a family history of cancer (Lerman et al., 1998; Lodder et al., 2001; Watson et al., 2004; Cameron & Muller, 2009; Hamilton et al., 2009; Caufield et al., 2010). It is imperative to note that even negative test results require the availability of counseling. Sobel and Cowan (2003) reveal that testing for Huntington's disease can be associated with feelings of helplessness, depression, guilt, and suicide attempts. Importantly, the authors emphasize that suicide attempts were reported not only among those who tested positive, but also in a woman who tested negative and felt disconnected from her siblings who tested positive (Sobel & Cowan, 2003). Positive BRCA test results can be associated with shock, fear, and guilt, and were shown to impact family relationships either positively or negatively (Douglas et al., 2009). A multi-center study that examined the long-term impact of BRCA testing on 193 participants reported that female carriers experienced higher distress levels 3 years after taking the test

(Foster et al., 2007). The involvement of health professionals is even more important, considering that many women from families with hereditary breast and ovarian cancer overestimate their risk to develop disease (Pasacreta, 2003). In addition, Croyle and colleagues (1997) found that women who never developed cancer, but have positive BRCA genetic test results, showed psychological distress levels similar to the ones observed in women at an average of 10 weeks after being diagnosed with cancer. It is important to remember that for DTC genetic testing that does not involve counseling, or does not involve the appropriate level of counseling, significant emotional and psychological damage may occur.

3. Sensitivity and Specificity

Like many other tests, genetic testing can give false-negative and false-positive results that, particularly in context of the DTC setting, complicate their interpretation and may cause substantial psychological and emotional harm. Wasson and collaborators (2006) point out that companies offering DTC genetic testing often do not provide information about the sensitivity and specificity of individual genetic tests either to consumers or to health professionals, making it impossible to estimate their reliability.

There are few reports about the frequency of errors during genetic testing. Honeywell et al. describe three patients who had initial false negative test results. A 5-year-old boy who presented with syncope after running was referred for a ryanodine receptor 2 gene (*RYR2*) testing, and an initial test that sequenced 33 of the 105 exons did not detect the mutation (Honeywell et al., 2008). A concurrent test for the same gene, conducted in another lab, detected a mutation in an exon that had previously been sequenced in the first lab. The mutation was subsequently confirmed after the first lab repeated the test. In a second patient, a 5-year-old child evaluated for a diagnosis of long-QT syndrome, sequence analysis of five genes linked to this condition did not detect any mutations, but another lab found a mutation in *KCNH2*, in an exon that was not tested in the first lab. In this case, the mutation was initially missed because it resided in an exon that had not been analyzed by the first laboratory. In a third case, several years after initial negative genetic test results from a research laboratory, a clinical laboratory detected a *KCNQ1* mutation in a patient.

Hofgärtner and Tait (1999) found significant problems in 0.33% of the genetic tests performed at 42 laboratories that they surveyed in the United States. From 3,977 tests conducted over a 5-year period in 39 Australian laboratories, Hertzberg et al. (2005) reported a 98.63% overall success rate, and revealed that 20 (51%) of the laboratories made at least one error, but three of them were responsible for 46% of all errors. A study conducted in 1997, that examined 47 centers in the United Kingdom, revealed errors in 3-6% of the laboratories (Preston et al., 1999), and an external quality assurance survey conducted over a 6-year period that enrolled 37 Australian laboratories testing two hemochromatosis gene mutations reported that a 0.53% error rate among the 3,016 responses received (Hertzberg et al., 2006).

Libby et al. (2006) mention the case of a young woman, with a history of deep venous thromobosis, who was tested for a factor V Leiden mutation in 4 different laboratories, two of which reported that she is heterozygous, and two did not identify the mutation. Tripodi and colleagues (2005) sent four DNA samples collected from patients with thromboembolic disorders to 41 laboratories to be tested for genetic factors conferring risk for thrombophilia, and noted that while most reports were accurate, both false-positive and false-negative results occurred. Palomaki et al. (2003) examined several laboratories in the United States offering genetic testing for cystic fibrosis between 1996-2001, and found that 0.5% and 1% of the results were false-positives and false-negatives, respectively.

An additional complication, specific for genetic testing, emerges from the fact that, often, many tests do not examine single mutations, but involve simultaneous genome-wide analyses of thousands or hundreds of thousands of mutations or single nucleotide polymorphisms in one individual. As Kohane and collaborators (2006) demonstrate, even assuming a sensitivity of 100% and a false-positive rate of 0.01%, when 10,000 independent tests performed, over 60% of the population will receive at least one false-positive result.

4. The Danger of Genetic Determinism

Sidney Wolfe (2002) quotes Canadian economist Stephen Leacock (1924) referring to the process of advertising as "the science of arresting the human intelligence long enough to get money from it". The appeal to consumers' emotions that is used in many DTC pharmaceutical product advertisements has been widely reported. Hollon (2005) emphasizes that most DTC advertisements for pharmaceutical products use vague and qualitative terms to describe the drug's benefits, and rely on emotional appeals, providing, at the same time, little health information. In a survey conducted between July 1998 and July 1999 to examine DTC advertisements conducted by pharmaceutical companies, Woloshin et al. (2001) reveal that 67% of the advertisements used emotional appeals, as opposed to describing the medication's expected benefits and adverse effects. Frosch and colleagues (2007) reported that 95% of television advertisements used emotional appeals, over half of them presented the products as medical breakthroughs, and many advertisements led consumers to believe that they could improve their health only by taking medications, or when they combined medications with lifestyle modifications, but not by lifestyle changes alone. Schickedanz and Herdman (2009) emphasize that the complexity associated with genetic testing is difficult to convey from either a web site or a 60-second commercial advertisement. One of the dangers associated with DTC genetic testing is that, with advertisements that appeal to emotions, and in the absence of adequate educational campaigns, patients and consumers might believe that their genetic predispositions will control their lives and that the consequences cannot be changed. This could not be further from the truth.

Gollust and collaborators (2002) discuss several examples of DTC advertisements that oversimplify the complexities surrounding genetic testing. For example, the authors mention a pregnancy magazine that depicts a newborn baby and informs readers how genetic screening would bring "peace of mind" for the baby and the parents. Many DTC campaigns use terms such as "hope" and "fear", which have the potential to manipulate consumers' emotions, behavior, and decisions. Liu and Pearson (2008) reveal that many DTC web sites make a powerful use of emotional appeals, often associating happiness or relief with taking a test, while portraying sadness or regret in individuals who did not get tested. The authors reveal that the most frequent emotional appeals were warmth, empowerment, and assurance, and found, on average, 2.52 emotional appeals on DTC web sites, as compared to 0.88 in the case of professional web sites (Liu & Pearson, 2008). Hull and Prasad (2001) describe their experience from the minutes preceding a performance they attended, when an advertisement encouraged women to undergo BRCA testing, telling them that the test would "dispel fear and provide hope". The authors discuss how such an advertisement is ethically flawed for several reasons. It is manipulative because it targets an audience that will witness, during the performance, a woman's struggle with ovarian cancer; it is misleading, because it does not adequately describe the product it advertises, fails to mention that results lack certainty, and targets the entire audience, instead of the small number of women who are at risk for BRCA-related tumors; and it is misguiding, because it encourages women to contact the company instead of discussing genetic testing with their health care providers (Hull & Prasad 2001). Despite the fact that BRCA genetic testing is recommended for only a small fraction of the population, a study conducted by Lowery et al. (2008) reveals that after a 5-month DTC advertisement for BRCA testing in Atlanta, GA and Denver, CO, over half of the women surveyed reported that they felt genetic testing benefits most women, and less than one third agreed that genetic testing is useful for only certain women, indicating that the participants overstated the benefits of genetic testing for this condition. In 2006, the United States Government Accountability Office (GAO) (2006) purchased DTC genetic tests from four web sites and submitted, for analysis, twelve DNA samples from a female and two DNA samples from an unrelated male, while indicating, in the description, that the material came from individuals with different lifestyles and personal characteristics. The investigation revealed that the "personalized" recommendations provided for these fictitious consumer profiles were misleading, and included claims that were medically unproven and ambiguous, stating, for example, that consumers "may" be at an increased risk for a specific condition. In addition, the investigation found that many "personalized" recommendations were not based on an individual's genetic profile, but were commonsense health advice – for example, if the lifestyle description mentioned that the consumer was a smoker, the recommendations indicated the need to stop smoking, and when the lifestyle description indicated that the consumer was not a smoker, the recommendation was to continue to avoid smoking. Two of the web sites recommended expensive personalized supplements that GAO reported to be similar to dietary supplements available from supermarkets at much lower prices, and one web site recommended an expensive supplement that can "repair damaged DNA", even though the term is very ambiguous, and no product with such an effect is known to exist based on medical and scientific findings to date (United States Government Accountability Office, 2006). Berg and Fryer-Edwards (2008) write about an advertisement for the D2 dopamine receptor gene, in which customers are informed that genetic testing can help them "understand and manage a child's behavior before it gets out of control", despite the fact that the involvement of this gene in addictive behaviors, such as alcoholism, is not supported by the recent scientific literature, and multiple genes, along with environmental influences, are thought to be involved. Advertisements that focus on the genetic factors, but fail to mention the complex gene-gene and gene-environment interactions that shape most medical conditions and phenotypic traits, could leave consumers with the impression that genetics is the only or the most important influence controlling those conditions. In fact, the scientific literature increasingly supports the idea that, even for diseases historically considered as having a strong genetic component, environmental influences are very important. For example, while several studies reported that the age of onset and behavioral manifestations are very similar in monozygotic twins with Huntington's disease living apart from birth, others found differences of 6 to 8 years in the age at onset, and in the nature of clinical manifestations, pointing towards the involvement of other factors, in addition to genetic influences (Sudarsky et al., 1983; Panas et al., 2008).

5. The Importance of Public Education

In an analysis describing several categories of information generated as a result of genetic testing, Evans (2008) points out that consumers are likely to learn about things that are "useful to know", along with things that they "already know", things that they "don't really want to know", some things that "aren't true", other things that they "don't want others to know" and, finally, some things that are "fun to know" (Evans, 2008).

When consumers respond to a DTC genetic testing campaign and undergo testing, the results will be provided to them directly, often without the involvement of healthcare professionals. Taking into account the complexity of gene-gene and gene-environment interactions that shape most medical conditions, it is questionable to what extent consumers and patients would benefit from learning about various disease predispositions under these circumstances. First and foremost, a thorough educational campaign, to identify and clarify the benefits and limitations of genetic testing, is essential every time a test is performed. It is important to realize that multiple factors, including diet, exercise, smoking, weight, environment, and ethnicity will determine whether or not an individual develops a specific disease (Farkas and Holland, 2009). More complex diseases, shaped by gene-environment interactions, exist as compared to single-gene conditions. Nevertheless, a 2005 survey of companies offering DTC genetic testing highlighted that many more tests are offered for complex diseases than for single-gene conditions, and six out of thirteen companies that were examined only offered tests for susceptibilities to complex traits (Berg & Fryer-Edwards, 2008).

Molster et al. (2009) conducted a telephone survey of 1,099 individuals in Australia, and noted that there was some misconception around the significance of the term "increased genetic risk". In their study, 75% of the participants concurred that an increased genetic risk means that they would develop the disease irrespective of what they did, and 68% indicated that a healthy lifestyle would not make a difference if they already had an increased genetic risk factors for a specific condition. In addition, 80% of those surveyed agreed that being a carrier for a mutation means that they have the disease (Molster et al., 2009).

As recently emphasized, genetic tests cannot be interpreted simply as "black or white" and they do not represent a diagnosis *per se* (Anonymous, 2008). Advertisements that urge consumers to get tested in order to relieve their fears, and messages that appeal to their emotions by depicting feelings of joy and relief when a specific mutation is absent, are often counterproductive, as they fail to portray the true complexity of genetic testing. For example, a woman carrying a *BRCA1* mutation could easily assume, in the absence of an educational campaign, that she will definitely develop cancer, despite the fact that 10-15% of women harboring this mutation never develop disease (Ford et al., 1994). Conversely, a negative test could provide a false sense of security, leading to a decreased appreciation of the importance of regular check-ups, and diminished surveillance and vigilance, despite the fact that breast cancer most often occurs without any known genetic risk factors. A recent study revealed that in families with breast cancer, even women who test negative for BRCA1 and BRCA2 have a higher risk to develop breast cancer than women in the general population, pointing towards the possibility that other, yet unidentified genetic factors, could be involved (Smith et al., 2007). In fact, the majority of the approximately 200,000 breast cancers and 25,000 ovarian cancers diagnosed annually in the United States cannot be explained solely by genetic factors (Lowery et al., 2008).

Liu and Pearson (2008) emphasize that consumers cannot always distinguish between accurate and inaccurate sources of information and they cannot always identify bias. This is particularly important in the case of genetic testing, where the amount of information becoming available is actively expanding. Public education emerges, therefore, as an intervention of utmost importance, which should empower consumers and patients. In a recent study, Gray et al. (2009) found that after women were informed about the risks associated with DTC BRCA testing, they expressed diminished interest in undergoing testing in this setting and showed, instead, a preference for hospital-based tests. Wilde and colleagues (2010) reported that from 36 participants enrolled in a study, 24 initially expressed an interest to undergo genetic testing for major depression, but after being informed about the positive and negative aspects related to predictive genetic testing, nine participants changed their mind. Nevertheless, all participants indicated a continuing interest to undergo testing, were the tests administered by medical professionals. Concomitant to consumer empowerment, an effective educational campaign targeting health professionals from various fields is essential. While in the past genetic counseling was a task restricted to a narrow group of medical and counseling professionals, increasingly, all health professionals and social workers will be confronted, on a daily basis, with ethical dilemmas related to genetic testing and counseling. A survey that was administered in 2006 to health professionals in Japan reported that, from the 1,124 general practitioners and 294 clinical geneticists who replied, 38% and 68.4%, respectively, were aware of DTC genetic testing (Ohata et al., 2009).

An important facet of educational initiatives should focus on understanding statistical risk. Gigerenzer et al. (2008) talk about a phenomenon that they call "collective statistical illiteracy" in health care, and describe among consumers, health professionals, and journalists. The authors explain how insufficient understanding of risk and uncertainty, in statistical terms, which sometimes is unintentional, but may also be the result of manipulation, has important health consequences (Gigerenzer et al., 2008). One such consequence is the increased susceptibility to emotional manipulation by advertisements that inform consumers about the need to undergo testing for specific conditions they might be at risk, undermining thus the concept of informed consent during medical decision-making processes. Understanding the benefits and risks of screening tests, and the need to learn about false-positive and false-negative results, emerge as cardinal aspects that should be incorporated into genetic testing educational initiatives.

6. Financial Considerations

The cost of DTC genetic tests ranges from under $100 to over $3000 and could increase when tests need to be repeated or when several family members have to be tested (Wasson et al., 2006; Berg and Fryer-Edwards, 2008). While discussing the ethical principle of justice, Wasson and collaborators (2006), in addition, point out that DTC genetic tests are biased towards individuals with more financial resources and, especially if a test that provides health benefits, this bias is unfair and inequitable.

7. Ownership of Submitted Samples

Fervent controversies surround the ownership of the DNA used for genetic tests, and the ownership of the information generated as a result of testing. Genetic information, just like medical information in general, is very private, and there are many debates on whether genetic tests are as private as medical tests, or more powerfully private. At the same time, genetic tests are informative not only about the risks for the individual being tested, but also about risks that others in the same family might face, and health care providers face a dilemma when they need to balance an individual's right to privacy with a family member's right to learn about predispositions that could affect their lives. By its very nature, genetic information always has certain items that are specific to the individual being tested, and other items that are shared with family members. Therefore, even when genetic tests are administered in a health care setting, under the supervision of health professionals, controversies emerge with respect to confidentiality and to individual family members' access to various components of the genetic test results (Clarke, 2007; Lucassen, 2007). These controversies are exacerbated when a company performs the testing upon direct request from consumers, and it is still unclear whether the DNA used for testing, and the information generated as a result of testing, belong to the company or to the individual being tested. Additional debates gravitate around how, for how long, and under what circumstances the test results can be retrieved, or the original DNA sample can be reused to either repeat the original test or to conduct additional ones. Gurwitz and Bregman-Eschet (2009) describe the example of a company that started providing genetic testing in 2007 and indicated, on its web site, that except for the saliva sample, which is required to conduct the test, it does not claim ownership over other materials that the consumer provides. The authors point out that one immediate implication is the ownership over biological samples, which could allow a company to treat these samples like any other property, possibly sell them, or use them for future tests, with or without notifying the individuals who provided the samples and/or obtaining informed consent from them. This remains an important topic to be addressed (Gurwitz & Bregman-Eschet, 2009). Berg and Fryer-Edwards (2008) report that from thirteen companies offering DTC genetic tests that they examined, only two clarified what happens to the DNA samples after the test is completed, and it is uncertain whether they would be immediately destroyed, stored, or used for future research projects, leaving several unsolved ethical and legal dilemmas.

8. Regulatory Considerations

Regulation of DTC genetic testing emerges as a very complex issue and involves several aspects that, as of now, represent mostly an unexplored territory for health practitioners, ethicists, and legislators alike. Several aspects that are part of the DTC genetic testing process may be regulated. One of them is the analytic validity of a test, which represents the ability of a laboratory to accurately and reliably determine the presence of the mutation examined, and depends on the sensitivity and the specificity of the laboratory methodology used (Palomaki et al., 2003). An additional factor that can be regulated is the clinical validity, which represents the ability of a genetic test to detect the presence or the absence of a medical condition or phenotype (Zimmern & Kroese, 2007). In addition, legislation that regulates false and deceptive advertisement becomes very important in this context (Hogarth et al., 2008).

In the United States, 25 states and the District of Columbia allow DTC genetic testing and 13 states explicitly prohibit it (Genetics and Public Policy Center, 2007; Hogarth et al., 2008). In Europe, seven countries (Austria, Belgium, France, Norway, Sweden, Switzerland and the Netherlands) have legislation to address DTC genetic testing. The Additional Protocol on Genetic Testing, passed on November 27, 2008 by the Council of Ministers of the Council of Europe, and ratified by several European Union member states, proposed that genetic testing for health reasons should only be conducted under "individualized medical supervision" (Council of Europe, 2008). However, an important controversy might revolve around the regulation of genetic tests that are not performed for health purposes (Hogarth et al., 2008; Gurwitz and Bregman-Eschet, 2009).

As some companies that offer genetic testing over the Internet have physical locations that are different from the countries where the tests can be purchased, regulatory attempts become even more difficult, if possible at all. In May 2002, Gollust and collaborators (2003) examined 105 sites that offer direct to consumer genetic services. Of these, 81 listed addresses in the United States, 17 listed international addresses, and 7 listed both domestic and international contact information. Developing effective regulatory guidelines to address DTC genetic testing becomes, therefore, a global priority.

THE INCIDENTALOME, MUTATIONS LINKED TO SEVERAL CONDITIONS, AND THE EXAMPLE OF APOE

The term *incidentalome* was coined in clinical medicine to describe situations when a medical or laboratory investigation unveils a medical condition that is distinct from the condition that prompted the original test. For example, an

X-ray may point towards findings, possibly of clinical significance, which were not the reason for performing the radiography in the first place. Incidental findings have been known for a long time in medicine. In up to half of the asymptomatic subjects, and in over half of the symptomatic ones, colonoscopy was found to reveal pathological findings in other organs (Wolf, 2008). A prospective evaluation of individuals undergoing chest computer tomography in a lung cancer screening program conducted in Italy revealed that approximately 4% of the individuals had adrenal masses (Bovio et al., 2006). Of 1,256 patients who underwent cervical computer tomography scans for traumatic injuries between January and June 2007, 230 (18.3%) had incidental findings, and a meta-analysis of 17 studies that involved 3,488 patients who performed computer tomography colonography showed incidental findings in 40% and extra-colonic cancers in 2.7% (Xiong et al., 2005; Barboza et al., 2009).

A similar phenomenon is being described in genetic testing. Results from genetic tests that are performed to examine a particular risk factor can often reveal predispositions to other diseases, that the individual did not intend to be tested for, or might not even want to know about. Sometimes, a specific mutation found in a patient can be linked, in subsequent studies, to additional conditions, while on other occasions, as new gene-disease associations are being discovered and characterized, re-analyzing existing genetic test results can point towards additional genetic predispositions. Kohane and colleagues (2006) talk about the "incidentalome" to describe the instances when mutations and predispositions are unveiled, but are different from the ones that represented the reason for performing the genetic test in the first place. As the number of mutations linked to medical conditions is increasing, the likelihood of revealing incidental findings during a genetic test is also expected to rise.

Incidental findings open an ethical dilemma. It is still unclear what duties and rights health care providers have in their relationship with patients, or with patients' families, when they learn about predispositions that they did not intend to test for, but were nevertheless revealed as part of a test. This dilemma becomes even more intense in context of the DTC genetic testing when, sometimes in the absence of counseling, consumers need to decide how to interpret multiple predispositions that they learn about, and how to communicate information that may also be relevant for other family members.

Apolipoprotein E provides an interesting example. The *APOE* gene, located on human chromosome 19, encodes a 299 amino acid plasma lipoprotein essential during cholesterol transport and proposed to transport lipids to sites of neuronal injury (Poirier, 1994; Samatovica, 2000). This protein has three isoforms, ApoE2, E3 and E4, which differ at positions 112 and 158. ApoE3, the most common isoform, contains cysteine and arginine, while ApoE2 has two cysteine residues and ApoE4 has two arginine residues at these positions (Weisgraber et al., 1981). Several studies revealed that the ApoE4 isoform confers increased risk for Alzheimer's disease, and the *APOE* genotype is thought to account for approximately 95% of the Alzheimer's disease cases in the United States (Saunders et al., 1993; Raber et al., 2004; Roses, 1996; Farrer et al., 1997). The same isoform is also associated with higher cholesterol levels and confers increased risk to develop coronary heart disease (Bennet et al., 2007; de Chaves and Narayanaswami, 2008). Moreover, the ApoE4 isoform was also linked to an increased risk of β-amyloid deposition in the brain subsequent to traumatic injury, and to more severe chronic neurologic deficit subsequent to head trauma in professional boxers (Nicoll et al., 1995). Patients with the ApoE4 isoform were more than twice as likely to have an unfavorable outcome 6 months after head injuries than patients without this isoform (Teasdale et al., 1997). It was reported that among children aged 15 and younger undergoing head injury, the impact of carrying the ApoE4 isoform could be viewed as being equivalent to aging 25 years (Teasdale et al., 2005). This association started controversial discussions over suggestions to test athletes involved in contact or concussion sports, in order to predict the ones at risk for more serious consequences after repeated head trauma. It was reported that in Australia, the Government of Victoria expressed an interest in using a genetic test for ApoE4 as a precondition for professional boxing (Spriggs, 2004; Savulescu, 2005), a proposal that opened fervent debates.

As we learn more about the human genome, and unveil more links between mutations and diseases, incidental findings are poised to become an increasingly common occurrence. Particularly as one mutation is often linked to several medical conditions, and can constitute a risk factor for some predispositions and a protective factor for others, interpreting genetic tests is becoming a challenging task. A recently identified SNP, *C1858T*, within the gene encoding an intracellular tyrosine phosphatase, *PTPN22*, substitutes an arginine with a tryptophan residue at position 620 in the resulting protein, and predisposes to several autoimmune diseases, including type 1 diabetes mellitus, rheumatoid arthritis, systemic lupus erythematosus, and Hashimoto thyroiditis (Criswell, 2005). An additional example is the methylenetetrahydrofolate reductase (*MTHFR*) *C677T* polymorphism, in which the C to T substitution in exon 4, which replaces alanine at position 225 of the resulting protein with valine, generates an enzyme with diminished activity (Gilbody et al., 2006; Timuragaoglu et al., 2006). When it is present in the heterozygous (CT) form, in which only one chromosomal copy of the gene harbors the mutation, MTHFR retains 65% of the enzymatic activity seen in individuals with both wild type copies (CC), and individuals harboring the mutation in its homozygous (TT) form have only approximately 30% of the enzymatic activity left (Frosst et al., 2005). This polymorphism was implicated in recurrent spontaneous abortion, esophageal cancer, certain types of migraine, peripheral artery disease, atherosclerosis among smokers, and relapse in pediatric acute lymphoblastic leukemia patients, and was, in addition, shown to confer a moderate, but statistically significant risk for depression, schizophrenia, and bipolar disorder (Lissak et al., 1999; Cherkas et al., 2004; Rothstein, 2004; Wang et al., 2004; Aplenc et al., 2005; Gilbody et al., 2006; Khandanpour et al., 2009; Rodríguez-Guillén Mdel et al., 2009).

Thus, a person undergoing genetic testing for the *APOE* isoforms will at the same time learn about being at higher risk to develop Alzheimer's disease and cardiovascular disease, and about the fact that head injuries could have more severe consequences. Similarly, testing for *MTHFR* polymorphisms may be informative about the risk for recurrent abortions or predisposition to neuropsychiatric conditions, which might not represent the reason somebody sought testing in the first place, and there is a certain chance that the person might not even want to know about them.

A special type of incidental finding, which represents a topic of great interest in medicine and anthropology, is misattributed parentage. Non-paternity rates of 10% and higher were reported in some studies (Stewart, 1989; Cervino & Hill, 2000), and there are arguments both in favor and against disclosing false paternity to parents, when it is revealed subsequent to genetic testing (Lucassen & Parker, 2001). It is still unclear what obligations health professionals have when confronted with this very complex issue and, as several authors pointed out, it is more advisable not to reveal this piece of information after it becomes available, unless the benefits outweigh the risks (Cho, 2008).

GENETIC TESTING IN MINORS AND INDIVIDUALS UNABLE TO CONSENT

Genetic testing in children increasingly emerges as the focus of intensive medical and ethico-legal debates. With parents always being informed about the outcome of genetic tests conducted in their children, the result is that children do not enjoy the same level of confidentiality as adults who are tested (Fryer, 1995; Duncan & Delatycki, 2006). Significant conflicts can thus arise between a parent's, or a guardian's interest to perform a genetic test, and the child's interest for genomic privacy. This opens a dilemma that promises to become even more pronounced with the availability of DTC genetic testing (Robertson, 2003; Duncan & Delatycki, 2006).

Several types of genetic tests were described in children (The Huntington's Disease Collaborative Research Group, 1993; Wertz et al., 1994; Gilbar, 2009). Some of them, such as the test for phenylketonuria, can diagnose a congenital disease and provide clear and immediate benefits when treatment, dietary intervention in this example, is available and immediately implemented. Of particular ethical interest are the genetic tests that do not provide an immediate health benefit. For example, testing for cystic fibrosis can identify carriers, but learning about one's carrier status only becomes relevant later in life, when reproductive decisions are made as young adults. Other tests can predict adult-onset medical conditions, many of which, such as adult-onset retinitis pigmentosa, or Huntington's disease, are untreatable, and raise even more significant ethical questions (Mezer & Wygnanski-Jaffe, 2009).

Genetic tests in children are surrounded by controversy, particularly when testing does not provide any immediate benefits to the child (Plass et al., 2009). Duncan and Delatycki (2006) point out that adolescence represents a vulnerable developmental stage, the stage at which identity is being shaped and the first intimate relationships are established, and predictive genetic testing during this period has a very different impact than during adulthood. Borry et al. (2008) argue that the benefits of receiving good news after genetic testing in minors might not be properly balanced against the risks associated with positive results and, in addition, the knowledge that despite not being sick they might develop disease later in life is processed differently in children as compared to adults. Bloch and Hayden (1990) underscore that the only instance that justifies genetic testing in children is when a clear benefit exists for the child, and Fryer (2000) emphasizes that, in addition to the breach of confidentiality that occurs when adults receive the results of genetic tests conducted in their children, the process of releasing test results directly to the parents is also in conflict with the policy of providing counseling to the individual undergoing testing.

Once a genetic test is performed, it is impossible to "undo" the process and, in a minor, this undermines the ability to autonomously decide whether or not to undergo testing later in life, upon reaching the age of consent. It was reported that only 10-15% of the adults with a family history of Huntington's disease decide to eventually undergo genetic testing (Borry et al., 2008). This, was pointed out, represents a good reason why one should never assume that children will later be interested in learning about their carrier status, irrespective of the nature of the specific condition. Instead, testing should be deferred until maturity, to provide children with the opportunity to later on exercise their "right to know" as adults (Duncan & Delatycki, 2006; Borry et al., 2008). However, each family, and probably every medical condition, open very unique circumstances, and Borry and collaborators (2007) argue that in certain families, overwhelming parental anxiety and uncertainty about the carrier status of their children could exert a more negative impact than testing itself. At the same time, Rhodes (2006) recommends that pediatricians encourage genetic testing in children at a young age, and points out that parents often authorize other medical procedures for their children without violating the principles of confidentiality, and genetic tests might be important in making decisions about children's health, well being, and future. While many genetic tests for adult-onset conditions do not provide immediate medical benefits to minors,

Rhodes (2006) emphasizes that uncertainty is often worse than even bad news, and stress was linked to well established psychological consequences, cardiovascular disease, poor immune function, endocrine dysfunction, and premature aging (Epel et al., 2004; Rhodes, 2006). Moreover, it is the parents who bear the emotional, social, moral burdens, and the responsibilities and the decision-making should, therefore, be theirs.

To examine how children's interests are protected in context of the DTC genetic testing, Borry et al. (2010) analyzed 29 web sites that offer DTC health related genetic tests and characterized four classes of companies. Thirteen companies did not address the issue of genetic testing in minors; eight allowed it with parental consent; four companies specified that they do not offer services to minors, but did not clearly indicate whether they would refuse to do so with parental authorization; and four other companies provided information clarifying that their tests are not for intended for minors. The study clearly revealed that some companies agree to provide genetic testing to the pediatric population, when testing is performed under parental request. Another study that examined five companies points out that policies often vary, with two companies refusing, and three accepting, DTC genetic testing in minors (Borry et al., 2009).

Genetic testing in minors can have severe psychological consequences. One of them is the feeling of worthlessness in those who test positive, and even in children who test negative, receiving the good news is sometimes associated with the development of "survivor guilt" (Wertz et al., 1994; Borry et al., 2008). Moreover, the knowledge that a child tests positive is, in many families, linked to the perception of increased link in other siblings (Wertz et al., 1994). As Wertz et al. so relevantly describe, siblings who are initially united by the "bond of risk" could suddenly find themselves "separated by the outcome of genetic testing" (Wertz et al., 1994).

In addition, just like in adults, false positive and false negative test results may have a long-lasting impact, which is even more complex when children are involved. In 1967, Rothenberg and Sills (1967) reported that, on a monthly basis, they saw approximately 2 to 4 families with initial false-positive PKU test results that, despite being followed by several negative tests, still left parents with acute and chronic anxiety. To help families cope with this situation, the authors implemented a two-step procedure in which a health care professional would first inform parents that a number of false-positive tests had previously been seen in the same hospital, and subsequently would provide support and reassurance until their anxiety was controlled (Rothenberg & Sills, 1967). Despite several negative tests following a false positive one, it was reported that anxiety often persisted (Hewlett & Waisbren, 2006). Gurian and colleagues (2006) reported a two-fold increased hospitalization rate during the first 6 months of life among children testing false positive for biochemical genetic conditions, and this was attributed to parental overprotection. In 1982, interviews with 31 parents whose children received false positive screening tests for congenital hypothyroidism in the Netherlands revealed that the suspicion of disease placed considerable stress both on the parents and on the parent-child relationships, and certain levels of anxiety persisted even after it was confirmed that the initial result was wrong (Tymstra, 1986). A study that examined 32 families that received several false positive neonatal screening tests for hypothyroidism reported, in 16 of them, persistent anxiety 6-12 months after the screening, and 13 families still exhibited stress 4 years later (Fyrö & Bodegård, 1987). Another report revealed that 6 months after false positive neonatal hearing screening tests, many parents were still concerned about hearing-specific conditions in their children (van der Ploeg et al., 2008). In light of all these issues, one of the most important considerations for DTC genetic testing will be the implementation of protective measures that intend to look out for the best interests of the pediatric population.

A topic with crucial implications for society, that so far has received little attention, is the genetic privacy of elderly individuals, whose decision-making capabilities might be impaired as a result of specific medical conditions. In light of DTC genetic testing, the importance of ensuring that the genetic privacy of this group of individuals is not infringed upon, either innocently or willingly, by family members or caretakers, cannot be overstated. Over 25 million individuals worldwide have dementia, and this number is anticipated to double every 20 years (Qiu et al., 2007; Qiu et al., 2009). In a study that examined individuals with mild Alzheimer's disease, Huthwaite and colleagues (2006) found impaired medical decision-making abilities at baseline, followed by an additional decline over the two years of the study, while Okonkwo and collaborators (2007) reported that patients with mild cognitive impairment have significant difficulties in their ability to consent to treatment. There is an urgent need to understand the implications of DTC genetic testing for individuals with impaired decision-making abilities, and for their families, and to establish and implement guidelines to protect their genetic privacy.

ACKNOWLEDGMENT

For critically reading the manuscript, and for providing insightful comments and helpful suggestions, I would like to express my appreciation and gratitude to H. James Birx, PhD, Department of Sociology, Anthropology, and Criminal Justice at Canisius College, and Barry Schwartz, PhD, Department of Psychology at Swarthmore College.

REFERENCES

Adeoye, S., & Bozic, K. J. (2007). Direct to consumer advertising in healthcare: history, benefits, and concerns. *Clinical Orthopaedics and Related Research*, *457*, 96–104.

Almasi, E. A., Stafford, R. S., Kravitz, R. L., & Mansfield, P. R. (2006). What are the public health effects of direct-to-consumer drug advertising? *PLoS Medicine*, *3*(3), e145. doi:10.1371/journal.pmed.0030145

Altman, R. B. (2009). Direct-to-Consumer Genetic Testing: Failure Is Not an Option. *Clinical Pharmacology and Therapeutics*, *86*(1), 15–17. doi:10.1038/clpt.2009.63

American College of Medical Genetics. (2004). ACGM statement on direct-to-consumer genetic testing. *Genetics in Medicine*, *6*(1), 60. doi:10.1097/01.GIM.0000106164.59722.CE

Anonymous,. (2008). Psychiatric genetic tests raise concerns. Home kits are now available, but they may not be testing for the right genes. *The Harvard Mental Health Letter*, *24*(11), 4.

Anonymous,. (2008). Direct-to-consumer genetic tests: flawed and unethical. *The Lancet Oncology*, *9*, 1113. doi:10.1016/S1470-2045(08)70288-2

Aplenc, R., Thompson, J., Han, P., La, M., Zhao, H., Lange, B., & Rebbeck, T. (2005). Methylenetetrahydofolate reductase polymorphism and therapy response in pediatric acute lymphoblast leukemia. *Cancer Research*, *65*(6), 2482–2487. doi:10.1158/0008-5472.CAN-04-2606

Badano, J. L., & Katsanis, N. (2002). Beyond Mendel: an evolving view of human genetic disease transmission. *Nature Reviews. Genetics*, *3*(10), 779–789. doi:10.1038/nrg910

Barboza, R., Fox, J. H., Shaffer, L. E., Opalek, F. M., & Faroki, S. (2009). Incidental findings in the cervical spine at CT for trauma evaluation. *AJR. American Journal of Roentgenology*, *192*(3), 725–729. doi:10.2214/AJR.08.1420

Basille, C., Frydman, R., & El Aly, A. (2009). Preimplantation genetic diagnosis: state of the art. *European Journal of Obstetrics, Gynecology, and Reproductive Biology*, *145*(1), 9–13. doi:10.1016/j.ejogrb.2009.04.004

Benn, P. A., & Chapman, A. R. (2010). Ethical challenges in providing noninvasive prenatal diagnosis. *Current Opinion in Obstetrics & Gynecology*, *22*(2), 128–134. doi:10.1097/GCO.0b013e3283372352

Bennet, A. M., Di Angelantonio, E., & Ye, Z. (2007). Association of apolipoprotein E genotypes with lipid levels and coronary risk. *Journal of the American Medical Association*, *298*(11), 1300–1311. doi:10.1001/jama.298.11.1300

Berg, C., & Fryer-Edwards, K. (2008). The ethical challenges of direct-to-consumer genetic testing. *Journal of Business Ethics*, *77*, 17–31. doi:10.1007/s10551-006-9298-8

Blauw, H. M., Veldink, J. H., & van Es, M. A. (2008). Copy number variation in sporadic amyotrophic lateral sclerosis: a genome-wide screen. *The Lancet Neurology*, *7*(4), 319–326. doi:10.1016/S1474-4422(08)70048-6

Bloch, M., & Hayden, M. R. (1990). Opinion: predictive testing for Huntington disease in childhood: challenges and implications. *American Journal of Human Genetics*, *46*(1), 1–4.

Blow, N. (2008). DNA sequencing: generation next-next. *Nature Methods*, *5*, 267–274. doi:10.1038/nmeth0308-267

Bolnick, D. A., Fullwiley, D., & Duster, T. (2007). Genetics. The science and business of genetic ancestry testing. *Science*, *318*(5849), 399–400. doi:10.1126/science.1150098

Borry, P., Goffin, T., Nys, H., & Dierickx, K. (2007). Attitudes regarding carrier testing in incompetent children: a survey of European clinical geneticists. *European Journal of Human Genetics*, *15*(12), 1211–1217. doi:10.1038/sj.ejhg.5201909

Borry, P., Goffin, T., Nys, H., & Dierickx, K. (2008). Predictive genetic testing in minors for adult-onset genetic diseases. *The Mount Sinai Journal of Medicine, New York*, *75*(3), 287–296. doi:10.1002/msj.20038

Borry, P., Howard, H. C., Sénécal, K., & Avard, D. (2009). Direct-to-consumer genome scanning services. Also for children? *Nature Reviews. Genetics*, *10*(1), 8. doi:10.1038/nrg2501

Borry, P., Howard, H. C., Sénécal, K., & Avard, D. (2010). Health-related direct-to-consumer genetic testing: a review of companies' policies with regard to genetic testing in minors. *Familial Cancer*, *9*(1), 51–59. doi:10.1007/s10689-009-9253-9

Bovio, S., Cataldi, A., & Reimondo, G. (2006). Prevalence of adrenal incidentaloma in a contemporary computerized tomography series. *Journal of Endocrinological Investigation*, *29*(4), 298–302.

Brookes, A. J. (1999). The essence of SNPs. *Gene*, *234*(2), 177–186. doi:10.1016/S0378-1119(99)00219-X

Brose, M. S., Rebbeck, T. R., & Calzone, K. A. (2002). Cancer risk estimates for BRCA1 mutation carriers identified in a risk evaluation program. *Journal of the National Cancer Institute*, *94*(18), 1365–1372.

Cameron, L. D., & Muller, C. (2009). Psychosocial aspects of genetic testing. *Current Opinion in Psychiatry*, *22*(2), 218–223. doi:10.1097/YCO.0b013e3283252d80

Cappuzzo, F., Varella-Garcia, M., & Shigematsu, H. (2005). Increased *HER2* gene copy number is associated with response to gefitinib therapy in epidermal growth factor receptor-positive non-small-cell lung cancer patients. *Journal of Clinical Oncology*, *23*(22), 5007–5018. doi:10.1200/JCO.2005.09.111

Caulfield, T., Ries, N. M., Ray, P. N., Shuman, C., & Wilson, B. (2010). Direct-to-consumer genetic testing: good, bad or benign? *Clinical Genetics*, *77*(2), 101–105. doi:10.1111/j.1399-0004.2009.01291.x

Centers for Disease Control and Prevention (CDC). (2001). Using tandem mass spectrometry for metabolic disease screening among newborns: A report of a work group. *MMWR. Morbidity and Mortality Weekly Report*, *50*(RR03), 1–22.

Centers for Disease Control and Prevention (CDC). (2004). Genetic testing for breast and ovarian cancer susceptibility: evaluating direct-to-consumer marketing--Atlanta, Denver, Raleigh-Durham, and Seattle, 2003. *MMWR. Morbidity and Mortality Weekly Report*, *53*(27), 603–606.

Cervino, A. C., & Hill, A. V. (2000). Comparison of tests for association and linkage in incomplete families. *American Journal of Human Genetics*, *67*(1), 120–132. doi:10.1086/302992

Cherkas, L. F., Oelsner, E. C., Mak, Y. T., Valdes, A., & Spector, T. D. (2004). Genetic influences on female infidelity and number of sexual partners in humans: a linkage and association study of the role of the vasopressin receptor gene (AVPR1A). *Twin Research*, *7*(6), 649–658. doi:10.1375/1369052042663922

Cho, M. K. (2008). Understanding incidental findings in the context of genetics and genomics. *The Journal of Law, Medicine & Ethics*, *36*(2), 280–285. doi:10.1111/j.1748-720X.2008.00270.x

Clague, A., & Thomas, A. (2002). Neonatal biochemical screening for disease. *Clinica Chimica Acta*, *315*(1-2), 99–110. doi:10.1016/S0009-8981(01)00716-1

Clarke, A. (2007). Should families own genetic information? No. *British Medical Journal*, *335*(7609), 23. doi:10.1136/bmj.39252.392940.AD

Claus, E. B., Schildkraut, J. M., Thompson, W. D., & Risch, N. J. (1996). The genetic attributable risk of breast and ovarian cancer. *Cancer*, *77*(11), 2318–2324. doi:10.1002/(SICI)1097-0142(19960601)77:11<2318::AID-CNCR21>3.0.CO;2-Z

Collins, F. S., & Guttmacher, A. E. (2003). Welcome to the genomic era. *The New England Journal of Medicine*, *349*(10), 996–998. doi:10.1056/NEJMe038132

Council of Europe. Additional Protocol to the Convention on Human Rights and Biomedicine, concerning Genetic Testing for Health Purposes. Strasbourg, November 27, 2008. Accessed March 4, 2010. http://conventions.coe.int/Treaty/en/Treaties/Html/203.htm

Criswell, L. A., Pfeiffer, K. A., & Lum, R. F. (2005). Analysis of families in the Multiple Autoimmune Disease Genetics Consortium (MADGC) Collection: the PTPN22 620W allele associates with multiple autoimmune phenotypes. *American Journal of Human Genetics*, *76*(4), 561–571. doi:10.1086/429096

Croyle, R. T., Smith, K. R., Botkin, J. R., Baty, B., & Nash, J. (1997). Psychological responses to BRCA1 mutation testing: preliminary findings. *Health Psychology*, *16*(1), 63–72. doi:10.1037/0278-6133.16.1.63

Cummings, S. (2000). The genetic testing process: how much counseling is needed? *Journal of Clinical Oncology*, *18*(Suppl 21), 60S–62S.

Danoff, T. M., Campbell, D. A., & McCarthy, L. C. (2004). A Gilbert's syndrome UGT1A1 variant confers susceptibility to tranilast-induced hyperbilirubinemia. *The Pharmacogenomics Journal*, *4*(1), 49–53. doi:10.1038/sj.tpj.6500221

de Chaves, E. P., & Narayanaswami, V. (2008). Apolipoprotein E and cholesterol in aging and disease in the brain. *Future Lipidology*, *3*(5), 505–530. doi:10.2217/17460875.3.5.505

Ding, Z., Harding, C. O., & Thöny, B. (2004). State-of-the-art 2003 on PKU gene therapy. *Molecular Genetics and Metabolism*, *81*(1), 3–8. doi:10.1016/j.ymgme.2003.09.010

Dorval, M., Patenaude, A. F., & Schneider, K. A. (2000). Anticipated versus actual emotional reactions to disclosure of results of genetic tests for cancer susceptibility: Findings from *p53* and *BRCA1* testing programs. *Journal of Clinical Oncology*, *18*(10), 2135–2142.

Douglas, H. A., Hamilton, R. J., & Grubs, R. E. (2009). The Effect of BRCA Gene Testing on Family Relationships: A Thematic Analysis of Qualitative Interviews. *Journal of Genetic Counseling*, *18*(5), 418–435. doi:10.1007/s10897-009-9232-1

Duncan, R. E., & Delatycki, M. B. (2006). Predictive genetic testing in young people for adult-onset conditions: where is the empirical evidence? *Clinical Genetics*, *69*(1), 8–16. doi:10.1111/j.1399-0004.2005.00505.x

Epel, E. S., Blackburn, E. H., & Lin, J. (2004). Accelerated telomere shortening in response to life stress. *Proceedings of the National Academy of Sciences of the United States of America*, *101*(49), 17312–17315. doi:10.1073/pnas.0407162101

Evans, J. P. (2008). Recreational genomics; what's in it for you? *Genetics in Medicine*, *10*(10), 709–710. doi:10.1097/GIM.0b013e3181859959

Farkas, D. H., & Holland, C. A. (2009). Direct-to-consumer genetic testing: two sides of the coin. *The Journal of Molecular Diagnostics*, *11*(4), 263–265. doi:10.2353/jmoldx.2009.090034

Farrer, L. A., Cupples, L. A., & Haines, J. L. (1997). Effects of age, sex and ethnicity on the association between apolipoprotein E genotype and Alzheimer disease: a meta-analysis. *Journal of the American Medical Association*, *278*(16), 1349–1356. doi:10.1001/jama.278.16.1349

Feuk, L., Carson, A. R., & Scherer, S. W. (2006). Structural variation in the human genome. *Nature Reviews. Genetics*, *7*(2), 85–97. doi:10.1038/nrg1767

Fields, C., Adams, M. D., White, O., & Venter, J. C. (1994). How many genes in the human genome? *Nature Genetics*, *7*(3), 345–346. doi:10.1038/ng0794-345

Ford, D., Easton, D. F., & Bishop, D. T. (1994). Risks of cancer in BRCA1-mutation carriers. Breast Cancer Linkage Consortium. *Lancet*, *343*(8899), 692–695. doi:10.1016/S0140-6736(94)91578-4

Foster, C., Watson, M., & Eeeles, R. (2007). Predictive genetic testing for BRCA1/2 in a UK clinical cohort: three-year follow-up. *British Journal of Cancer*, *96*(5), 718–724. doi:10.1038/sj.bjc.6603610

Freeman, J. L., Perry, G. H., & Feuk, L. (2006). Copy number variation: new insights in genome diversity. *Genome Research*, *16*(8), 949–961. doi:10.1101/gr.3677206

Frosch, D. L., Grande, D., Tarn, D. M., & Kravitz, R. L. (2010). A decade of controversy: balancing policy with evidence in the regulation of prescription drug advertising. *American Journal of Public Health*, *100*(1), 24–32. doi:10.2105/AJPH.2008.153767

Frosch, D. L., Krueger, P. M., Hornik, R. C., Cronholm, P. F., & Barg, F. K. (2007). Creating demand for prescription drugs: a content analysis of television direct-to-consumer advertising. *Annals of Family Medicine*, *5*(1), 6–13. doi:10.1370/afm.611

Frosst, P., Blom, H. J., & Milos, R. (1995). A candidate genetic risk factor for vascular disease: a common mutation in methylenetetrahydrofolate reductase. *Nature Genetics*, *10*(1), 111–113. doi:10.1038/ng0595-111

Fryer, A. (1995). Genetic testing of children. *Archives of Disease in Childhood*, *73*(2), 97–99. doi:10.1136/adc.73.2.97

Fryer, A. (2000). Inappropriate genetic testing of children. *Archives of Disease in Childhood*, *83*(4), 283–285. doi:10.1136/adc.83.4.283

Fyrö, K., & Bodegård, G. (1987). Four-year follow-up of psychological reactions to false positive screening tests for congenital hypothyroidism. *Acta Paediatrica Scandinavica*, *76*(1), 107–114. doi:10.1111/j.1651-2227.1987.tb10424.x

Genetics and Public Policy Center. (2007). *Survey of direct-to-consumer testing statutes and regulations*. Berman Institute of Bioethics, Johns Hopkins University. http://www.dnapolicy.org

Geransar, R., & Einsiedel, E. (2008). Evaluating online direct-to-consumer marketing of genetic tests: informed choices or buyers beware? *Genetic Testing*, *12*(1), 13–23. doi:10.1089/gte.2007.0024

Gigerenzer, G., Gaissmeier, W., Kurz-Milcke, E., Schwartz, L. M., & Woloshin, S. (2008). Helping doctors and patients make sense of health statistics. *Psychological Science in the Public Interest*, *8*(2), 53–96.

Gilbar, R. (2010). Genetic testing of children for familial cancers: a comparative legal perspective on consent, communication of information and confidentiality. *Familial Cancer*, *9*(1), 75–87. doi:10.1007/s10689-009-9268-2

Gilbody, S., Lewis, S., & Lightfoot, T. (2006). Methylenetetrahydrofolate reductase (MTHFR) genetic polymorphisms and psychiatric disorders: a HuGE review. *American Journal of Epidemiology*, *165*(1), 1–13. doi:10.1093/aje/kwj347

Goddard, K. A., Robitaille, J., & Dowling, N. F. (2009). Health-related direct-to-consumer genetic tests: a public health assessment and analysis of practices related to Internet-based tests for risk of thrombosis. *Public Health Genomics*, *12*(2), 92–104. doi:10.1159/000176794

Gollust, S. E., Hull, S. C., & Wilfond, B. S. (2002). Limitations of direct-to-consumer advertising for clinical genetic testing. *Journal of the American Medical Association*, *288*(14), 1762–1767. doi:10.1001/jama.288.14.1762

Gollust, S. E., Wilfond, B. S., & Hull, S. C. (2003). Direct-to-consumer sales of genetic services on the Internet. *Genetics in Medicine*, *5*(4), 332–337. doi:10.1097/01.GIM.0000076972.83711.48

Gonzalez, E., Kulkarni, H., & Bolivar, H. (2005). The influence of *CCL3L1* gene-containing segmental duplications on HIV-1/AIDS susceptibility. *Science*, *307*(5714), 1434–1440. doi:10.1126/science.1101160

Government Accountability Office. (2006) Nutrigenetic testing: tests purchased from four Web sites mislead consumers. Available from http://www.gao.gov/products/GAO-06-977T. Accessed February 21, 2010.

Graham, D. S., Akil, M., & Vyse, T. J. (2007). Association of polymorphisms across the tyrosine kinase gene, TYK2 in UK SLE families. *Rheumatology (Oxford, England)*, *46*(6), 927–930. doi:10.1093/rheumatology/kel449

Gray, S. W., O'Grady, C., Karp, L., Smith, D., Schwartz, J. S., Hornik, R. C., & Armstrong, K. (2009). Risk information exposure and direct-to-consumer genetic testing for BRCA mutations among women with a personal or family history of breast or ovarian cancer. *Cancer Epidemiology, Biomarkers & Prevention*, *18*(4), 1303–1311. doi:10.1158/1055-9965.EPI-08-0825

Grünhage, F., Nattermann, J., & Gressner, O. A. (2010). Lower copy numbers of the chemokine *CCL3L1* gene in patients with chronic hepatitis C. *Journal of Hepatology*, *52*(2), 153–159. doi:10.1016/j.jhep.2009.11.001

Gurian, E. A., Kinnamon, D. D., Henry, J. J., & Waisbren, S. E. (2006). Expanded newborn screening for biochemical disorders: the effect of a false-positive result. *Pediatrics*, *117*(6), 1915–1921. doi:10.1542/peds.2005-2294

Gurwitz, D., & Bregman-Eschet, Y. (2009). Personal genomics services: whose genomes? *European Journal of Human Genetics*, *17*(7), 883–889. doi:10.1038/ejhg.2008.254

Hamilton, J. G., Lobel, M., & Moyer, A. (2009). Emotional distress following genetic testing for hereditary breast and ovarian cancer: a meta-analytic review. *Health Psychology*, *28*(4), 510–518. doi:10.1037/a0014778

Helgadottir, A., Thorleifsson, G., & Manolescu, A. (2007). A common variant on chromosome 9p21 affects the risk of myocardial infarction. *Science*, *316*(5830), 1491–1493. doi:10.1126/science.1142842

Hemmings, S. M., Kinnear, C. J., & Lochner, C. (2006). Genetic correlates in trichotillomania. A case- control association study in the South African Caucasian population. *The Israel Journal of Psychiatry and Related Sciences*, *43*(2), 93–101.

Hertzberg, M., Neville, S., Favaloro, E., & McDonald, D. (2005). External quality assurance of DNA testing for thrombophilia mutations. *American Journal of Clinical Pathology*, *123*(2), 189–193. doi:10.1309/09F827BC91M3D91C

Hertzberg, M., Neville, S., & McDonald, D. (2006). External quality assurance of molecular analysis of haemochromatosis gene mutations. *Journal of Clinical Pathology*, *59*(7), 744–747. doi:10.1136/jcp.2005.026005

Hewlett, J., & Waisbren, S. E. (2006). A review of the psychosocial effects of false-positive results on parents and current communication practices in newborn screening. *Journal of Inherited Metabolic Disease*, *29*(5), 677–682. doi:10.1007/s10545-006-0381-1

Hofgärtner, W. T., & Tait, J. F. (1999). Frequency of problems during clinical molecular-genetic testing. *American Journal of Clinical Pathology*, *112*(1), 14–21.

Hogarth, S., Javitt, G., & Melzer, D. (2008). The current landscape for direct-to-consumer genetic testing: legal, ethical, and policy issues. *Annual Review of Genomics and Human Genetics*, *9*, 161–182. doi:10.1146/annurev.genom.9.081307.164319

Hollon, M. F. (2005). Direct-to-consumer advertising. A haphazard approach to health promotion. *Journal of the American Medical Association*, *293*(16), 2030–2033. doi:10.1001/jama.293.16.2030

Hollox, E. J., Armour, J. A. L., & Barber, J. C. K. (2003). Extensive normal copy number variation of a beta-defensin antimicrobial-gene cluster. *American Journal of Human Genetics*, *73*(3), 591–600. doi:10.1086/378157

Honeywell, C. R., Gollob, M. H., Rutberg, J., Gow, R. M., & Geraghty, M. T. (2008). Discrepant DNA analysis in three patients with inherited arrhythmia: molecular genetic test results deserve a second glance. *American Journal of Medical Genetics. Part A*, *146A*(11), 1466–1469. doi:10.1002/ajmg.a.32336

Hudson, K., Javitt, G., Burke, W., & Byers, P., & American Society of Human Genetics Social Issues Committee. (2007). ASHG Statement on direct-to-consumer genetic testing in the United States. *Obstetrics and Gynecology*, *110*(6), 1392–1395.

Hull, S. C., & Prasad, K. (2001). Reading between the lines: direct-to-consumer advertising of genetic testing in the USA. *Reproductive Health Matters*, *9*(18), 44–48. doi:10.1016/S0968-8080(01)90089-8

Human Genetics Commission. (2003). Genes Direct: Ensuring the effective oversight of genetic tests supplied directly to the public. London: Department of Health. Available at: http://www.dh.gov.uk/en/index.htm, Accessed March 07, 2010.

Huthwaite, J. S., Martin, R. C., Griffith, H. R., Anderson, B., Harrell, L. E., & Marson, D. C. (2006). Declining medical decision-making capacity in mild AD: a two-year longitudinal study. *Behavioral Sciences & the Law*, *24*(4), 453–463. doi:10.1002/bsl.701

Iervolino, A. C., Perroud, N., Fullana, M. A., Guipponi, M., Cherkas, L., Collier, D. A., & Mataix-Cols, D. (2009). Prevalence and heritability of compulsive hoarding: a twin study. *The American Journal of Psychiatry*, *166*(10), 1156–1161. doi:10.1176/appi.ajp.2009.08121789

Imelfort, I., Batley, J., Grimmond, S., & Edwards, D. (2009). Genome sequencing approaches and successes. *Methods in Molecular Biology (Clifton, N.J.)*, *513*, 345–358. doi:10.1007/978-1-59745-427-8_18

International Human Genome Sequencing Consortium. (2004). Finishing the euchromatic sequence of the human genome. *Nature*, *431*(7011), 931–945. doi:10.1038/nature03001

Ioannidis, J. P. (2009). Personalized genetic prediction: too limited, too expensive, or too soon? *Annals of Internal Medicine*, *150*(2), 139–141.

Ionita-Laza, I., Rogers, A. J., Lange, C., Raby, B. A., & Lee, C. (2009). Genetic association analysis of copy-number variation (CNV) in human disease pathogenesis. *Genomics*, *93*(1), 22–26. doi:10.1016/j.ygeno.2008.08.012

Jameson, J. L. (1998). *Principles of Molecular Medicine* (1st ed.). Totowa, NJ: Humana Press.

Janssens, A. C., Gwinn, M., Bradley, L. A., Oostra, B. A., van Duijn, C. M., & Khoury, M. J. (2008). A critical appraisal of the scientific basis of commercial genomic profiles used to assess health risks and personalize health interventions. *American Journal of Human Genetics*, *82*(3), 593–599. doi:10.1016/j.ajhg.2007.12.020

Kaklamani, V. G., Wisinski, K. B., & Sadmi, M. (2008). Variants of the adiponectin (ADIPOQ) and adiponectin receptor 1 (ADIPOR1) genes and colorectal cancer risk. *Journal of the American Medical Association*, *300*(13), 1523–1531. doi:10.1001/jama.300.13.1523

Kaput, J. (2008). Nutrigenomics research for personalized nutrition and medicine. *Current Opinion in Biotechnology*, *19*(2), 110–120. doi:10.1016/j.copbio.2008.02.005

Katsanis, S. H., Javitt, G., & Hudson, K. (2008). Public health. A case study of personalized medicine. *Science*, *320*(5872), 53–54. doi:10.1126/science.1156604

Kawanishi, C., Lundrgen, S., Agren, H., & Bertilsson, L. (2004). Increased incidence of *CYP2D6* gene duplication in patients with persistent mood disorders: ultrarapid metabolism of antidepressants as a cause of nonresponse. A pilot study. *European Journal of Clinical Pharmacology*, *9*(11), 803–807.

Kaye, J. (2008). The regulation of direct-to-consumer genetic tests. *Human Molecular Genetics*, *17*(R2), R180–R183. doi:10.1093/hmg/ddn253

Kehrer-Sawatzki, H. (2007). What a difference copy number variation makes. *BioEssays*, *29*(4), 311–313. doi:10.1002/bies.20554

Khandanpour, N., Willis, G., & Meyer, F. J. (2009). Peripheral arterial disease and methylenetetrahydrofolate reductase (MTHFR) C677T mutations: a case-control study and meta-analysis. *Journal of Vascular Surgery*, *49*(3), 711–718. doi:10.1016/j.jvs.2008.10.004

Kilpivaara, O., & Mukherjee, S., S., Schram, A.M., et al. (2009). A germline JAK2 SNP is associated with predisposition to the development of JAK2 (V617F)-positive myeloproliferative neoplasms. *Nature Genetics*, *41*(4), 455–459. doi:10.1038/ng.342

Kohane, I. S., Masys, D. R., & Altman, R. B. (2006). The incidentalome: a threat to genomic medicine. *Journal of the American Medical Association*, *296*(2), 212–215. doi:10.1001/jama.296.2.212

Kuhn, L., Schramm, D. B., & Donninger, S. (2007). African infants' *CCL3* gene copies influence perinatal HIV transmission in the absence of maternal nevirapine. *AIDS (London, England)*, *21*(13), 1753–1761. doi:10.1097/QAD.0b013e3282ba553a

Kyogoku, C., Morinobu, A., & Nishimura, K. (2009). Lack of association between tyrosine kinase 2 (TYK2) gene polymorphisms and susceptibility to SLE in a Japanese population. *Modern Rheumatology*, *19*(4), 401–406. doi:10.1007/s10165-009-0173-1

Lakusta, L., Dessalegn, B., & Landau, B. (2010). Impaired geometric reorientation caused by genetic defect. *Proceedings of the National Academy of Sciences of the United States of America*, *107*(7), 2813–2817. doi:10.1073/pnas.0909155107

Lander, E. S., Linton, L. M., & Birren, B. (2001). Initial sequencing and analysis of the human genome. *Nature*, *409*(6822), 860–921. doi:10.1038/35057062

Leacock, S. (1924). *The garden of folly* (pp. 122–131). New York: Dodd Mead.

Lee, H. S., Korman, B. D., & Le, J. M. (2009). Genetic risk factors for rheumatoid arthritis differ in Caucasian and Korean populations. *Arthritis and Rheumatism*, *60*(2), 364–371. doi:10.1002/art.24245

Lee, S. S., & Crawley, L. (2009). Research 2.0: social networking and direct-to-consumer (DTC) genomics. *The American Journal of Bioethics*, *9*(6-7), 35–44. doi:10.1080/15265160902874452

Lerman, C., Hughes, C., & Lemon, S. J. (1998). What you don't know can hurt you: adverse psychologic effects in members of BRCA1-linked and BRCA2-linked families who decline genetic testing. *Journal of Clinical Oncology*, *16*(5), 1650–1654.

Levy, H. L., & Albers, S. (2000). Genetic screening of newborns. *Annual Review of Genomics and Human Genetics*, *1*, 139–177. doi:10.1146/annurev.genom.1.1.139

Libby, E. N., Booker, J. K., Gulley, M. L., Garcia, D., & Moll, S. (2006). False-negative factor V Leiden genetic testing in a patient with recurrent deep venous thrombosis. *American Journal of Hematology*, *81*(4), 284–289. doi:10.1002/ajh.20543

Lissak, A., Sharon, A., Fruchter, O., Kassel, A., Sanderovitz, J., & Abramovici, H. (1999). Polymorphism for mutation of cytosine to thymine at location 677 in the methylenetetrahydrofolate reductase gene is associated with recurrent early fetal loss. *American Journal of Obstetrics and Gynecology*, *181*(1), 126–130. doi:10.1016/S0002-9378(99)70447-3

Little, J., Higgins, J. P., & Ioannidis, J. P. (2009). STrengthening the REporting of Genetic Association studies (STREGA): an extension of the STROBE Statement. *Annals of Internal Medicine*, *150*(3), 206–215.

Liu, Y., & Pearson, Y. E. (2008). Direct-to-consumer marketing of predictive medical genetic tests: assessment of current practices and policy recommendations. *Journal of Public Policy & Marketing*, *27*(2), 131–148. doi:10.1509/jppm.27.2.131

Lobo, D. S., & Kennedy, J. L. (2009). Genetic aspects of pathological gambling: a complex disorder with shared genetic vulnerabilities. *Addiction (Abingdon, England)*, *104*(9), 1454–1465. doi:10.1111/j.1360-0443.2009.02671.x

Lodder, L., Frets, P. G., & Trijsburg, R. W. (2001). Psychological impact of receiving a BRCA1/BRCA2 test result. *American Journal of Medical Genetics*, *98*(1), 15–24. doi:10.1002/1096-8628(20010101)98:1<15::AID-AJMG1014>3.0.CO;2-0

Lowery, J. T., Byers, T., Axell, L., Ku, L., & Jacobellis, J. (2008). The impact of direct-to-consumer marketing of cancer genetic testing on women according to their genetic risk. *Genetics in Medicine*, *10*(12), 888–894. doi:10.1097/GIM.0b013e31818de6d7

Lucassen, A. (2007). Should families own genetic information? Yes. *British Medical Journal*, *335*(7609), 22. doi:10.1136/bmj.39252.386030.AD

Lucassen, A., & Parker, M. (2001). Revealing false paternity: some ethical considerations. *Lancet*, *357*(9261), 1033–1035. doi:10.1016/S0140-6736(00)04240-9

Maddox, B. (2003). The double helix and the 'wronged heroine'. *Nature*, *421*(6921), 407–408. doi:10.1038/nature01399

Manolio, T. A., Brooks, L. D., & Collins, F. S. (2008). A HapMap harvest of insights into the genetics of common disease. *The Journal of Clinical Investigation*, *118*(5), 1590–1605. doi:10.1172/JCI34772

Marietta, C., & McGuire, A. L. (2009). Currents in contemporary ethics. Direct-to-consumer genetic testing: is it the practice of medicine? *The Journal of Law, Medicine & Ethics*, *37*(2), 369–374. doi:10.1111/j.1748-720X.2009.00380.x

Maxam, A. M., & Gilbert, W. (1977). A new method for sequencing DNA. *Proceedings of the National Academy of Sciences of the United States of America*, *74*(2), 560–564. doi:10.1073/pnas.74.2.560

McCarthy, M. I., Abencasis, G. R., Cardon, L. R., Goldstein, D. B., Little, J., Ioannidis, J. P., & Hirschhorn, J. N. (2008). Genome-wide association studies for complex traits: consensus, uncertainty and challenges. *Nature Reviews. Genetics*, *9*(5), 356–369. doi:10.1038/nrg2344

Metzker, M. L. (2010). Sequencing technologies - the next generation. *Nature Reviews. Genetics*, *11*(1), 31–46. doi:10.1038/nrg2626

Mezer, E., & Wygnanski-Jaffe, T. (2009). Ethical issues in ocular genetics. *Current Opinion in Ophthalmology*, *20*(5), 382–386. doi:10.1097/ICU.0b013e32832f7feb

Mitchell, P. B., Meiser, B., Wilde, A., Fullerton, J., Donald, J., Wilhelm, K., & Schofield, P. R. (2010). Predictive and diagnostic genetic testing in psychiatry. *The Psychiatric Clinics of North America*, *33*(1), 225–243. doi:10.1016/j.psc.2009.10.001

Molster, C., Charles, T., Samanek, A., & O'Leary, P. (2009). Australian study on public knowledge of human genetics and health. *Public Health Genomics*, *12*(2), 84–91. doi:10.1159/000164684

Morozova, O., & Marra, M. A. (2008). Applications of next-generation sequencing technologies in functional genomics. *Genomics*, *92*(5), 255–264. doi:10.1016/j.ygeno.2008.07.001

Moser, K. L., Kelly, J. A., Lessard, C. J., & Harley, J. B. (2009). Recent insights into the genetic basis of systemic lupus erythematosus. *Genes and Immunity*, *10*(5), 373–379. doi:10.1038/gene.2009.39

Mouchawar, J., Hensley-Alford, S., & Laurion, S. (2005). Impact of direct-to-consumer advertising for hereditary breast cancer testing on genetic services at a managed care organization: a naturally-occurring experiment. *Genetics in Medicine*, *7*(3), 191–197. doi:10.1097/01.GIM.0000156526.16967.7A

Murphy, W. J., Larkin, D. M., & Everts-van der Wind, A. (2005). Dynamics of mammalian chromosome evolution inferred from multispecies comparative maps. *Science*, *309*(5734), 613–617. doi:10.1126/science.1111387

Nakajima, T., Kaur, G., Mehra, N., & Kimura, A. (2008). HIV-1/AIDS susceptibility and copy number variation in CCL3L1, a gene encoding a natural ligand for HIV-1 co-receptor CCR5. *Cytogenetic and Genome Research*, *123*(1-4), 156–160. doi:10.1159/000184703

National Human Genome Research Institute. (n.d.). Retrieved from www.genome.gov

Nicoletto, M. O., Donach, M., De Nicolo, A., Artioli, G., Banna, G., & Monfardini, S. (2001). BRCA-1 and BRCA-2 mutations as prognostic factors in clinical practice and genetic counselling. *Cancer Treatment Reviews*, *27*(5), 295–304. doi:10.1053/ctrv.2001.0233

Nicoll, J. A. R., Roberts, G. W., & Graham, D. I. (1995). Apolipoprotein E4 allele is associated with deposition of amyloid beta-protein following head injury. *Nature Medicine*, *1*(2), 135–137. doi:10.1038/nm0295-135

Ohata, T., Tsuchiya, A., Watanabe, M., Sumida, T., & Takada, F. (2009). Physicians' opinion for 'new' genetic testing in Japan. *Journal of Human Genetics*, *54*(4), 203–208. doi:10.1038/jhg.2009.11

Okada, Y., Mori, M., & Yamada, R. (2008). SLC22A4 polymorphism and rheumatoid arthritis susceptibility: a replication study in a Japanese population and a metaanalysis. *The Journal of Rheumatology*, *35*(9), 1723–1728.

Okonkwo, O., Griffith, H. R., & Belue, K. (2007). Medical decision-making capacity in patients with mild cognitive impairment. *Neurology*, *69*(15), 1528–1535. doi:10.1212/01.wnl.0000277639.90611.d9

Palomaki, G. E., Bradley, L. A., Richards, C. S., Richards, C. S., & Haddow, J. E. (2003). Analytic validity of cystic fibrosis testing: a preliminary estimate. *Genetics in Medicine*, *5*(1), 15–20. doi:10.1097/00125817-200301000-00003

Panas, M., Karadima, G., Markianos, M., Kalfakis, N., & Vassilopoulos, D. (2008). Phenotypic discordance in a pair of monozygotic twins with Huntington's disease. *Clinical Genetics*, *74*(3), 291–292. doi:10.1111/j.1399-0004.2008.01036.x

Pasacreta, J. V. (2003). Psychosocial issues associated with genetic testing for breast and ovarian cancer risk: an integrative review. *Cancer Investigation*, *21*(4), 588–623. doi:10.1081/CNV-120022380

Paul, D. B. (2008). Patient advocacy in newborn screening: continuities and discontinuities. *American Journal of Medical Genetics. Part C, Seminars in Medical Genetics*, *148C*(1), 8–14. doi:10.1002/ajmg.c.30166

Pena, S. D., & Chakraborty, R. (1994). Paternity testing in the DNA era. *Trends in Genetics*, *10*(6), 204–209. doi:10.1016/0168-9525(94)90257-7

Pettersson, E., Lundeberg, J., & Ahmadian, A. (2009). Generations of sequencing technologies. *Genomics*, *93*(2), 105–111. doi:10.1016/j.ygeno.2008.10.003

Pfister, E. L., Kennington, L., & Straubhaar, J. (2009). Five siRNAs targeting three SNPs may provide therapy for three-quarters of Huntington's disease patients. *Current Biology*, *19*(9), 774–778. doi:10.1016/j.cub.2009.03.030

Plass, A. M., Baars, M. J., & Cornel, M. C. (2009). Testing the children: do non-genetic health-care providers differ in their decision to advise genetic presymptomatic testing on minors? A cross-sectional study in five countries in the European Union. *Genetic Testing and Molecular Biomarkers*, *13*(3), 367–376. doi:10.1089/gtmb.2008.0119

Poirier, J. (1994). Apolipoprotein E in animal models of CNS injury and in Alzheimer's disease. *Trends in Neurosciences*, *17*(12), 525–530. doi:10.1016/0166-2236(94)90156-2

Preston, F. E., Kitchen, S., Jennings, I., & Woods, T. A. (1999). A UK National External Quality Assessment scheme (UK Neqas) for molecular genetic testing for the diagnosis of familial thrombophilia. *Thrombosis and Haemostasis*, *82*(5), 1556–1557.

Qiu, C., De Ronchi, D., & Fratiglioni, L. (2007). The epidemiology of the dementias: an update. *Current Opinion in Psychiatry*, *20*(4), 380–385. doi:10.1097/YCO.0b013e32816ebc7b

Qiu, C., Kivipelto, M., & von Strauss, E. (2009). Epidemiology of Alzheimer's disease: occurrence, determinants, and strategies toward intervention. *Dialogues in Clinical NeuroSciences*, *11*(2), 111–128.

Raber, J., Huang, Y., & Ashford, J. W. (2004). ApoE genotype accounts for the vast majority of the AD risk and AD pathology. *Neurobiology of Aging*, *25*(5), 641–650. doi:10.1016/j.neurobiolaging.2003.12.023

Ramus, S. J., & Gayther, S. A. (2009). The contribution of BRCA1 and BRCA2 to ovarian cancer. *Molecular Oncology*, *3*(2), 138–150. doi:10.1016/j.molonc.2009.02.001

Rhodes, R. (2006). Why test children for adult-onset genetic diseases? *The Mount Sinai Journal of Medicine, New York*, *73*(3), 609–616.

Ries, N. M., & Castle, D. (2008). Nutrigenomics and ethics interface: direct-to-consumer services and commercial aspects. *OMICS: A Journal of Integrative Biology*, *12*(4), 245–250. doi:10.1089/omi.2008.0049

Risch, N., Spiker, D., & Lotspeich, L. (1999). A genomic screen of autism: evidence for a multilocus etiology. *American Journal of Human Genetics*, *65*(2), 493–507. doi:10.1086/302497

Robertson, J. A. (2003). The $1000 genome: ethical and legal issues in whole genome sequencing of individuals. *The American Journal of Bioethics*, *3*(3), W35–W42. doi:10.1162/152651603322874762

Rodríguez-Guillén Mdel, R., Torres-Sánchez, L., & Chen, J. (2009). Maternal MTHFR polymorphisms and risk of spontaneous abortion. *Salud Pública de México*, *51*(1), 19–25.

Roses, A. D. (1996). Apolipoprotein E and Alzheimer's disease. A rapidly expanding field with medical and epidemiological consequences. *Annals of the New York Academy of Sciences*, *802*, 50–57. doi:10.1111/j.1749-6632.1996.tb32598.x

Rothenberg, M. B., & Sills, E. M. (1967). Iatrogenesis-PKU anxiety syndrome. *The American Journal of Psychiatry*, *124*(1), 109.

Rothstein, M. A. (2004). *Genetics and Life Insurance: Medical Underwriting and Social Policy*. Boston: MIT Press.

Samatovicz, R. A. (2000). Genetics and brain injury: apolipoprotein E. *The Journal of Head Trauma Rehabilitation*, *15*(3), 869–874. doi:10.1097/00001199-200006000-00002

Sanger, F., & Coulson, A. R. (1975). A rapid method for determining sequences in DNA by primed synthesis with DNA polymerase. *Journal of Molecular Biology*, *94*(3), 441–448. doi:10.1016/0022-2836(75)90213-2

Sanger, F., Nicklen, S., & Coulson, A. R. (1977). DNA sequencing with chain-terminating inhibitors. *Proceedings of the National Academy of Sciences of the United States of America*, *74*(12), 5463–5467. doi:10.1073/pnas.74.12.5463

Saunders, A. M., Strittmatter, W. J., & Schmechel, D. (1993). Association of apolipoprotein E allele epsilon 4 with late-onset familial and sporadic Alzheimer's disease. *Neurology*, *43*(8), 1467–1472.

Savulescu, J. Compulsory genetic testing for APOE Epsilon 4 and Boxing. In Tamburrini CM, Tännsjö T. (Ed.) *Genetic Technology and Sport: Ethical Questions (Ethics and Sport)*. 1st Edition, Routledge.

Schickedanz, A. D., & Herdman, R. C. (2009). Direct-to-consumer genetic testing: the need to get retail genomics right. *Clinical Pharmacology and Therapeutics*, *86*(1), 17–20. doi:10.1038/clpt.2009.56

Scriver, C. R., & Waters, P. J. (1999). Monogenic traits are not simple: lessons from phenylketonuria. *Trends in Genetics*, *15*(7), 267–272. doi:10.1016/S0168-9525(99)01761-8

Service, R. F. (2006). Gene sequencing: The Race for the $1000 Genome. *Science*, *311*, 1544–1546. doi:10.1126/science.311.5767.1544

Shastry, B. S. (2002). SNP alleles in human disease and evolution. *Journal of Human Genetics*, *47*(11), 561–566. doi:10.1007/s100380200086

Shastry, B. S. (2007). SNPs in disease gene mapping, medicinal drug development and evolution. *Journal of Human Genetics*, *52*(11), 871–880. doi:10.1007/s10038-007-0200-z

Shendure, J., Mitra, R. D., Varma, C., & Church, G. M. (2004). Advanced sequencing technologies: methods and goals. *Nature Reviews. Genetics*, *5*(5), 335–344. doi:10.1038/nrg1325

Sigurdsson, S., Nordmark, G., & Göring, H. H. (2005). Polymorphisms in the tyrosine kinase 2 and interferon regulatory factor 5 genes are associated with systemic lupus erythematosus. *American Journal of Human Genetics*, *76*(3), 528–537. doi:10.1086/428480

Smith, A., Moran, A., & Boyd, M. C. (2007). Phenocopies in BRCA1 and BRCA2 families: evidence for modifier genes and implications for screening. *Journal of Medical Genetics*, *44*(1), 10–15. doi:10.1136/jmg.2006.043091

Sobel, S., & Cowan, C. B. (2003). Ambiguous loss and disenfrenchized grief: the impact of DNA predictive testing on the family as a system. *Family Process*, *42*(1), 47–57. doi:10.1111/j.1545-5300.2003.00047.x

Sobrino, B., Brión, M., & Carracedo, A. (2005). SNPs in forensic genetics: a review on SNP typing methodologies. *Forensic Science International*, *154*(2-3), 181–194. doi:10.1016/j.forsciint.2004.10.020

Sobrino, B., & Carracedo, A. (2005). SNP typing in forensic genetics: a review. *Methods in Molecular Biology (Clifton, N.J.)*, *297*, 107–126.

Spriggs, M. (2004). Compulsory brain scans and genetic tests for boxers-or should boxing be banned? *Journal of Medical Ethics*, *30*, 515–516. doi:10.1136/jme.2003.003541

Stewart, A. D. (1989). Screening for cystic fibrosis. *Nature*, *341*(6244), 696. doi:10.1038/341696b0

Sudarsky, L., Myers, R. H., & Walshe, T. M. (1983). Huntington's disease in monozygotic twins reared apart. *Journal of Medical Genetics*, *20*(6), 408–411. doi:10.1136/jmg.20.6.408

Teasdale, G. M., Murray, G. D., & Nicoll, J. A. (2005). The association between APOE epsilon4, age and outcome after head injury: a prospective cohort study. *Brain*, *128*(Pt 11), 2556–2561. doi:10.1093/brain/awh595

Teasdale, G. M., Nicoll, J. A., Murray, G., & Fiddes, M. (1997). Association of apolipoprotein E polymorphism with outcome after head injury. *Lancet*, *350*(9084), 1069–1071. doi:10.1016/S0140-6736(97)04318-3

Thakur, M., Grossman, I., & McCrory, D. C. (2007). Review of evidence for genetic testing for CYP450 polymorphisms in management of patients with nonpsychotic depression with selective serotonin reuptake inhibitors. *Genetics in Medicine*, *9*(12), 826–835. doi:10.1097/GIM.0b013e31815bf98f

Thangadurai, S. (2004). The Human Genome Project: the role of analytical chemists. *Analytical Sciences*, *20*(4), 595–601. doi:10.2116/analsci.20.595

The Huntington's Disease Collaborative Research Group. (1993). A novel gene containing a trinucleotide repeat that is expanded and unstable on Huntington's disease chromosomes. *Cell*, *72*(6), 971–983. doi:10.1016/0092-8674(93)90585-E

The International HapMap Consortium. (2003). The International HapMap Project. *Nature*, *426*, 789–796. doi:10.1038/nature02168

The International HapMap Consortium. (2007). A second generation human haplotype map of over 3.1 million SNPs. *Nature*, *449*, 851–861. doi:10.1038/nature06258

The International Schizophrenia Consortium. (2008). Rare chromosomal deletions and duplications increase risk of schizophrenia. *Nature*, *455*(7210), 237–241. doi:10.1038/nature07239

Timuragaoglu, A., Dizlek, S., Uysalgil, N., Tosun, O., & Yamac, K. (2006). Methylenetetrahydrofolate reductase C677T polymorphism in adult patients with lymphoproliferative disorders and its effect on chemotherapy. *Annals of Hematology*, *8*(12), 863–868. doi:10.1007/s00277-006-0175-4

Tripodi, A., Chantarangkul, V., Menegatti, M., Tagliabue, L., & Peyvandi, F. (2005). Performance of clinical laboratories for DNA analyses to detect thrombophilia mutations. *Clinical Chemistry*, *517*, 1310–1311. doi:10.1373/clinchem.2005.049981

Tsoutsman, T., Bagnall, R. D., & Semsarian, C. (2008). Impact of multiple gene mutations in determining the severity of cardiomyopathy and heart failure. *Clinical and Experimental Pharmacology & Physiology*, *35*(11), 1349–1357. doi:10.1111/j.1440-1681.2008.05037.x

Tymstra, T. (1986). False positive results in screening tests: experiences of parents of children screened for congenital hypothyroidism. *Family Practice*, *3*(2), 92–96. doi:10.1093/fampra/3.2.92

UK Human Genetic Commission. (2007). *More Genes Direct: A Report on Developments in the Availability, Marketing and Regulation of Genetic Tests Supplied Directly to the Public*. (UK Human Genetics Commission, London; http://www.phgfoundation.org/news/3933/)

van Dellen, A., & Hannan, A. J. (2004). Genetic and environmental factors in the pathogenesis of Huntington's disease. *Neurogenetics*, *5*(1), 9–17. doi:10.1007/s10048-003-0169-5

van der Ploeg, C. P., Lanting, C. I., Kauffman-de Boer, M. A., Uilenburg, N. N., de Ridder-Sluiter, J. G., & Verkerk, P. H. (2008). Examination of long-lasting parental concern after false-positive results of neonatal hearing screening. *Archives of Disease in Childhood*, *93*(6), 508–511. doi:10.1136/adc.2007.129320

Venter, J. C., Adams, M. D., & Myers, E. W. (2001). The sequence of the human genome. *Science*, *291*(5507), 1304–1351. doi:10.1126/science.1058040

Wang, X. P., Lin, Q. D., Ma, Z. W., & Zhao, A. M. (2004). [C677T and A1298C mutation of the methylenetetrahydrofolate reductase gene in unexplained recurrent spontaneous abortion]. *Zhonghua Fu Chan Ke Za Zhi*, *39*(4), 238–241.

Wasson, K. (2008). Consumer alert: ethical issues raised by the sale of genetic tests directly to consumers. *The American Journal of Bioethics*, *8*(6), 16–18. doi:10.1080/15265160802248351

Wasson, K., Cook, E. D., & Helzlsouer, K. (2006). Direct-to-consumer online genetic testing and the four principles: an analysis of the ethical issues. *Ethics & Medicine*, *22*(2), 83–91.

Watson, J. D., & Crick, F. H. (1953). Molecular structure of nucleic acids; a structure for deoxyribose nucleid acid. *Nature*, *171*(4356), 737–738. doi:10.1038/171737a0

Watson, M., Foster, C., & Eeles, R. (2004). Psychosocial impact of breast/ovarian (BRCA1/2) cancer-predictive genetic testing in a UK multicentre clinical cohort. *British Journal of Cancer*, *91*(10), 1787–1794. doi:10.1038/sj.bjc.6602207

Weisgraber, K. H., Rall, S. C. Jr, & Mahley, R. W. (1981). Human E apoprotein heterogeneity. Cysteine–arginine interchanges in the amino acid sequence of the apo-E isoforms. *The Journal of Biological Chemistry*, *256*(17), 9077–9083.

Wertz, D. C., Fanos, J. H., & Reilly, P. R. (1994). Genetic testing for children and adolescents. Who decides? *Journal of the American Medical Association*, *272*(11), 875–881. doi:10.1001/jama.272.11.875

Wheeler, D. A., Srinivasan, M., & Egholm, M. (2008). The complete genome of an individual by massively parallel DNA sequencing. *Nature*, *452*(7189), 872–876. doi:10.1038/nature06884

Wilde, A., Meisner, B., Mitchell, P. B., & Schofield, P. R. (2010). Public interest in predictive genetic testing, including direct-to-consumer testing, for susceptibility to major depression: preliminary findings. *European Journal of Human Genetics*, *18*(1), 47–51. doi:10.1038/ejhg.2009.138

Williams, R. A., Mamotte, C. D., & Burnett, J. R. (2008). Phenylketonuria: an inborn error of phenylalanine metabolism. *The Clinical Biochemist. Reviews / Australian Association of Clinical Biochemists*, *29*(1), 31–41.

Wolf, S. M. (2008). Introduction: the challenge of incidental findings. *The Journal of Law, Medicine & Ethics*, *36*(2), 216–218. doi:10.1111/j.1748-720X.2008.00265.x

Wolfe, S. M. (2002). Direct-to-consumer advertising--education or emotion promotion? *The New England Journal of Medicine*, *346*(7), 524–526. doi:10.1056/NEJM200202143460713

Woloshin, S., Schwartz, L. M., Tremmel, J., & Welch, H. G. (2001). Direct-to-consumer advertisements for prescription drugs: what are Americans being sold? *Lancet*, *358*(9288), 1141–1146. doi:10.1016/S0140-6736(01)06254-7

Xiong, T., Richardson, M., Woodroffe, R., Halligan, S., Morton, D., & Lilford, R. J. (2005). Incidental lesions found on CT colonography: their nature and frequency. *The British Journal of Radiology*, *78*(925), 22–29. doi:10.1259/bjr/67998962

Yang, Y., Chung, E. K., & Wu, Y. L. (2007). Gene copy-number variation and associated polymorphisms of complement component C4 in human systemic lupus erythematosus (SLE): low copy number is a risk factor for and high copy number is a protective factor against SLE susceptibility in European Americans. *American Journal of Human Genetics*, *80*(6), 1037–1054. doi:10.1086/518257

Zeggini, E., & Ioannidis, J. P. (2009). Meta-analysis in genome-wide association studies. *Pharmacogenomics*, *10*(2), 191–201. doi:10.2217/14622416.10.2.191

Zimmern, R. L., & Kroese, M. (2007). The evaluation of genetic tests. *Journal of Public Health (Oxford, England)*, *29*(3), 246–250. doi:10.1093/pubmed/fdm028

Chapter 6
The Applications of Omics Technologies and the Challenges of Ethics in Nutritional Sciences

Minakshi Bhardwaj
Cardiff University, UK

ABSTRACT

During the past two decades, there have been numerous developments in the genetic and genomic technologies enabling us to understand complex biological systems in an integrative manner through holistic approaches in research. Since the sequencing of the human genome, efforts are made to identify the number of the genes and their functions. The tools for determining the functionality of the genes are just beginning to appear. Initially the methodologies to identify functionality of the genes were largely based on comparative studies between model organisms. The very high number of genes with unknown functions demanded the need to develop new methods and technologies that may be helpful in assigning functions to the identified genes. Advancements in computing techniques and software opened the door for new technologies to be able to take an applied approach by studying biomolecules needed for proper functioning of the cell and take a holistic approach in biomedical research. Besides genomics, several other technologies are developed in the last decade that take an 'omics' approach, i.e., an integrated approach in the study of cell function. It is hoped that the applied integrative omics approaches may be helpful in establishing cause and effect relationships between genotype and phenotype. These 'omics' approaches include the integration of genomics, proteomics, transcriptomics, metabolomics and other omic technologies to do the non-targeted studies of biomolecules involved in the proper functioning of the cells and their responses to environmental changes. The applications of these technologies have been also utilized in the field of nutrition for studies on how nutrients and other metabolites effect the proper functioning of the cell. With these emerging techniques to understand the molecular functioning of the body, it is envisaged that they might be helpful to give personalized medical care and dietary advice to people based on their individual genotypes in the future. Whilst nutritional genomics is a rapidly growing field in the nutritional sciences focusing on the diet-gene relationships, there is an increasing understanding that other technologies will also be crucial in understanding the whole biological processes

DOI: 10.4018/978-1-61692-883-4.ch006

involved in metabolism of food. In this chapter I wish to outline the use of contemporary technologies that are involved in establishing the intricate linkages between diet and the genes, and the ethical challenges they raise in their applications.

HUMAN GENOME TO APPLIED OMICS

Although sequencing of the human genome revolutionized biomedical research, the fundamental limitation of the human genome is that it just provides the blueprint and potentiality of the organism. But it doesn't tell what actually takes place, what genes are expressed and how proteins are encoded. Even if we are able to identify all the genes, inactivating all genes and then studying them for different properties and phenotypes is an enormous task which is not possible to be completed in the near future. Furthermore, the mutations in important genes are lethal, meaning that such mutations can't be obtained or studied. Therefore new techniques are emerging such as transcriptomics, metabolomics, and proteomics to study the functionality of the genes in response to changing conditions. Thus the functionality of the genes can be established by linking it to the expression of the known genes that are co-expressed.

Whilst genomics and genomic variation has taken the forefront of the biomedical research and drug development, it is important to realize that the genomic approaches are not necessarily dependent on the sequenced human genome. Human genome is dynamic and the genetic variation basically is seen in two aspects, one in the responsiveness to drugs and food, second in genetic predisposition to diseases. Most of the applied omics technologies used in nutritional science studies focus on the first aspect. Nevertheless, in the case of nutrition, human genetic variation is not as easy as it seems. Genetically determined inborn metabolic and biochemical factors need to be differentiated from the variations that are also caused by life cycle (growth, pregnancy and old age etc) and lifestyle factors (socio-economic conditions, environmental stressors, etc.).

The use of applied genomics has two main deliverables, one that result in the identification and selection of specific biomolecules that control or represent a biological process, and second that use the complete set of biomolecules to assess the response or the quality of the transcriptome, proteome or metabolome. Another aspect of new omics technologies lies in their design and applications. Some of the technologies are developed for disease specific applications, identifying surrogate biomarkers for late onset diseases for example, and some are food specific applications, such as identifying toxicants or micronutrients which upon exposure may be lethal to life.

FUNCTIONAL FOODS

Functional foods were developed at a particular sensitive time in the eighties when genetic modification in foods was beginning to bring distrust and discontent among ordinary public. The development of functional foods was based on the premise of providing additional health value to the foods through alteration of physical structure and chemical composition of food products in order to achieve particular effects in the body functions. Functional foods encapsulated public health agenda as the supporters suggested that they not only will satisfy individual consumer needs for healthy food but also will contribute to the reduction in the food related illnesses (TAB1999). There is no legal definition of functional foods, except for their claims about enhancing nutritional value of food, such that when consumed will have positive effects on the body. The claims made by functional foods are seen both in terms of health and nutrient content. The health claims include enhanced function claims (improving digestive system, concentration) and disease risk claims (will reduce the risk of getting cancer or heart disease). The nutrient claims mention nutrient contents (such as high in fibre), comparative claims (less fat) and nutrient function claims (lycopene in tomatoes reduce the risk of prostate cancer).

Functional foods form a borderline between food and medicine; hence they need a different way of thinking about their intake and consumption. However, one can argue that their market has increased over the years in parallel with the increase in the food and lifestyle related conditions, e.g., obesity, cardiovascular diseases, and diabetes. Health foods shops are found in most countries and supermarkets have developed special sections for "healthy foods." They range from processed foods for allergies such as gluten free food as well as over target groups such as age and gender ones. Nevertheless, it is estimated that functional foods are bought by healthy adults more than the people who actually need them. Although the market for functional foods has significantly grown over the years with the increasing health conscious consumers, they are equally challenged in ethical, legal and social domains.

Availability and consumerism do not necessarily equate to their benefits. An ordinary consumer is not equipped with verifying health claims, which has a dimension of trust. Trust is a slow process, developed overtime with experience. Trust in the products also comes from the information sources and regulations (Poulsen, 1999). Principle of utility demands that the claims about benefits and safety of functional foods be verified before it is brought to the market and that there be post market surveillance. Functional foods are expensive and unaffordable to the people who need them the most; for example it is projected that diet related conditions such as obesity, heart attacks are found more in the less privileged low income groups. Functional foods are also not prescribed and recommended in the health system. Hence there are issues related to its accessibility and justice in health care. The flood of healthy foods in the supermarket shelves is overwhelming and the abundant choices of food products represent a real challenge for the consumers, who encounter dilemma in decision making in their food choices.

NEW BIOMEDICAL RESEARCH AND IMPACT ON FUNCTIONAL FOODS

Nutritional Genomics

Nutrition science traditionally focuses on the prevention and optimization of health rather than curing a well characterized disease; hence it has always been difficult to prove the efficacy of functional foods. The completion of the sequencing of the human genome in the year 2000 brought a paradigm shift in the biomedicine to research and future applications. This led to development of the concepts and research on genetic variation and dietary response; looking into the evolutionary aspects of diet and the role of nutrients in gene expression. With the possibility of estimating changes in the gene expression in nutritional sciences, hopes are increased to answer different questions related to genetic susceptibility in response to dietary changes.

Nutritional genomics is a developing science which looks into gene-nutrient relationship and in the future specific functional foods can be created based on genetic disposition. There are two dynamic ways in which this diet and gene relationship is explored. *Nutrigenetics* is concerned with how individual genetic variation determines personal risk to disease, nutritional requirements and metabolic response to the diet. On the other hand *nutrigenomics* explores genome wide influence of nutrition, looking into the functional effects of nutrients on the genome, patterns of gene expression and metabolic responses in response to dietary interventions. It is hoped that with the nutrigenomics will be crucial in understanding the polygenic, diet related diseases that require a complex approach in understanding multiple genetic variants interacting with multiple environmental factors. Though their individual contribution is small, their combined affect on whole body metabolism results in disease and sickness. Diet related conditions develop over a long period of time and even several decades. Therefore finding accurate biomarkers as clinical endpoints of diseases such as stroke, cancers or cardiovascular diseases is a big challenge. The challenge also lies in identifying biomarkers that can be used for exposure of an individual to certain food ingredients or biomarkers that will measure the effect of exposure and probable consequences.

Transcriptomics

The central dogma of biology involves DNA transcription to mRNA and then translation to proteins which are building blocks of life. The transcriptome is the complete set of RNA that can be produced from DNA and transcriptomics is the study of transcriptome, i.e., gene expression at the level of mRNA. Nutrients including vitamins, minerals, etc. can modify and affect the gene transcription and translation which in turn can alter metabolism and growth, resulting in disease. Currently DNA microarrays are used for genome-wide monitoring of gene expression that simultaneously assess the transcriptions and expression of all genes. Using one array the expression level of up to 50,000 transcripts can be analyzed (Mariman E 2006). Microarray may be able to identify the differences between normal cells and diseased cells before and after the exposure to different dietary components which will be helpful in identifying different biomarkers for prognosis, disease diagnosis and new therapeutic tools. In particular it is envisaged that transcriptomics will be crucial in understanding the intricate relationship between diet and aging, and how caloric restriction over a prolonged period may result in slowing of aging process (Lee et al, 1999).

The use of microarrays also has a number of technical difficulties and drawbacks. Microarrays require large logistical support and they are an expensive technology. The complexity of identifying and analyzing nutritional biomarkers using microarrays may compromise the kinds of samples that can analyzed. The main issue with microarray technology is that it is less sensitive to assays and credible measurements can only be obtained for most abundant gene expressions and smaller expressions get undetected. The tests need to be repeated several times to obtain small results, making studies potentially very expensive. Other issues such as technical variations between different arrays and analytical procedures may lead to differences in different transcription processes. There is a danger of misinterpretation of data leading to resulting in false positives (Corthesy-Theulaz et al., 2005). Hence the issue of statistical validity of outputs using microarrays in transcriptomics is critical. Significant care will be needed in designing and analysing microarray studies in transcriptomics. However there is an integrated database of microarray expressions for nutritional science is available (Saito et al., 2005). The database serves as a portal for storing, managing and sharing of gene expression in nutritional sciences.

Proteomics

Proteins are structural and functional elements of the body. Proteins are encoded by genes, and are indispensable for life as they carry out most biological functions. In order to understand the functions of cells and organs, it is important study which proteins are present in cells, how they interact with each other and what they do. Proteins serve as mediators between genes and phenotypes. The number of proteins is significantly higher compared to the number of genes in any organism. It is estimated that the human proteome encompasses several 100,000 proteins and its variants, compared to entire human genome which is said to comprise approximately 25,000 genes. The term proteomics was first coined in 1996 and its roots conceptually lie in genomics (Wilkins et al., 1996). The proteome is defined as the study and measurement of all the proteins in a cell or organ in a given time, reflecting the expression of genes in a given time. The success of proteomics depends on application of several technologies that will help to identify and analyze each protein and protein products in a cell or an organ. In nutritional research, the impact of dietary interventions in the regulations of proteins or new expressions of proteins can be studied through proteomics, but it is important to recognize that the main objective of the proteomics is not necessary to study proteins but changes in proteins structures and functions through diet.

It is hoped that proteomics may be important in studying the effects of nutrients on proteome and explain physiological changes occurring when food is eaten and nutrients give stimulus to gene expression. Diet related proteomic changes may serve as very good biomarkers for nutritional adequacies, allergic proteins and allergic responses to the environmental factors; such as food allergies. Allergies are becoming increasingly common responses to complex diets that most people rely on, and proteomics will be helpful in understanding allergic responses with immunologically oriented approaches. Hence proteomics can be powerful tool for food safety assessment, including human responses to genetically modified foods.

The technologies applied in proteomics have been constantly developing, beginning with 2-dimensional gel electrophoresis that proved to have considerable drawbacks such as inability to detect proteins with extreme properties (very small, very large, very hydrophobic, etc.) and difficulty in identification of proteins and expensive. Other new technologies applied in proteomics include Mass Spectrometry (MS) and Isotope Coded Affinity Tags (ICAT) (Zhang et al., 2008). All these technologies are in their experimental stages and not established. There are not only technical issues related with outputs of all of these technologies, but also the costs involved. Nevertheless, there is optimism in the nutritional science community about the possibility of using protein microarrays to explore novel insights into nutrient-gene or nutrient-drug interactions, such as high level cholesterol diet and resulting protein expressions that may be useful in understanding gene expressions with such diets.

Metabolomics

Metabolites serve as the end product of any cellular regulatory process. They are the by-products of biological responses occurring as a result of genetic changes and environmental influence. It is widely understood that chronic diseases have both dietary and genetic components and metabolomics hopes to provide tools to investigate this complex relationship between nutrition and metabolism. Metabolomics is defined as comprehensive analysis of the whole metabolome under a given set of conditions (Goodacre et al., 2004). Unlike genomics, proteomics and transcriptomics where the focus of analysis is generally on specific compounds, metabolomics strategies focus on a spectrum of molecules with diverse properties (Weckworth & Fiehn, 2002). Metabolomics is about small molecules and one of the aims of metabolomics is to identify those small molecules that make difference between effects of different diets in order to help us better understand the interaction between human health and regulatory role of nutrients.

Conventional methods in the identification of metabolites focused on specific compounds and used for diseases related to inborn errors of metabolism. However, the target of metabolomics is to avoid such bias and be able to detect every single metabolite. The two most popular technologies applied in metabolomics are using Nuclear Magnetic Resonance (NMR) and Mass

Spectrometry (MS). There are both advantages and disadvantages in both the techniques. The advantages with NMR lie in requiring no or very little sample preparation in order to generate a good profile of small metabolites, and it is inherently quantitative. However NMR is not as sensitive as MS. But MS requires detailed sample preparations and there is a chance of losing some compounds as a result. It is hoped that the profiling of entire set of metabolites existing in biological samples will generate a wealth of information related to physiological status of an organism. However, bioinformatics tools needed for identification and analysis of data are still in their embryonic stage. And unlike genomics and proteomics where investigators can actually access databases such as Genbank or Swiss-Prot, there is no database available for metabolomics comparison and analyses (Whitfield et al., 2004).

Toxicogenomics

Toxicogenomics is the study of response of whole genome toxic substances and environmental stressors. It aims to identify relationship between environmental stressors and disease susceptibility, identify biomarkers of disease and exposure to toxic substances and elucidate the molecular mechanisms of toxicity (Waters & Fostel, 2004). There are similarities between toxicogenomics and acute-toxicity studies, but they differ in the methodologies used. Toxicogenomics cannot be studied in isolation to other technologies mentioned above, and an integrative approach of transcriptomics, proteomics and metabolomics is needed to contribute to development of toxicogenomics along with contributions from traditional toxicology and pathology studies.

The inherent challenge of toxicogenomics lies in its objective. It aims to integrate data from different studies in order to produce more refined understanding of toxicological responses at molecular level and their phenotypic expressions on organisms. Although these studies can be performed at laboratory level, large scale integrative databases do not exist that combine all these data together as biological end points from different studies; however, efforts are initiated to establish databases such as Chemical Effects in Biological Systems (CEBS) knowledge base. Scattered databases need common profiling in order to elucidate toxicogenomics results. Another concern lies in the diversity of chemicals and stressors present in the environment, their time and amount/dose that defines relationship between chemicals and diseases. The genetic diversity of human populations adds to the complexity of genetic susceptibility and of elucidating adverse affects of toxicants on individual genetic makeup.

ETHICAL ISSUES

New technologies bring new ethical issues. The development of biomedical technologies are directly associated with human well being and social welfare. Ethical issues become apparent both in applications of technologies and its implications on society. Beside the technological concerns raised at the level of research and use in the above sections, the use of omics technologies in nutrition also raises ethical issues at the level of observation and intervention. Some of these concerns are expressed below in detail.

Theory to Practice

Omics approaches provide opportunities and considerable challenges in terms of their applications. Despite the understanding that these technologies are in their embryonic stages, hopes are raised about their potential use in the clinical applications. The rapidity of innovation in genomics is overwhelming and the issue is that with technologies, large amounts of data can be obtained in much less time and more efficiently. However, converting data into knowledge is a serious concern even today, for example the studies on the properties of vitamin A took two decades to establish its biological plausibility because of lack of rigorous field data to go forward and questions about its biological plausibility (Sommer & Davidson, 2002). Similarly the functional aspects of a lot of genes are still not sorted and scientists are not trained to take a holistic approach of looking into large quantities of data and interpreting it with other suitable applications. Although the need for integrating omics technologies to study factors effecting human health is well recognized, it will be crucial to follow if they can accurately identify and elucidate multiple factors in a given environment and condition. Large epidemiological studies are based on questionnaires and personal health information provided by the subjects; hence there are also concerns about confounding factors such as sample size and the level of statistical significance needed for making a useful correlation to draw any meaningful information.

Screening for Diet

Beyond laboratory technical difficulties, challenges also lies in the clinical practice, in terms of promoting healthy eating patterns and preventing risk in light of the emerging epidemic of food and lifestyle related conditions. Whereas drug interventions primarily deal with therapy, nutrition advice will significantly take a role in the prevention of the diseases. Screening for diet may affect the daily choices of people. Under which conditions genetic testing and screening should be provided and how to protect information access is critical to acknowledge. Population screening programs for diabetes, obesity and cardiovascular diseases are not known and the causes for these conditions may not be all genetic. Further research is needed to analyze the significance both in terms of frequencies and penetrance of mutations in populations. Therefore determining the criteria for diseased or at risk groups and individuals needs careful consideration. Genetic counseling strategies at present focus on "disease related options" and nutrigenomics may not necessarily be "disease related." A distinction for nutrigenomics counseling is needed, as it may involve asymptomatic and otherwise healthy population, and it may be regarded as dictation and imposition of ordinary choices on normal lives (Bhardwaj, 2007). The screening methods also have implications by age, group and condition. Currently neonatal metabolic disorder screening is being done, and screening methods for women are also categorized to include average healthy female, pregnant women, postmenopausal women in terms of dietary recommendations. Other groups include young adults, males and old age people. It is hoped that the screening methods will aid in developing new bioactive molecules in functional foods that may be targeted for specific groups and thereby increasing the bioefficacy of the functional foods; for example older population may have specific needs that may require different dosages or strength of nutrient content in the foods (Raats et al., 2008).

Screening particularly for dietary recommendations raises issue of choice and personal autonomy. Food is associated with social status, cultural phenomenon, a lifestyle choice and has an inherent aesthetic value. Medicalization of food will impact these values and it is difficult to rationalize at this stage if screening methods will actually impact overall health at public level or widen the gap between different social strata.

Personalised Dietary Advice

The purpose of screening is not only to identify susceptible groups, but also to gain information about individual differences between individuals so that a personalized dietary advice can be given based on their genome profile. Hence the genetic testing for individually tailored diet will have a significant impact on the ways in which future dietary advice will be given. However, the relationship between individuals and food they eat is different than medicine. Unlike medicine, food decisions are not always based on oriented around health, and food related decisions are part of a "normal daily activity." Food is not prescribed like drugs and it is questionable to what extent individuals will want personalized diet.

The field of nutrigenomics is gradually developing; however, the rapidity with which Over the Counter (OTC) genetic tests for individualized diet have appeared in the market has raised serious concerns about their validity, authenticity of the tests and the possibilities of misleading the ordinary public. Commercial companies are exploiting "individual variation" as the basis for marketing their products. Diet-related genetic testing is increasingly available through internet and over the shelves in pharmacies in some parts of the world. Personalized diet related testing is commercialized both as test-kits and home brew tests for dietary regimes. There is already hype for 'tied-selling' of the products in the markets, providing testing and treatment as supplements or nutriceuticals designed specifically for individual needs. It does not recognize the issues of access and affordability of such tests (Green et al., 2003). The individualized genetic testing for dietary advice is yet to be scientifically proven. Though some may argue for individual autonomy of choice, responsibility and empowerment that personalization can offer; yet knowledge and information doesn't necessarily reflect healthy preferences and rational choices (Chadwick, 2009).

Quality Assurance of Foods

It is necessary to acknowledge that nutrigenomics and associated technologies will not only focus on human genome and human health. In the case of nutrition, it is expected that these technologies will have a critical impact on the food industry. In particular by understanding the complete proteome or metabolome of food and food ingredients will be helpful in understanding specific quality, authenticity of foods including their origin For example by examining the proteome of wheat the quality of the bread can be predicted.

Proponents of the new technologies argue that it will also help in the production of desired molecules for food ingredients by using metabolic engineering. Metabolomics and genomics might help to determine the parameters needed to induce secondary metabolites such as functional foods. Taking a holistic approach, it is possible not only to determine amount of metabolites but also nature, for example to use it in flavor formation and flavor profile, making food more palatable and approachable for a range of customers. The practical benefits of these technologies are for food processing industry in the short term, for example complex fermentation related food products: The quality of the starter culture may be used to predict the end product such as beer.

The promises of these new technologies are invaluable for commercial food industry. One can, however, question the promissory nature of these technologies and its value for customers. Traditionally, food development and improvement has been based upon trial and error method that involved experimentation with available ingredients. The new technologies are building upon the premise of "identifying" the "nutritious value" of the existing molecules and develop new ones as "required." The issue of "requirement" is associated with identifying accurate markers for individual health. That is where the major challenges of these technologies lie. Linking markers with nutrient value of biomolecules is in its infancy; however, at the level of food industry one can argue that the commercialization of nutrigenomics and applied genomic technologies is building upon the fundamentals of functional foods and is a response to the scientific advancements in nutritional research that identify the linkages between poor diet, malnutrition and illness, for example n-3 PUFA and cardiac function, calcium and osteoporosis, fibre and gut health.

DEVELOPING COUNTRIES

Asian traditions of medicine, such as Ayurveda, Chinese, Unani medicine are based on recognizing the medicinal powers of food. Traditional knowledge is applied in daily activities and choices about food. Whereas five fruits and vegetable a day concept is newly developed in the past ten years in Britain, Japanese traditional outlook on food focused on more than 10 different varieties of food in a day. Developing on the traditional wisdom, functional foods were first engineered in Japan for *general use* to supplement micronutrients and some macronutrients that may be "lacking" due to some dietary practices, and to improve normal health through reducing the risk and enhancing immune system. In other parts of the developing world, functional foods are just beginning to be commercialized; this is unlike the developed world, where it was introduced around two decades ago. In many parts of developing world, commercialization of medicinal properties of foods was generally focused on developing traditional medicine than "novel foods."

Developing countries suffer from *double burden* of diseases, infectious diseases such as malaria, HIV, hepatitis C etc., and fairly recent diet related chronic conditions such as such as obesity, cardiovascular diseases, diabetes and so on. Although genomic technologies have been successfully implemented in the case of infectious diseases, developing countries are struggling to develop disease prevention strategies for genetic and diet related conditions. One reason is the lack of attention received prior to the epidemic and underdeveloped facilities for research when compared to the industrialized countries. The field of nutrigenomics is yet to come to front at public level in developing countries. The science of nutrigenomics and other omics technologies is still evolving; therefore it is too early to predict the implications of omics technologies for developing countries. One can argue for the possibilities of screening for susceptibility biomarkers for populations for high prevalent diseases such as hypertension, diabetes, coronary heart disease and certain cancers. However, nutrigenomics and other omics services do not have a place in developing countries as yet because as they do not help to contribute towards meeting most urgent needs that involve prevention of infectious disease and providing basic nutrition to masses.

CONCLUSION

Currently most of the omics driven technology are based on promises; one can expect the huge investments in research and clinical trials for nutrition and diet related conditions, but the clinical applications of these omics technologies will not be available soon. The promissory nature of these technologies is a fertile ground for its commercialization without much consideration for immediate implications on society.

There is a need to understand food as being independent from technological advancements. With climate change becoming critical, agriculture production will inevitably change. The changes coming with climate change will also impact the food availability and food choices. As much as there are arguments in favour of using technologies for "health" purposes, the availability of basic food to large populations will still remain a primary concern at the global level.

ACKNOWLEDGMENT

The support of the Economic and Social Research Council (ESRC) is gratefully acknowledged. This work is part of the Research Programme of the ESRC Genomics Network at Cesagen (ESRC Centre for Economic and Social Aspects of Genomics).

REFERENCES

Bhardwaj, M. (2007). From farm to pharma: public health challenges of nutrigenomics. *Personalised medicine*, 4(4), 423-430.

Chadwick, R. (2009). Nutrigenomics. In P. Atkinson (Ed.), *Handbook of Genetics and Society* (94-104). Routledge.

Corthesy-Theulaz, I. (Ed.). (2005). Nutrigenomics: the impact of biomics technology on nutrition research. *Annals of Nutrition & Metabolism*, *49*, 355–365. doi:10.1159/000088315

Goodacre, R. (Ed.). (2004). Metabolomics by numbers: acquiring and understanding global metabolite data. *Trends in Biotechnology*, *22*, 245–252. doi:10.1016/j.tibtech.2004.03.007

Green, M. R. (Ed.). (2003). Nutrigenetics: where next for food industry? *The Pharmacogenomics Journal*, *3*, 191–193. doi:10.1038/sj.tpj.6500180

Lee, C. K. (Ed.). (1999). Gene expression profile of aging and its retardation by caloric restriction. *Science*, *285*, 1390–1393. doi:10.1126/science.285.5432.1390

Mariman, C. M. E. (2006). Nutrigenomics and nutrigenetics: the 'omic' revolution in nutritional sciences. *Biotechnology and Applied Biochemistry*, *44*, 119–128. doi:10.1042/BA20050112

Poulsen, J. B. (1999). Danish consumers attitudes towards functional foods. *MAPP working paper no. 62.2.*

Raats, M. M. (Ed.). (2008). *Food for the Aging Population*. Woodhead Publishing.

Saito, K. (Ed.). (2005). A Nutrigenomic database-integrated repository for publications and associated microarray data in nutrigenomics research. *The British Journal of Nutrition*, *94*, 493–495. doi:10.1079/BJN20051536

Sommer, A., & Davidson, F. R. (2002). Assessment and control of vitamin A deficiency: the annecy accords. *The Journal of Nutrition*, *132*, 2845S–2850S.

van der Werf, M. J. (Ed.). (2001). Nutrigenomics: application of genomics technologies in nutritional sciences and food technology. *Journal of Food Science*, *66*(6), 772–780. doi:10.1111/j.1365-2621.2001.tb15171.x

Waters, M. D., & Fostel, J. M. (2004). Toxicogenomics and systems toxicology: aims and prospects. *Nature Reviews. Genetics*, *5*, 936–948. doi:10.1038/nrg1493

Weckworth, W., & Fiehn, O. (2002). Can we discover novel pathways using metabolomics analysis? *Current Opinion in Biotechnology*, *12*, 156–160. doi:10.1016/S0958-1669(02)00299-9

Whitfield, P. D. (Ed.). (2004). Metabolomics: an emerging post genomic tool for nutrition. *The British Journal of Nutrition*, *92*, 549–555. doi:10.1079/BJN20041243

Wilkins, M. R. (Ed.). (1996). Progress with proteome projects: why all proteins expressed by genomes should be identified and how to do it. *Biotechnology & Genetic Engineering Reviews*, *13*, 19–50.

Zhang, X. (Ed.). (2008). Novel omics technologies in nutrition research. *Biotechnology Advances*, *26*, 169–176. doi:10.1016/j.biotechadv.2007.11.002

Chapter 7
Beyond Informed Consent:
A Model of Collective Guardianship for Ethical Genetic Research

Leonardo de Castro
National University of Singapore, Singapore

Chin Leong Teoh
National University of Singapore, Singapore

ABSTRACT

This chapter affirms the continuing relevance of requiring informed consent for health research in a context consisting of evolving genetic research methodologies and non-paradigmatic ways by which human beings become subjects of genetic research. The chapter also recognizes the special status of genetic materials and genetic data as subjects of research, as well as the different ways in which genetic materials and genetic data may be "owned." Different senses of ownership necessitate variable ways of implementing informed consent and these have to be clarified and carefully matched. Taking into account the specific interests expressed by human participants in human tissue research,the authors can see that these can be best promoted by a kind of oversight function delegated to ethics committees. The idea of a "one-time" or absolute consent given at the time of recruitment sounds appealing in that it minimizes inconveniences to many stakeholders, including researchers and human subjects. However, there remain valid reasons to be wary lest the system allow some types of research (or use of human research materials) that subjects would disapprove of unless sufficient pertinent information could be provided at the moment of recruitment. Thus the authors present an option for something close to "one-time" or absolute consent with safety nets in the form of oversight functions "delegated" to oversight ethics committees. The exercise of oversight function should involve flexibility to negotiate specific instructions given by the subject(s), such as those that may have something to do with uses that could have a particular religious or cultural significance.

DOI: 10.4018/978-1-61692-883-4.ch007

INFORMED CONSENT PARADIGM FOR RESEARCH ETHICS

Notwithstanding the problems and issues that have been encountered in the implementation of informed consent, the concept continues to have an indispensable relevance to research involving human subjects. The idea of informed consent for research has served a very useful purpose. Human subjects of research have come from very vulnerable populations and it has proven to be necessary to require their informed consent for participation as one of the safeguards against their being exploited. There is a highly uneven relationship between investigators and research subjects, notwithstanding efforts at research ethics education and training. For many participants, the starting point is utter ignorance and the situation is made worse by other factors such as physician dependence or economic bondage.

In international research, the consequences of the subjects' geographical distance from originators of research are being magnified. Those responsible for preparing protocols in international research are likely to be worlds apart from those who become research subjects and the latter are unable to communicate directly with the sponsoring team if they have any questions, misgivings, or apprehensions. Communication always requires intermediaries and local investigators have been known to resist making any modifications to protocols that require having to go back to the originating countries in a process that requires a large amount of time.

On the other hand, the strict implementation of informed consent for all research involving human subjects has resulted in the inability of many prospectively useful research projects to proceed or to be carried out expeditiously even when the risks to participants were not great. It is necessary to explore alternative paradigms of informed consent that can continue to protect human subjects at the same time that they enable important research projects to proceed and respect the efforts (even responsibility) of researchers to extend the frontiers of human knowledge. This is especially true in matters of genetic research where research methodologies defy paradigmatic implementation of informed consent procedures. The challenge is to find ways of safeguarding human subjects' interests while recognizing the freedom of scientists to undertake essential genetic research.

CHALLENGES TO THE INFORMED CONSENT PARADIGM

One of the reasons why informed consent for genetic research is especially problematic is that there are a number of ways by which humans could become research subjects. Not all of these ways are recognized as involving research, especially when we consider the initial starting point when an individual's genetic materials or information become candidates for use in a particular study. When we consider the opportunities for sharing of research resources among various laboratories, we realize how difficult it is even merely to determine the specific point when consent for the use of genetic materials for a specific study could and should have been secured.

In general, to become a research subject is to have oneself–or one's bodily parts or information–become the subject of investigation or examination that could yield generalizable information, where "generalizable" means that which could be inferred to be true also of others in a similar situation, or that which could be inferred to apply on a broader scale, e.g., to an individual at other times in his or her life. Thus, there are a number of possible entry points for someone (or something) becoming a research subject, including the following:

1. Having tissues/samples taken for diagnostic examination
2. Having one's records filed in a health care institution

3. Participating in non-genetic research that requires tissues to be drawn and stored
4. Submitting oneself for genetic testing
5. Having a genetic relative participate in research in any of the above ways

By "entry point" we mean the moment when control over materials or information belonging to an individual starts to be vested in another, explicitly or otherwise. At this point of entry, someone other than the individual source/owner assumes control through possession. There may be no intention at this entry point to engage in research using the materials or information. The specific study may come, as it usually does, at a later point in time. Thus, if someone provides bodily material for diagnostic examination, research on those materials may come much later, possibly even years after the entry point. This could happen also when someone's medical records are filed in an institution and the information is used later on in a study, or when people provide materials for genetic testing or neonatal screening and the materials are subsequently used for research. Indirectly, the situation could also arise when genetic relatives provide materials for testing and the attached information could be said to "belong" not only to the individual source but to those with whom pertinent information is shared on account of their shared genotype.

The idea is that these possible entry points for genetic research indicate situations that are not amenable to paradigmatic ways of implementing informed consent for these reasons: possession or control of the materials or information that become the subject of research is usually transferred long before a particular research project could be conceived or initiated; sufficient information concerning possible research projects could not be provided to the prospective subjects at the point of entry. The situation is very different from those cases where there is a clearly delineated point of recruitment and enough information about a research project is available and ready to be passed on to those who are being recruited to serve as subjects. In the latter cases, fully informed consent can be sought before a decision is made to transfer possession or control of human research materials.

CROSS-MATCHING OF DATABASES

Additional consent issues arise because of increased collaboration among laboratories and other research institutions. Genetic materials and databases that are in the possession of one institution could be shared with an entire network in a way that might not have been part of information provided to human subjects during the time of recruitment or donation. The cross-matching of databases from collaborating institutions indicate research studies that could not be adequately covered by the usual procedures for taking individual informed consent. To be sure, there are guidelines such as UNESCO's International Declaration on Human Genetic Data (2003) serving to remind us that:

Consent should be essential for the cross-matching of human genetic data, human proteomic data or biological samples stored for diagnostic and health care purposes and for medical and other scientific research purposes, unless otherwise provided for by domestic law for compelling reasons and consistent with the international law of human rights (UNESCO, 2003).

However, this provision obviously presents compliance problems where databases have been kept for long periods already, thus making access to owners understandably difficult. It is not easy to convince researchers that they should hold back on studies when very little risk could be identified and there is hardly any imaginable reason why the owners of the samples would have refused consent. Requiring consent in the usual form poses what many will regard as an unreasonable impediment to scientific research. On the one hand, the concerns on the part of the research community have to be supported. On the other hand, the misgivings of research subjects have to be respected. The seemingly incompatible positions can be reconciled provided we are willing to adopt a flexible approach to informed consent. Essential genetic research can be advanced without necessarily sacrificing the interests and safety of human subjects.

GENETIC DATABASES AND CONSENT

Those who favor loosening long-established informed consent procedures in the use of genetic databases have focused on two main arguments in order to support their position. The first argument holds that modification of traditional informed consent protocols is necessary if one is to pursue public benefits that can be obtained from the responsible use of genetic databases or biobanks. Some versions of this argument point out that too much emphasis has been given to the importance of informed consent and that its value has been held unquestioningly and unjustifiably to trump all other values at stake in the biobanking equation. A correction is needed in order to tone down this overemphasis and allow the promotion of public good through the conduct of essential genetic research. On this view, strict adherence to prevailing informed consent standards in relation to genetic research and biobanks may not only be impractical, inconvenient or costly, but even morally questionable. It would be morally questionable because it could entail the suppression of research that is necessary for finding solutions to pressing health concerns.

A second argument invokes public opinion studies showing a willingness on the part of respondents to provide some form of "blanket (or absolute) consent" in the collection and use of genetic materials and samples, hence indicating that rigorous standards of informed consent ought to be relaxed. The point is that if there is general acceptance of, if not outright support for, "blanket consent" amongst the population at large, then that ought to be taken as empirical evidence that adopting such a policy may not be as morally problematic as it first appears.

These two arguments are discussed in this paper and their implications examined in relation to a proposed formula for dealing with informed consent for research involving biobanks or genetic databases. The arguments are also discussed in connection with the notion of property and with the significance that can be attached to the results of public opinion surveys insofar as they are taken to reflect considered views regarding the use of genetic samples for research.

OWNERSHIP AND CONSENT

Individual informed consent seems to imply that genetic samples are a "property" and are "owned" only by the individuals concerned in that these individuals have ultimate authority for deciding on their use or disposal. This approach also treats research on genetic information essentially as equivalent to research on the human subject herself in that the usual requirement is for an identical form of consent. However, the nature of genetic materials and genetic data points to notions of multiple "ownership" or custodianship indicating multiple loci for seeking and providing consent. If we take this into account, we can see that the conventional paradigm for informed consent is inappropriate not only because it fails to recognize the weight of "public good" arising from essential genetic research but also because it ignores legitimate claims for collective guardianship and decision-making regarding highly valuable resources.

The concept of property needs to be clarified. We do not use the words "property" or "ownership" here to situate the discussion in a legal context. What we aim to do is to distinguish ordinary language notions that are akin, but not necessarily equivalent to ownership. A claim to ownership is only one of the foundations that may be invoked when one wishes to make a claim to some form of control. There are a number of other foundations that one may invoke in order to assert a right to use, or dispose of something. What we mean is illustrated by the ways in which we use the words "my," "mine" or "ours." To be sure, there are many things that are "ours" in the sense that we own them or that they are our property, and that therefore we have the right to use them or dispose of them as we please. Thus, we may own some pairs of shoes and, by virtue of such ownership, have the right to determine their usage or disposal without the interference of others. We may wear them or destroy them and others should not be allowed to prevent the implementation of our wishes since control over such property is solely vested in the owners.

However, there may be other moral claims to control that are not founded on the notion of property and may even override claims to ownership. For example, a person may own a valuable painting after having paid millions of dollars for it at an auction. That painting may have a value that goes beyond the monetary equivalent that has been used to pay for it. If a private individual had bought the Mona Lisa, for example, we would not readily admit that that individual has a right to do with that painting whatever he pleases. That claim to have control over the disposal of the painting could be overridden by a public interest that does not stem from property ownership. In this sense, there is a responsibility to exercise a form of control over the disposal of the painting that relates to the notion of custodianship. This custodianship is distinct from property ownership and can even come into conflict with the latter.

In a sense, there are many people who are vested with the custodianship of the Mona Lisa. They may not own the painting in that they cannot be rightfully regarded as its owners. However, that does not take away from them some interest in how the Mona Lisa is used or disposed of.

Similarly, there may be situations where a number of people can be regarded as having a claim to be custodians of an individual's genetic samples. The claim can arise in relation to different possible forms of attachment to the particular samples. For example, one may relate to the samples donated by a sibling because of the way in which information derived from those samples may apply to him and have a bearing on his privacy. The samples may also be the source of information that could give rise to discrimination against him.

Members of a closely-knit ethnic group may have a similar basis for making a claim to custodianship regarding samples derived from one of their kind. The claim is not based on ownership of the samples as a kind of property. However, it is a strong claim that arises because of what can be done with the information associated with the genetic materials. That information could be used to discriminate against them or to harm them in other ways.

In other situations, the claim to some kind of custodianship can arise from non-kinship forms of attachment. For instance, researchers themselves – on their own behalf or on behalf of the public, may make a case for the use of specific genetic materials in a situation where such use is foreseen to lead to, or be urgently needed by, therapeutic or preventative measures. The point is that the members of the group can have a morally legitimate basis, not grounded on the concept of property ownership, for making a claim to veto, or to support, the use of the materials for research.

This clarification has a bearing on how we understand informed consent and should be seen to inform the discussion in the rest of this paper.

INFORMED CONSENT AND PUBLIC OPINION SURVEYS

The arguments identified above indicate important challenges that need to be unpacked because of their implications for policy-making. In general, it is no cause for surprise that policy-making takes public opinion surveys into account. In regard to genetic research, the question may then be posed: Should we regard public opinion surveys as relevant to policy formulation? Can we make decisions regarding the disposition of genetic materials and biobanks on the basis of public opinion surveys?

One starting point is provided by Caulfield's position (2007) about the use of public opinion polls for policy-making. He observes that, contrary to views presented by commentators like Wendler, public perception data are at best mixed, and do not offer conclusive grounds for preferring "one-time general consent" or indeed any particular approach to consent (p. 222). Indeed, a lot depends on the actual survey questions that are posed and data can be found even among the studies cited by Wendler that do not conclusively support his policy proposal. For instance, a Swedish study that was cited by Wendler revealed that "48% of respondents 'feel respected and involved' by 'repetitive informed consent procedures', [while] only 11.3% felt that they received 'superfluous information,' and only 9.2% felt 'bothered' by the receipt of more information" (Caulfield, 2007, p. 219). In addition, Wendler appeared not to give due significance to studies showing that "when given the choice, the public opt for the specific, or fresh, consent approach" (Caulfield, 2007, pp. 219-220). Thus, the notion of a blanket consent or a one-time consent with no adequate safety nets for the protection of individual interests would not be able to provide them the level of assurance that research subjects would usually seek at the beginning of a project.

One thing that some surveys show is that the public are suspicious of commercial interests in the use of biobanks. Caulfield argues that this can lead to public distrust of biobanks, with specific informed consent being the only way to bolster public confidence in the handling of genetic databases. While some studies show that the public acknowledges the special nature of genetic information and may accept the use of research ethics boards as a "surrogate decision-maker," this acceptance has also been compatible with a preference for specific informed consent (Caulfield, 2007, p. 222). Indeed, extrapolating exactly what the public finds acceptable or not across time and cultures is difficult when different sets of questions are posed under differing contexts and conditions.

Caulfield himself has offered this piece of analysis as a way of neutralizing either approach to consent based on the preferences of the public (Caulfield, 2007, p. 222). The lack of a categorical message from public opinion surveys does not, by itself, imply that these cannot be used to guide policy. Although the equivocal nature of currently available data prevents us from endorsing the use of either blanket or one-time consent, a more carefully conceptualized and executed survey with specific details regarding issues that touch the particular sensitivities of participants could yield more conclusive results. Moreover, available public opinion surveys serve the purpose of indicating what information could be relevant to the subjects when their consent is sought.

Some writers have expressed skepticism concerning the legal and ethical value of public opinion surveys in this respect. In particular, Caulfield asks whether "consent policies [should] really follow public opinion data" (Caulfield, 2007, p. 222) when the main issue (apart from concerns about privacy and confidentiality) is one involving autonomy. For him, autonomy implies control over genetic material because it "implicates personal integrity", the latter being aligned with notions of human dignity and rights. Caulfield thus concludes that if the crux of the issue is about "a fundamental human right, then one legitimately can question whether the assertion of that right can ever be supplanted by an assertion indicated through the sampling of public opinion" (Caulfield, 2007, p. 223). It can be argued that populist or majority-oriented policies in such cases represent not the promotion of preferences and desires (and hence autonomy) but rather the tyranny of democratic rule, where minority voices are silenced and rendered insignificant for the purpose of determining public policy. The possibility of such a scenario should certainly make us cautious about using only public perception data in the formulation of policies. Political legitimacy should not be confused with moral legitimacy if individual rights are taken seriously.

There are several ideas here that need to be clarified. Firstly, appeal to public perception data does not have to result in a policy that merely favors the majority opinion. It can provide useful insights into the general mood and even specific concerns of the population so that policy can take these into account. Caulfield accepts this (p. 218) as he refers to the importance of appreciating public support of research and a general distrust for commercial uses of biobanks.[1] Certainly such data will be helpful in formulating consent guidelines (whether blanket or specific or somewhere in between). On the other hand, one mustn't confuse (even) blanket consent with presumed consent or no consent.

In the case of blanket consent, there is at least one point of explicit contact providing an opportunity to establish protection for the rights of potential donors since permission from them is needed in order to proceed in a particular manner. Potential donors must be fully informed of the nature of the consent that is being asked for, and it is necessary for them to know why some people may reject such a general consent regime. If they reject that arrangement themselves, no disadvantage should be allowed to accrue to them as a result of their decision. In addition, there should be no reason for them to fear any form of reprisal. In fact, we may go so far as to actively encourage donors who are uncertain or hesitant about donation to simply refuse general consent. In this way, the negative impact on those who do not form part of the majority can be minimized and their pertinent rights can be protected, even if a regime of general consent were in place. The purported administrative convenience of general consent need not undermine one's moral right of refusal, as it might in the case of dissenting minorities in a democracy who may experience the tyranny of the majority once a particular law is passed and implemented uniformly and universally. The proposed analogy with repressed minorities in politics doesn't have to carry over to the case of consent with respect to biobanks. Perhaps because of the concern expressed by commentators, issues of this kind can be addressed when the details of consent implementation are laid out.

GENETIC INFORMATION AS INTIMATE TO ONE'S PERSONHOOD

It is useful to clarify the notion that genetic information is "intimate to one's personhood" in relation to the view that one has the right to make decisions about it (Caulfield, 2007, p. 224). In the first place, one may ask what notion of "personhood" is being invoked. This ambiguity has implications for the ethical concepts that one should appeal to in examining proposals for informed consent. Genetic information may be deeply personal in the sense that it may reveal various things about an individual, her family and relatives. For this reason, the concepts of privacy and confidentiality seem most relevant to safeguarding one's "personhood." We can call this respect for "privacy-based personhood." If autonomy applies in this regard, it does so derivatively as an enabling concept or faculty to allow for the realization of the primary values and goals of respecting privacy and confidentiality.

On the other hand, appealing to autonomy in the way Caulfield does seems to imply a different notion of "personhood," perhaps one linked to the beliefs and values of an individual, where these may be influenced by one's culture or religion in forming one's sense of identity. In the latter case, it is the socio-cultural or religious milieu that one is immersed in that determines, in part, one's personal sense of *who* they are–their existential sense of self. It sets the stage and provides the resources for self-definition and understanding, where particular elements are incorporated and others contrasted against surrounding peoples, cultures and sub-cultures. Someone becomes who they are, and imbibes the values that are central to their sense of being or belonging, by virtue of their participation in this relational nexus of socio-cultural interaction. We may call this "existential personhood." But it is unclear how genetic information may be "intimate to one's personhood" in the social or relational sense of constituting one's identity and the sense that one may have of oneself. If the concern is that knowledge of one's genetic profile can have existential implications, this concern is already addressed by the privacy provision. Hence, appealing to respect for autonomy in such cases adds nothing to the analysis. Of course, appeal to autonomy does not have to be articulated in those terms. One can have a right to decide what to do with one's genetic samples if it is assumed that one owns that information in a sense that can be interpreted to entail the need to maintain and assert control. Here, a case can be made for the use of genetic information that may infringe on one's sense of existential personhood. There are two notions here that require further clarification. The first is the notion of ownership, which we touched on previously. We noted that there can be a morally legitimate interest in genetic samples that could come into conflict with, and sometimes override, traditional ownership claims based on the notion of property. These include appeals based on the greater good as well as claims made on the basis on privacy-based personhood. However, the idea of appropriate interest, and hence custodianship, of genetic samples can also be defined in communal terms that help protect or further one's sense of existential personhood.

The nature of genetic goods and their potential use to further the wellbeing of mankind means that it is a public good in a special sense of the term. It doesn't merely refer to a good that could potentially benefit everyone. There are also implications for co-ownership of genetic materials. Hence, "owning" genetic goods is not similar to owning a shirt or a car. The latter are examples of "private property." When one gives up ownership of a shirt or car (perhaps when it is discarded), it is not normally seen as becoming the property of the community at large. However, public genetic goods are different. They are more akin to goods which express the collective valuation of a society. Perhaps they are more akin to great works of art or the value placed on places of historic or cultural significance. They express collective value judgments of worth of a kind different from material possessions like shirts and cars. As such, they are worthy of protection and preservation in line with communal interests, hopes and aspirations. This is one way of interpreting genes as being the "common heritage of humanity," a reference made in the Universal Declaration on the Human Genome and Human Rights (1997) (UNESCO, 1997) and subsequently reaffirmed in HUGO Ethics Committee's (2002) Statement on Human Genomic Databases (HUGO Ethics Committee, 2002). The heritage is not just genetic or biological; it is also socio-cultural because of its acquired meaning and value in relation to health and the reduction of suffering. If this is correct, then it is not unreasonable to expect communities to offer shared ownership of or co-guardianship for the use of this precious resource, in ways that align with communal values, when individuals have agreed to contribute or donate their genetic samples. Hence, there may be a time after donation when the individual's actual and ongoing views on the use of genetic samples have been merged into that of the community and are inextricable from the latter.

Subject to certain conditions that are outlined below in the Proposed Broad Consent Regime, ethics review boards can be well placed to decide in a way that respects the donor's prior wishes while protecting the responsible use of genetic samples for the betterment of mankind, without specific informed consent every step of the way. This point goes beyond appeals to altruism or beneficence as the motivating factor for donation. In addition, there is a direct appreciation of the human condition and the bonds that bind us, which emerge from a spirit of humanitarian solidarity, especially in combating suffering and disease. Such an appeal holds both donors and co-guardians accountable to the same criterion of ethical judgment and militates against overemphasis on the respect for autonomy defined narrowly as obtaining specific informed consent. Given the social meaning of genetic samples, and the unlikelihood of harm caused by research on the samples themselves (a point we return to below), it perhaps becomes puzzling in such a context why specific informed consent ought to be seen as the sole or overriding research ethics criterion. Ethical decisions which emerge from the concept of solidarity appear more appropriate, adequate and encompassing, given the nature of genetic public goods.

Solidarity does not imply prioritizing the community over individual interests since communities themselves can show solidarity over the importance of individual rights. Instead, it acts as a motivational corrective to overly-narrow interpretations of the respect for autonomy and situates the individual within the interests, hopes and aspirations of the community. It is a principle which seeks to give due respect for, and to balance different values and valid interests that can be found in society. If such an approach to considering ethical problems seems reasonable then it only shows us the limits of a one-dimensional emphasis on autonomy as the leading bioethical principle.

In addition, the principle of solidarity can also imply that individuals have an ethical duty to support research that may be of benefit to others. Hence, Chadwick and Berg argue that "(i)t is not obvious…why a right to refuse to participate in genetic research, when it could be to the benefit of others, should be overriding" (Chadwick and Berg, 2001, p. 320). This is especially the case for certain genetic diseases, where the cooperation of particular families or communities can make all the difference to obtaining the needed genetic samples for research and medical progress that can lead to the reduction of human suffering.

Such arguments give us reason to reconsider ordinary notions of "ownership" and "property" which seem to give autonomy arguments their force. We have seen that even preserving the individual's primary right to control her own genetic materials does not foreclose a notion of co-ownership or guardianship of genetic samples after (some may argue, even before) they have been donated. The foregoing analysis brings out some elements crucial to the debate.

The second notion that requires clarification involves the observation that the goal of respecting one's "personhood" or "personal integrity" in the biobank debate is not the same as the notion as it applies to classic cases in research ethics. Cases like the Nazi medical experiments, Willowbrook and Tuskegee, for instance, are cases that emphasize the importance of specific informed consent because they involve "personal integrity" and "personhood" in a way that affects the physical and psychological well-being of the subjects themselves. The experiments were literally conducted *on* the subjects and caused considerable harm to them without their being able to refuse it.[2] Chadwick and Berg argue that "(p)resent thinking [in research ethics] is at least in part the result of a response to crimes against humanity in totalitarian states, which occurred more than half a century ago - a situation with little similarity to present research or the medical uses of databases. There might be reasons to question the transferability of rules and principles developed in one context, to the problems of today and tomorrow" (Chadwick and Berg, 2001, p. 320).

The kind of "personal integrity" and "personhood" that we are seeking respect for in the case of tissues and samples donated to biobanks is of a different order, with a different rendering of what constitutes risk or benefit to potential donors. Firstly, there is no direct physical harm that can be inflicted on the subjects themselves when research is done on genetic samples. The samples are physically separated from the body of the source person and damage to the samples is not manifested in the form of damage to the source in the same way that damage to one's attached face is necessarily damage to the person-owner of the face. Secondly, there can only be a kind of harm done if the samples are used in a manner that would offend the personal values or *beliefs* of the donor – for instance if her samples were used directly or indirectly in embryonic stem cell research and she finds this kind of research objectionable because of her religious beliefs about the status of the embryo.[3] This distinction between different notions of personhood is important because it allows us to take a closer look at what is really at stake in the biobank debate. The nature of genetic information, personal though it may be, does not invoke respect for personhood (and autonomy) in the first sense of avoiding harm to the self. However, it may invoke the second sense of personhood (and autonomy) in respecting one's "personal integrity" or dignity in cases where one can determine the *uses* to which her donation of genetic materials will be put to.[4] This involves primarily being sensitive to the kind(s) of research he or she would want to support on *moral* grounds.

Disagreement exists about whether respect for autonomy extends to one determining the use of one's genetic samples, and if so, whether this should override all other values at stake. Hence, Wendler states that "respect for autonomy does not determine whether people should provide one-time general consent or should specify which diseases may be studied using their samples" (Wendler, 2006, p. 546). One can argue that there is nothing in the principle of autonomy that specifies its manner of application. Hence, the provision of one-time general consent can be seen as fulfilling the autonomy requirement, especially if individuals know ahead of time what they are signing up for, and agree to it. Chadwick and Berg also wonder if "individuals should be free, from an ethical point of view, to refuse to help in an effort to relieve suffering for what could be regarded as trivial reasons, such as refusing to allow samples to be reused for research on drug abuse because of the disapproval of drug users" or in research on alcohol abuse because of a disdain for alcoholics (Chadwick and Berg, 2001, pp. 319 and 321). In other words, it is not clear why the individual's concerns (which may not even be distinctly *moral* concerns) should trump all other considerations in the use of donated samples when more people could potentially benefit by allowing its use. It may be thought to be a puzzling feature of a proposed *ethical* regime of consent that it would allow for the irrational, non-moral or even immoral (e.g., discriminatory) wishes and fancies of individuals to trump the use of genetic samples in research that could potentially reduce suffering for many. But this is just what follows if autonomy is seen to be the primary value. If one cannot accept this consequence, then it would seem to imply that unqualified autonomy cannot be the primary value on which to ground an ethical regime of consent.

It is important not to confuse the issue by bringing in issues of privacy and confidentiality. These important values relate more to the handling, storage and safeguarding of genetic databases rather than the issue of consent. This is not to say that privacy and confidentiality are not relevant when considering situations of the sort mentioned. Regardless of whether specific or broad consent is taken, the issue of respecting the privacy and confidentiality of genetic information still stands, and exists as a distinct ethical challenge. Issues of consent, however, revolve centrally around conceptions of autonomy. The point here is that even a more expansive notion of autonomy, like the one Caulfield prefers, does not rule out all forms of general consent.

ELEMENTS OF A PROPOSED BROAD CONSENT REGIME

Given the analysis above, the relevant notion of personal integrity that we are seeking to protect is consistent with any broad consent regime[5] that:

a. Details a list of potential uses of the samples at the outset and allows for subsequent uncoerced decision-making according to the donors' values,
b. Allows for withdrawal of consent at any time for any reason (or even no reason) without adverse repercussions,
c. Accepts the need to update the list of potential uses of the donated samples periodically, and seeks to inform donors of the updated list and whether they wish to withdraw consent,
d. Allows for individuals to be informed about or counseled on what is at stake before agreeing to donation,
e. Allows for individuals to be informed about or counseled on what is at stake, up to the level at which they feel their autonomy has been respected[6]
f. Gives potential donors the option of not having to reconsent in the future if they so choose,
g. Provides a mechanism for donors to keep themselves informed about developments with respect to research materials if they want to, without being forced to make decisions they have not indicated a prior wish to make. (For instance, they can be told in advance that updates concerning donated material and pertinent research will be available on a website or through regular newsletters that they can choose to read or ignore), and
h. Accepts co-guardianship of the donated samples by a research ethics board on specific uses of the donated material.[7]

Non-specific (e.g., broad) varieties of consent need not imply a less rigorous process of engagement with the potential donor so that they understand what lies ahead. Multiple ownership or custodianship of genetic materials means that research ethics boards provide an additional layer of consent after the donor has given her consent. What is referred to here is something more that the usual ethics review done by committees prior to the start of research. Instead, what is envisioned is a process by which designated bodies are tasked to conduct reviews at critical junctures to determine whether new research protocols are in keeping with conditions set by donors of genetic materials. Why is this needed? For one thing, additional consent can have the virtue of safeguarding the interests of communities who, in general, possess identity-defining characteristics. Hence, future research that may involve the use of embryos may be rejected by an ethics committee in a predominantly Catholic community, if it has been entrusted with the task of ensuring that research undertaken on donated materials complies with Catholic values. Even if a particular donor did not actually object to research involving embryos, no real harm to the donor is brought about by a research ethics committee refusing to use donated samples for such research on the grounds that they are respecting the shared moral interests of the community at large. Indeed such an approach can promote "personal integrity" in the second sense mentioned above, in ensuring that local sensitivities are appreciated and respected.[8]

In addition, ethics review boards can protect the interests of donors by reviewing "coding measures, information security, and other potential risks for the donor that might arise from, for example, changes in legal status, principal investigators, or organization of the original biobank" (Hanson et al., 2006, p. 269). Certainly the context within which research ethics committees operate can be reconceptualized to include greater community participation and consultation in the setting of research agenda priorities, use of genetic samples, public education, levels of consent deemed adequate and respectful of autonomy, and indeed any other issues which may be of interest or concern to the public (Lipworth, Ankeny & Kerridge, 2006, p. 126). One doesn't have to label such approaches feminist or communitarian to appreciate that long-term sustainable scientific research can only occur within a community that understands and supports such research. Research ethics boards would do well to keep abreast of such public sentiments, which is another way for public opinion polls and dialogue sessions to inform policy making.

At the same time, broad consent does not rule out a system of initial codification that would indicate in greater or lesser degree whether, for instance, the potential donor would object to their samples being used in relation to research tied to abortion, stem cell research, animal experimentation or which may have an immediate commercial or profit intent. Even though not every objection can be taken into account in such codification, common ethical concerns or flash points can certainly be accommodated in any broad consent regime, and similar concerns can be added as befitting local cultural sensitivities. If the cost of using genetic samples is thought to be too high with detailed codification of individual preferences, broad consent can still accommodate such preferences if potential donors simply opt out on the basis of information given at the outset.

Hansson also argues that a system of broad consent is more respectful of autonomy than prevailing guidelines that do not seek any form of consent, favoring instead consent that is given by research ethics boards:

In the UK, Estonia, and Sweden, ethics-review boards have a mandate to approve a biobank study without requiring informed consent; in Iceland, the national bioethics committee can grant this permission. Regulations in Canada, Germany, Norway, the Netherlands, and the USA permit biobank research without consent provided that samples are not identifiable. Thus, when risks of harm to donors in a research project are low (i.e., when protected sample codes prevent the spread of information), or when obtaining consent is impractical—a provision stated in several ethical guidelines—research without consent is usually permitted. Given that ethics-review boards might grant biobank research without consent, it seems odd that participants themselves should not be allowed to give broad consent to future biobank research (Hansson et al., 2006, pp. 267-268).

However, there are those who hold that the special nature of genetic information means that anonymity, privacy and confidentiality can never be completely ensured. Even if we accept that specific or broad consent by themselves cannot resolve the problem of the security of genetic information, it is argued that consent (and, in turn, respect for autonomy) cannot be valid until this feature of genetic research is acknowledged and incorporated into consent procedures. Hence, Lunshof et al. (2008) argue for a system of open consent that, in a sense, privileges truth-telling or veracity over the ensurance of complete privacy, confidentiality or secrecy in the handling of genetic samples. Here is how they describe the importance of open consent as it applies to the Personal Genome Project:

Open consent means that volunteers consent to unrestricted redisclosure of data originating from a confidential relationship, namely their health records, and to unrestricted disclosure of information that emerges from any future research on their genotype–phenotype data set, the information content of which cannot be predicted. No promises of anonymity, privacy or confidentiality are made. The leading moral principle is veracity—telling the truth—which should precede autonomy. Although, in clinical medicine, veracity is the legal norm in many jurisdictions, physicians may try to justify the withholding of information by invoking the 'therapeutic privilege.' In research, there is no such privilege, and when seeking informed consent from research subjects, distorted or incomplete information could undermine trust in researchers and in science (Lunshof et al., 2008, p. 409).

The following features of the Personal Genome Project can certainly be added to any broad consent regime, where research participants are aware that:

- Their data could be included in an open-access public database.
- No guarantees are given regarding anonymity, privacy and confidentiality.
- Participation involves a certain risk of harm to themselves and their relatives.
- Participation does not benefit the participants in any tangible way.
- (Where necessary) compliance with monitoring of their well-being through (e.g. quarterly) questionnaires may be required.
- Withdrawal from the study is possible at any time.
- Complete removal of data that have been available in the public domain may not be possible (Lunshof et al., 2008, p. 409).

Lunshof et al. conclude that "(t)he moral goal of open consent is to obtain valid consent by effectuating veracity as a precondition for valid consent and effectuating voluntariness through strict eligibility criteria, as a precondition for substantial informed consent" (Lunshof et al., p. 409).

Again, it is not clear if specific consent will do any better than broad consent in building trust if the key concern is to openly acknowledge that the confidentiality of information is not totally guaranteed. Here, the notion of autonomy at stake has more to do with risk assessment in relation to harm to the self than one's right (if there is indeed such a right) to determine the use of one's genetic samples. It would seem that if the goal is to build trust, then potential donors should be assured that there are institutional and process-related safeguards in place to protect their genetic information and possible legal avenues for redress should someone intentionally and maliciously breach privacy arrangements to the detriment of the donor. The important goal of building public trust in research need not imply anything about the superiority of specific consent over broad consent. Indeed, in order to bolster public confidence in genetic research, we should ensure not just that veracity should precede autonomy but also that competence in the setting up of genetic security processes and protocols should precede research. Consent, even when fully informed, cannot take the place of thoroughness and accountability in the setting up of security or coding systems to protect genetic privacy and confidentiality, if public trust is to be maintained and enhanced. This implies the prior need to develop, adopt and maintain best practice standards in the protection of personal information. It also requires appropriate training and education for researchers, administrators and ethics review boards on the proper handling of sensitive genetic data. An ethics review board seeking to determine if genetic samples should be shared or used in other research platforms should also satisfy themselves that these new collaborating or participating institutions have adequate security systems and protocols to ensure genetic information is not leaked. In the fast-paced world of scientific research, such protections first have to be ensured before permission could be given to share genetic samples. Such measures will help increase public trust in genetic research since due diligence is observed before research approval. Scientists will then not be accused of sacrificing individual interests and rights in the name of scientific progress. In addition, adequate secrecy laws with punitive measures for security offenses or lapses and avenues for redress, especially as they pertain to the use of genetic databases, will help boost public confidence in genetic research. In this regard, access to genetic databases for criminal identification or profiling purposes will have to be seriously curtailed if not forbidden altogether. Depending on the legal decisions adopted at the local level, such decisions should be flagged out in broad consent regimes so that potential donors can give valid and informed consent on this particular issue.[9]

CONCLUSION

The proposed broad consent regime, which was crafted on the basis of a detailed look at notions of "ownership" and "personhood," and how these affect personal integrity, shows one way that support for genetic research need not be inconsistent with the protection of individual autonomy. Changes in the research landscape and different rationales for consent mean that paradigmatic ways of respecting autonomy via specific informed consent may not always hold. However, a broad consent regime like the one outlined seems sufficient for the purposes of balancing both individual and research interests.

REFERENCES

Caulfield, T. (2007). Biobanks and blanket consent: the proper place of the public good and public perception rationales. *King's Law Journal*, *18*(2), 209–226.

Chadwick, R., & Berg, K. (2001). Solidarity and equity: new ethical frameworks for genetic databases. *Nature Reviews. Genetics*, *2*(4), 318–321. doi:10.1038/35066094

Hansson, M. G., Dillner, J., Bartram, C. R. J., Carlson, A., & Helgesson, G. (2006). Should donors be allowed to give broad consent to future biobank research? *The Lancet Oncology*, *7*(3), 266–269. doi:10.1016/S1470-2045(06)70618-0

HUGO Ethics Committee. (2002). Statement on Human Genomic Databases. Retrieved from http://www.hugo-international.org/img/genomic_2002.pdf. Accessed 15 February 2010.

Levitt, M. (2003). Public consultation in bioethics: what's the point of asking the public when they have neither scientific nor ethical expertise? *Health Care Analysis*, *11*(1), 15–25. doi:10.1023/A:1025381828650

Lipworth, W., Ankeny, R., & Kerridge, I. (2006). Consent in crisis: the need to reconceptualize consent to tissue banking research. *Internal Medicine Journal*, *36*(2), 124–128. doi:10.1111/j.1445-5994.2006.01020.x

Lunshof, J. E., Chadwick, R., Vorhaus, D. B., & Church, G. M. (2008). From genetic privacy to open consent. *Nature Reviews. Genetics*, *9*(5), 406–411. doi:10.1038/nrg2360

UNESCO. (1997). Universal Declaration on the Human Genome and Human Rights. Retrieved from http://portal.unesco.org/en/ev.phpURL_ID=13177&URL_DO=DO_TOPIC&URL_SECTION=201.html. Accessed 15 February 2010.

UNESCO. (2003). Article 22 International Declaration on Human Genetic Data. Retrieved from http://portal.unesco.org/en/ev.phpURL_ID=17720&URL_DO=DO_TOPIC&URL_SECT ION=201.html. Accessed 15 February 2010.

Wendler, D. (2006). One-time consent for research on biological samples. *British Medical Journal*, *332*, 544–547. doi:10.1136/bmj.332.7540.544

ENDNOTES

1 Levitt argues that public consultation can be of benefit because public concerns or their lack of knowledge can provide feedback for the nature of public education that is needed. It can also be of importance to gauge public sentiment that may provide a kind of mandate for specific proposals. At the same time public consultation can be a forum for the public to educate and inform policy

makers and scientists of issues in the moral landscape of emerging technologies that the latter might have missed or discounted. While public feedback (a survey of what is the case) should not unquestioningly lead to policy proposals (which should be based instead on what ought to be the case), Levitt (2003) argues that no policy proposals have been successful that haven't taken public sentiments into account. Wendler's (2006) own conclusion can be seen as an extension of this last point. He states that public support for one-time general consent shows that it is "socially acceptable and should lead to high rates of donation." Furthermore, offering general consent forms to potential donors satisfies the "reasonable person" test of how people generally and reasonably wish to be ethically treated.

2 Even in such classic cases, it can be argued that respect for autonomy is not the primary value, since even if someone agreed to be a subject of Nazi medical experiments, one would feel an obligation not to allow it to proceed.

3 At this point, the paper is not focusing on what has been called informational harm, which may arise when subjects learn of things they would rather not, or when genetic information is used by third parties for discriminatory purposes or when that information leads to stigmatization. These are certainly important issues and call for protections given on the grounds of privacy and confidentiality. However, research on genetic samples does not directly compromise the principle of autonomy as long as privacy and confidentiality arrangements remain secure or subjects are aware of the extent to which they *can* be secure. Any ethically viable biobank should ensure the latter, irrespective of the kind of consent that was given.

4 Some of the available public data in this regard tells us that the *use* of genetic samples is not the key concern for members of the public. Instead, "(r)espondents who were unwilling to donate samples tended to be concerned with the method of obtaining samples, not the possible use of the samples for research. In one study 38% of unwilling respondents cited a fear of needles or injections, and in another study many people who were unwilling to donate cited the time required for the consent discussion." (Wendler, 2006). In the latter case, more detailed consent is obviously not the solution, and broad consent could help bring about not only higher donation rates but also better respect people's autonomy by treating them in a way that *they* want, instead of how academics *think* they should be treated in consent proceedings. The relevance of public data here can thus lead to an enhancement, not diminution, of autonomy, especially when the other guidelines discussed in the text are adhered to.

5 Following Hansson et al., 2006, we will state our preference for a regime of broad or general consent that is at a point between "blanket" consent (which does not specify any restrictions) and "specific" consent understood in the traditional manner.

6 Indeed Hansson, et al. argue that "(a)cceptance of broad consent and future consent implies a greater concern for autonomy than if such consents are prohibited" (Hansson et al., 2006, p.267). But this is just another way of saying that respect for autonomy implies that people should be able to give consent at the level or in the way they feel is sufficient, given initial counseling of the facts, risks and benefits of participation.

7 Wendler (2006, pp. 546-547) suggests the following 6 aspects to be included in one-time general consent: "request to obtain samples for future research; risks, if any;

absence of direct benefits; information, if any, to be provided by individuals; reliance on ethics committees to review and approve future research provided it finds the research is ethical and poses no greater than minimal risk; solicitation of individual questions." He also acknowledges that additional elements may have to be incorporated depending on "individual studies."

8 Wendler found that "(i)n the six studies that examined the issue, most people (79-95%) were willing to provide one-time general consent and rely on ethics committees to determine the studies for which their samples would be used." In addition, "(n)ine studies found that many people would like some information on the projects for which their samples will be used, although the type of information desired was not specified" (Wendler, 2006, p. 546). Our proposed guidelines so far have been consistent with the preferences indicated by such public data.

9 To be sure, this issue complicates the sharing of genetic data across legal jurisdictions which may allow for the use of such data in criminal investigations. There should perhaps be international agreement that biobanks should only be used for medical or scientific research, and not for other questionable purposes. This will help maintain public trust in the use of biobanks.

Chapter 8
Genomics and Genetic Engineering: Playing God?

Varghese M. Daniel
Monash University, Australia

ABSTACT

The hurricane growth of genomics and genetic engineering poses challenging ethical questions pertaining to the technological application in human life. Many secular and religious bioethicists observe that the new proposals of genetic engineering are described as "playing God." The metaphor has evoked both optimistic and pessimistic perspectives among the scholars in bioethics. The American President's Advisory Commission for Bioethics describes the ethical arguments in relation with this metaphor in many volumes. The negative renditions of "playing God" conclude that even though human beings are God's creation they might still be able to play God, which could lead human beings and the entire cosmos to disaster. This perceptive proposes that modern genetic technologies and the researches in genomics could lead humanity into such a disaster. Contrary to this urging, some other bioethicists endorse that as an image of God, humans are called to play God. This chapter critically analyse the rationality of these arguments and its milieu in the context of Christian theology and verify its universal relevance in the context of bioethics.

THE NEW DEVELOPMENTS IN GENETICS

The dawn of the new millennium has witnessed the announcements of two major scientific breakthroughs. One is that the researchers have assembled the entire genome (gene map) of living organism, a bacterium, which they consider to be the second step in a three-step process to create the first synthetic organism, which would possess an artificial life (Borman, 2008a, 2008a; Dragnea, 2008; Pennisi, 2008). The second amazing news is that British scientists have created human embryos with the DNA from three people (Cree et al., 2008; Krishnan et al., 2008). Academic literature and media recurrently draw

DOI: 10.4018/978-1-61692-883-4.ch008

our attention to the new discoveries in the field of human genetics and medical technology, and its proposed benefits to humanity. The ethical problems related to genetic engineering are great subjects of concern for bioethicists. 'Playing God' is one of the very popular metaphors, which is used by many in relation with ethics, medicine and genetic engineering. Surprisingly, both secular and religious bioethicists quite often describe the discoveries, inventions and technological applications in genetics using the metaphor 'playing God.' The American President's Advisory Commission for Bioethics reports about different genetic technologies and interventions, where it describes the ethical arguments in relation with this metaphor in many volumes (National Bioethics Advisory Commission, 2003, 1997; President's Commission for the Study of Ethical Problems in Medicine and Biomedical and Behavioral Research, 1982). In bioethics literature almost all genetic technologies like gene therapy (Sullivan and Salliday, 2007), pre-implantation genetic diagnosis (Kalfoglou et al., 2005), modern tissue engineering and germline interventions (Peters, 1995), gender selection (Dahl et al., 2004; O'Keefe, 2004), genetic nanotechnologies (Ebbesen and Jensen, 2006), genetic screening (Mallia and ten Have, 2005), synthesizing a minimal genome (Cho et al., 1999), human-animal hybrids (Robert and Baylis, 2005), stem cell research, and cloning (Maienschein, 2003, pp. 129-236; Winston, 2003) are attributed to as 'playing God.' It has been discussed as a 'state of ethical principle' and as 'invoking perspective' (Verhey, 1995).

The repercussions of new medical interventions and discoveries are subjected to humanity and people of faith. They influence the relationships between God, humans and nature. Most of the religions believe that human beings are created by God, principally Christianity. Both the Bible and patristic literature affirm that God is the Almighty Sovereign and that human beings are created in the image and likeness of God, but they are not God. Both Eastern and Western theologians concretely characterize and distinguish the limitations of being human from the ultimate and boundless power of Almighty God. The believers acknowledge the domains of life, birth and biological make-up to be under the unsurpassable authority of the Creator. Hence any human attempt to control or regulate life functions within the above domains is interpreted to be 'playing God.' This paper aspires to analyse the meaning and implications of the phrase 'playing God' attributed to genetic discoveries and interventions, in the light of Oriental Christian theology.

PLAYING GOD: THE MEANING

Generally, the metaphor 'playing God' refers to criticise any action undertaken by human beings, outside of the 'traditional' definition of what human beings ought to do. Recently it has also been used to cover a range of human actions, which aspires to create radical changes in the scientific or medical world. In 1982, President's Commission for Study of Ethical Problems in Medicine and Biomedical and Behavioural Research Report portrayed 'playing God' as an expression of 1) a sense of awe and concern 2) an arrogant interference with nature and 3) an opposition to create new form of life (President's Commission, 1982, pp. 53-57). In order to work through this reading of human action or hubris, theologian Ted Peters attempts to locate three overlapping meanings for 'playing God' (Peters, 2003, p. 11). Firstly, human action is undertaken in order to learn God's awesome secrets. Some secrets regarding the existence of life or of the world are not revealed to humans, and they remain 'enigmatic.' This means that when human beings are trying to learn the secrets which only God knows, then it is as though we are "playing God." Secondly, "playing God has to do with the actual wielding of power over life and death" (Peters, 2003, p. 11). Thus any action or decision-making which touches the aspects of human life or death is 'playing God.' The third meaning is attributed to the use of science to alter life and influence human evolution. So any action or decision capable of turning human nature from one purpose to another is 'playing God.' Even though the phrase 'playing God' was initially 'foreign to theologians' (Peters, 2003, p. 13), it has now become a term frequently referred to in theological and even secular circles when dealing with genetic and medical technologies. The Jewish and Christian emphasis regarding the place of human in the entire cosmos is evident in the words of Rabbi Elliot (Dorff, 2001). He describes that humans are not the owners of the body; but the body is the property of God. Humans safeguard their life on earth using commandments of God. There is equality in creation where all humans are created in the image and likeness of God and hence it is impossible to reject the personhood of any human beings, whoever who has body and life. However, he states the limitation of humans as 'finite and fallible beings.' Audrey Chapman observes another meaning for "playing God," a phrase which has been employed to refer to the concern that people is using genetics as an inappropriate way to change nature scientifically.[1] Courtney Campbell also recognizes this phrase as a moral stop sign to scientific pursuit and medical practice, which attempt to explore and understand the secrets, which are beyond human capacity (National Bioethics Advisory Commission, 1997, pp. 15-16). Leon Kass articulates the term 'playing God' as an expression from the fear of super humanism (Kass, 2003). Bart Hansen and Paul Schotsmans preserve the metaphor to explain human's role as created co-creator (2001). However they distinguish God's work of creation from human creativity.[2]

All these descriptive approaches recognise the necessity of the metaphor in the bioethics debates. The above meaning of 'playing God' reflects a theological ambivalence, regarding whether human beings should feel free to alter the genetic make up, or to make life and death decisions, or should refrain from trespassing on a prerogative that belongs to God alone. It is worthwhile to explore that ambivalence of the meaning of 'playing God' in the context of genetic engineering and bioethics.

PLAYING GOD: WE SHOULD NOT DO IT

History of science and medicine has witnessed that the new discoveries and technological inventions are received by humanity with great surprise and astonishment. Some of the scientists were persecuted and their theories have been condemned, because the proposal was found to be opposed to the existing believes and practices.[3] Many scientific revolutions and technological inventions have been faced by many objections (Floris Cohen, 1994, pp. 308-313).

Generally, any new technology which tends to control or manipulate human lives may invite severe objections (Wilmut, Campbell and Tudge, 2000, p. 298). However, once the technology comes into practice, it is slowly being adopted as normal in the society. Nevertheless, some scholars assert that certain inventions and scientific proposals enable us to foresee various ethical challenges which stand as the "sword of Damocles" on the solidarity and harmony of the future humanity. According to Lori B. Andrews, the genetic interventions may produce unpredictable consequences, which could affect human life perilously (Andrews, 2001, pp. 31-170). She predicts that it would affect the definitions and concepts of various aspects of life such as parenthood, personhood, doctor-patient relationship, gender balance, uniqueness of a newborn etc. The present world's solidarity is indebted to natural selection and evolution. The author of this process is considered as God and hence playing God refers to the human intervention in the territories that traditionally have been regarded as provinces for God's acts (Smith and Cohen, 2003, p. 49).

Sometimes our scientific knowledge of certain phenomena is limited. Isaac Newton for example resorted to a notion of 'God's direct intervention,' which he believed to be much above the law of gravity (Verhey, 1995, p. 353). When human beings are in a situation where they fail to understand certain basic truths, regardless of however hard they try, a 'God of the Gaps' may be portrayed to act as a source of help (Verhey, 1995, p. 353). Human beings often stand dumbfounded in the face of certain questions answerable only by God and which reflect our powerlessness or ignorance as creatures. Verhey points out that human beings have defined certain aspects of life or of the world, which are considered as the mysteries of life or of the world to be inaccessible by human beings (Verhey, 1995, p. 353). The absence of perfect knowledge in human beings may lead to unpredictable results attributed to divine omniscience (National Bioethics Advisory Commission, 1997). The fallible state of human beings to evaluate actions according to self interest and his/her own powerlessness to control the outcomes indicates that adopting technologies like genetic screening, stem cell research, new proposed reproductive technologies or gene therapy will be human ventures to play God (Zoloth, 2001, p. 223). In discussions about the manipulations of genetic material and creating new life, the phrase 'playing God' has entered as a continuation of the discussion in medical ethics started by Paul Ramsay and Joseph Fletcher four decades before. In 1970s these scholars employed this metaphor to deliver their ethical arguments in the context of bioethics. Ramsay, therefore, warns, "Men ought not to play God before they learn to be men and after they have learnt to be men, they will not play God" (Ramsay, 1970, pp. 138, 143). Here Ramsay is not just evaluating the limitation of the scientific research but also indicates the weakness of human being.

Ruth Chadwick (1989) also suggests that the decision makers in genetic engineering are not infallible beings, and that they are incapable of predicting unforeseen consequences and possible disasters. This presumes that the age of technological triumphs would seem to bring any scientific fiction into reality. The charge of playing God reminds that we are prone to error, bias, arrogance, ignorance, game of risk, and going even beyond human limitation (Smith and Cohen, 2003, p. 50).

All these negative renditions of "playing God" conclude that even though human beings are God's creation *they might still be able to play God,* which could lead human beings and also the entire cosmos to disaster. In this way of understanding, many people assume that the above said genetic technologies could lead humanity into such a disaster (Resnik and Langer, 2001; Ebbesen and Jensen, 2006).

This concept that humans are prohibited to enter the mysteries of the nature and world comes from the two ancient philosophical-theological teachings: 1) The Augustinian thought of 'Original Sin' (human being's inability to not to sin "*necessitas peccandi*") (Harnack, 2009, p. 117) and 2) the Aristotle-Aquinas thought of Natural (Moral) Law (Aquinas, 1947, I-II, q 94, a 2).[4] Augustinian theology describes human's inherited ontological sinful nature, which is transmitted through every procreative act, starting with Adam, and makes all humans to be of sinful nature.[5] Human nature is in other words predisposed towards wrongdoing. Being human, one has no natural power to do what is right out of free will. In other words the state of sinfulness causes the act of sinfulness. Augustine emphasizes the huge gap between God and man in nature and establishes God's sovereignty. The explanation and its depth of the gap between God and humans have compelled Augustine to neglect the freedom of the human.

The second dominant thought behind this negative connotation of the metaphor playing God, is the theory of Natural Law. The 'novelty' of the every new technology persuades to rethink the laws of nature. The historical development of Natural Law theory starts with Stoicism (Long, 1974, p. 147). They believed in the existence of harmony between nature and life. In accordance to this harmony humans will be able to do what is right and good. Aristotelian philosophical understanding construed natural law as a moral law. This is identified by many scholars, in stating that Aristotle's theory of nature and technology leaves no place for authentic human creativity, for a 'human world', and thus precludes technological innovations (Schummer, 2001). In thirteenth century, Thomas Aquinas has interpreted this Aristotelian theory of natural law as God's law, and that God is the origin and source of natural law. *Morality is hence understood as going along with natural law and immorality has been conceived as what is going against natural law* (Takala, 2004). This thought has predisposed in the oppositional attitude towards the rapid developments in modern medicine and technology, which bring may artificial methods and lead to many scholars to observe it as 'playing God'.[6] William S Bainbridge observes:

God is the supreme wisdom, whose will is the source of natural law: Man should not be creating life. Leave it to God, he knows how we should be made. Man is not able to control himself well enough to avoid the temptation to do something wrong with his technology (Bainbridge, 2003).

Rifkin and Perlas write:

Plato, St. Thomas Aquinas, Charles Darwin... there were not evil men. Their cosmologies were not product of intrigue. These learned gentlemen were merely trying to express, as best they could, the working of nature... (now) humanity is abandoning the idea that the universe operates by ironclad truths because it no longer feels the need to be constrained by such fetters. Nature is being made anew, this time by human beings. We no longer feel ourselves to be guest in someone else's home and therefore obliged to make our behavior conform with a set of pre-existing cosmic rules. It is our creation now... (Rifkin and Perlas, 1983, pp. 242ff).

They observe that the rapid development and ubiquitous impact of modern technology has the power to redefine humanity and nature. They are however terrified that this hurricane journey will result in great disaster because it breaks the natural law and makes human-made laws. Here the normative nature of 'naturalness' and its necessity for the world, which has been proposed by Aristotle-Aquinas-Darwin era, is questioned.

PLAYING GOD: WE MUST DO IT

The second position is that 'playing God' implies that human beings are called to play God, or it is our vocation and hence we have to do it. The proponents of this theory in the contemporary bioethics literature are Allan Verhey, Andrew Dutney and Gareth Jones.[7] They actually follow the perspective of Joseph Fletcher. Fletcher argues early on that we are called to play God through genetic engineering and other technological inventions (Fletcher, 1970). Quoting Fletcher, Allan Verhey writes that the metaphor 'playing God' is not only a warning and reflection of human limitation, but is an invitation to imitate God's power and care for His creation. It is a vocation to replicate God's healing ministry. He advocates that in such a way genetic therapy could be useful to reduce the sufferings of the poor and powerless (Verhey, 1995, p. 355). Dutney upholds that we humans are called to play God and that "modern Christians play God as much as any other modern people" (Dutney, 2001, p. 12). He argues that modern human beings inevitably play God in their daily life one way or another. He perceives cloning as one more scientific invention. More importantly, according to Dutney it reflects the call for human beings to "play God," the virtue or vice of which is dependent on scientific developments.

Gareth Jones also believes that the divine grace and creativity are evident in all these genetic and scientific realms, and human creativity is to follow suit (Gareth Jones, 2005). He says:

I have retained this term 'playing God' for the sake of simplicity, in order to draw a contrast between its negative connotations and what I take to be far more positive ones. It may not be the best term from a Christian angle, but it has the merit of immediacy. It has the connotations of fulfilling a God-given mandate to serve others, to care for the vulnerable, and to heal. It acknowledges that humans are to participate in the process of transforming the world, by sustaining, restoring and improving what has been temporarily entrusted to us (Gareth Jones, 2009).

He continues that not only scientists and doctors but even ordinary people are playing God. The image of God in humans inspires them to exercise responsibility, to improve the world for ourselves and others and to control what can be controlled (Gareth Jones, 2005, p. 208).

These positive attitudes of playing God have a close relation with 'promethean determinism,' as Ted Peters explains, in which humans assume the origin and development of technology as a way to control nature, giving a genetic freedom to take decision to our progress (Peters, 2003, pp. 8-9). This is contrary to 'puppet determinism, in which viewing our genes as the key to what we are and presumes that we are victims of our genes and we would have no choice. In the positive way of conception of playing God, puppet determinism is replaced by promethean determinism, which allows human to play God. Though Ted Peter undoubtedly not argues that playing God is a positive metaphor, he upholds that as co-creators humans can engage in the activities of the genetic science with a responsible attitude (Peters, 2003, pp. 14-16).

The positive theory of playing God identifies basic Christian anthropological notion of human as the creation of God. However, when they explain humanity has a vocation to play God or has the capability to play God, they assume that humans ontologically participate in the essence (*ousia*) of God. This philosophical reflection is in close proximity to the *Advaita*[8] philosophy in classical Hinduism.[9] This Hindu philosophy of non-duality equates human beings as identical with *Brahman* (God). *Atman* (human self) itself is *Brahman*. In this philosophical context playing God is meaningful, because *atman's* (self) action, which leads to truth, is considered as Brahman's activities. A*vidya* (ignorance) causes man to move further away from the Self and obscure his / her knowledge of the truth. Hence human actions in connection with *Avidya* are not considered as true actions (*Dharma*) of *Atman* (Deussen, 1908, pp. 85ff).

The positive theory of playing God reflects the absolute power of human beings over the created world as being an image or being acting god of the world. This mode of approach passes through the concept of natural law because according to the Natural Law theory the most general precepts are unchanging and if the created beings change at all, it would be the lower levels of generalization (Durbin, 1984, pp. 212-213).[10]

PLAYING GOD: AN ORIENTAL CHRISTIAN RESPONSE

The above discussion based on the metaphor 'playing God' brings contradictive conclusions in bioethics literature. It is noted that both interpretations appear mainly from the Western theological and philosophical tradition. I would aim to analyse the significance of the phrase 'playing God' on the basis of Oriental Orthodox anthropology and inquire whether it should be used in bioethics as neither a positive nor a negative point of view. I use the theological-anthropological principles pioneered by Paulos Mar Gregorios[11] as the tool for this purpose. His anthropological approach is far different from that of the Western thought based on Augustinian–Aquinas theology.[12]

Mar Gregorios employs three Greek words to illuminate the relationship between God and human beings, which he borrows from Gregory of Nyssa, a fourth century Father in the Eastern world. First word is *'Diastema'* which explains the ontological distance between God and his creation. He quotes Gregory of Nyssa, as follows:

For being by nature invisible, He (God) becomes visible only in His operations, and only when He is contemplated in the things that are external to Him. But the meaning of the Beatitude (Matthew 5.8) does not merely indicate that we can infer the nature of the cause from its operation, for in that case even the wise of the world might gain a knowledge of transcendent wisdom and power through the harmonic structure of the universe (Gregorios, 1980, p. 70).

Mar Gregorios emphasizes unequivocally God's transcendence and human limitation. "God is beyond the reach of humans both ontologically and epistemologically" (Gregorios, 1980, p. 99). Human beings are only a creation of the eternal God. Although human have the image and likeness of God, they are not their own creators. God - the Creator, and human beings - the creation are different in "*ousia.*" This distance is not the 'God of Gaps,' which Fletcher and Verhey have described and buried (Verhey, 1995, pp. 352-357). This *Diastema* is the distance between the Creator and the creation in ontological nature. Even the above protestant theologians have not denied it. Verhey says "The world and its order are not God, but they are God's. They are the work of God" (p. 353). Even though God's *ousia* is distinguished from human, still human beings are called to participate in God's energy. Mar Gregorios describes this participation with another Greek word "*Metaousia.*" It is a way to experience God.

Without that participation, nothing can exist... What we can participate in is the being, life and goodness of God as it is given to us in God's energia which brought us in to being, sustains us in life and leads us in the good (Gregorios, 1980, p. 128).

This participation proves that 'matter' is not alien to God. All created beings reflect the will and wisdom of God and hence they can produce good. This human existence and the participation in the *energia* of God guide them to another concept in the God-Man relationship, which is called *epinoia.* It is the method of discovering things unknown. Mar Gregorios summarises Gregory of Nyssa's ideas about *epinoia* in the following way:

a) It is the faculty of mind that goes beyond sense impressions to relate them to each other in the mind and thus forms concepts. b) All human advance and progress is dependent on the proper exercise of this faculty. c) This faculty, however, is not autonomous, but given by God and functions properly only under God's guidance. d) This faculty can err; but this is essentially the character of all human faculties that can be used for opposite purposes that is for the truth and for falsehood, for good or for evil. The faculty is implanted in human beings for good but man has the moral freedom to use it for the opposite (Gregorios, 1980, p. 44).

This idea of *epinoia* shows that humans have the capacity to create and discover new things, which may be unrevealed to them. This ability can be interpreted to represent the role of human beings as co-creators. *Epinoia* is the capacity of human beings to participate in God's art of creation. God's purpose for human beings is not just for them to be 'puppets' in this created world, but to be an active instrument of God's energy in this creative world. This faculty of *epinoia* can produce error. It does not deny the ability of a doctor to heal and to kill through his medical knowledge. The faculty is meant to produce good; however, human free will can also lead to evil. Here human quality as co-creator reflects more a priestly stewardship along with liturgical characteristics.[13] In this approach humans do not stand as rebels but rather as participators in the *energia* of God through *metaousia*. These three core concepts of *diastema, metaousia* and *epinoia* in Eastern anthropology establish human goodness, limitation, authority and freedom.

Can Humans Really 'Play God'?

As discussed above, the Western theological and philosophical reflections in bioethics assert that humans are competent to play God either positively or negatively. Before making an Oriental response to the above question, it is better to analyse the meaning of the word 'play' or 'playing.' There are many denotations found in the English dictionaries for the word 'play' (verb) or 'playing' (present participle). I found three meanings of this word which could be relevant in the context of the metaphor playing God. Playing means 'to engage in recreational activities like sports and games' (Webster's Third, 1976, p. 1736) or 'to perform musical instrument' (Collins, 1998, p. 1187) or 'to represent a character in a theatrical performance or a film' (New Oxford, 1998, p. 1421). Other words like 'imitating,' 'pretending' and 'representing' could also be attributed to this meaning in the bioethical context. Is a human being able to be or to do any of these activities as God is or as He does? Western Christian theological reflection on bioethics asserts that humans are able to 'play God' either positively or negatively. Contrary to this conclusion, Oriental anthropology and its reflection on bioethics utterly reject this analysis. Firstly, the above said Eastern theological thought asserts that human beings cannot recognize fully or participate in the *ousia* of God,[14] and hence denies on the basis of this Eastern anthropology the argument of 'playing God.' If we employ the above meaning for playing we need to accept that those who *do play* must take control over the object. For example, if a human being wants to play the guitar well, he must master that object. Also, if any one wants to imitate or to pretend or to represent a character in the theatre setting, they must 'understand' what character and personality he / she is pretending to be. The actor should recognize or identify what nature he or she encompasses. When we employ the metaphor 'playing God,' it conveys this impression to their readers. Based on the above Eastern theological and anthropological perspective human beings are unable to 'understand' or to recognize God's beingness and nature. It is impossible to limit God within the human knowledge.[15] The attributes of God in the Bible, patristic literature and in the liturgical texts, help humans to understand who God is.[16] But that understanding is only partial. We know God only through God and his *energia* is the pathway to guide us to the knowledge of God.

Secondly, indeed the concept of the image of God has acquired much importance in Eastern Christian anthropology. This image is the uniqueness of human beings but is never a license or authority to play God. God's image is not to be understood as an ontological part of God's *ousia* but a function of energy which was graciously given to the human beings by the creator during the process of creation.[17] This image of God is a reflection of the responsibility given to human beings from God to maintain, sustain and care for the rest of the creation.[18] This image of God reflects the human free will and helps us in entering into new regimes of knowledge and to make new discoveries, but it is *our responsibility* to ensure that the consequences of freedom of will leads to the betterment or uplifting of humanity and not the reverse. It is possible to lead oneself to vice or to an error through the use of his/her freewill. Genetic engineering is viewed by Oriental Christian theology from this point of view. Through genetic interventions humans have the potentiality and freedom to lead humanity to either good or bad. It is not God's wish to deprive them of this freedom and make them a puppet in God's hands but they are expected to assume responsibility for his actions and hence required to take wise decisions at the right time.

Thirdly, all human actions and decisions cannot be considered to come from the fallen state of humanity or would be the reflection of the original sin. Augustinian and Ambrose notion of 'original sin,' transmitted from generation to generation from parents to child through the act of procreation, is absolutely rejected by the Oriental Christian Church (Varghese, 1974, p. 91). Mar Gregorios denies the Augustinian view that "being human is not anything of such kind that, having come into being, he can as of himself do anything rightly" and Mar Gregorios calls this is as 'childish dependence' of humans on God (Varghese, 1974, p. 46). Mar Gregorios argues:

It is an affront to human dignity, and certainly not the view which Christ and apostles hold about man. The 'world come of age' cannot brook this insult to mankind. It is not the Christian gospel, which undermines man in order to exalt God. It is too petty a God who can have glory at only the expense of the glory of man (Varghese, 1974, pp. 46-47).

In Oriental anthropology, a human being who grows into the image and likeness of God (*theosis*) is capable of making right decisions through rightly interacting with his/her Creator and by understanding one's responsibility as the crown of creation. Thus irrespective of whether genetic technologies leads to good or bad, it could never be pictured as 'playing God.' Thus the medical and genetic interventions are not a threat to the concept of God; rather they reveal how great is human, the God's creation, and his freedom as a human.

Fourthly, humans are called to utilize their capability, which is the gift of God. Scientific technology itself is not against the will of God, because God blessed human beings to preserve or sustain the world. The contingent universe can be changed or can be extended (Gregorios, 2007). The idea of contingency implies an active role for human beings. God's imminence through His *energia* facilitates the extension of His presence over creation. Nevertheless human beings need to understand their spiritual bond with nature. Thus the contingent world and the image of God in humans allows us to invent, innovate and discover new ways with our natural resources, which we can find in the *energia* of God.[19] Here we require a *sense of responsibility* with our technological expedition. The thought of responsibility will be from a holistic-*agapic* approach rather than individualistic-self-centred approaches. To take such a holistic-*agapic* approach, we need open dialogs and a broadminded approach with a prayerful mind. Basil, a fourth century (AD) Eastern Father and the Bishop of Caesarea started the first Christian hospital open to the public on a regular basis, and Basil's rationale for health and medical works proves that God created a contingent world that has the possibility of essential changes, which we need to triumph over the effects of the Fall. Christ's incarnation and His healing ministry have to be continued. Basil found the technological advances in the field of medicine promising for the benefit of human beings especially in the healing ministry. But at the same time he rejected the irresponsible exercise of medical technology.[20] However this creativeness and participation in the *energia* cannot be considered as playing God as Fletcher and Dutney argue, rather they (scientists and doctors) are simply human beings exercising their human nature.

Fifthly, we need to recognize the difference between God's creation and scientific and technological progresses. According to Bible, God creates the world from *ex nihilo* (out of nothingness). Humans, however, create or discover new technologies not from the *ex nihilo*, but from the existing and living organisms, which have been originally created by God.[21] For instance, the most recent genetic technological claim–that researchers have assembled the entire genome (gene map) of living organism, a bacterium, is considered to be the second step of a three-step process in their effort to create the first synthetic organism, a creation of an artificial life. Even when scientists are able to create artificial life, the feat cannot be considered equal to God's creation *ex nihilo* because humans use materials already existing in nature, which is not alien from God (Gregorios, 2007, p. 107). Cloning technology also would be considered in this way. Definitely cloning technology is a scientific breakthrough in the field of genetic engineering and it may produce good or bad consequences in the later stage. The acquaintance of advanced cell biology and natural reproductive cloning in single cellular organisms has led to experiments on animals. The successful (to some extent) attempts to produce stem cells and to reproduce offspring in animals have to a certain extent persuaded some that it is possible to apply this technology for human beings. There are enormous arguments for and against human cloning. Nobody knows what will be the outcome when we open the door to human cloning. As pointed out before, both proponents and opponents argue that human cloning is 'playing God.' Based on the above Oriental anthropological understanding these positive and negative elucidation of playing God are meaningless, since we must note that the clone is generated from a living cell and not *ex nihilo*.[22] At this juncture, the basic difference between a human creation and God's creation can be identified. God created the world out of nothing through His *energia*. His *energia* provides, from this theological perspective, the biochemicals for the formation of the first cell. Through mitosis and meiosis organisms grow in the world, which is the gracious gift of God. The cell culture as a whole could be understood as an expression of *epinoia*, irrespective of whether cloning may be bring about new ethical, theological and social problems. Hence, those technologies cannot be attributed as playing God and cannot be equated with God's creation out of nothingness.

Sixthly, the human experiments and technological revolutions may be unnatural, but that cannot be considered as playing God or against God's commandments. The 'novelty' of genetic engineering and technology persuades many people to think that they are against the laws of nature. Oriental Christianity rejects the Aristotle-Aquinas natural law. The introduction of artificial methods to enhance human life has a long history at least from the Neolithic age, according to the scientific human history. Cooking food, using clothes to cover human nakedness, using motor vehicles, etc. are also not traditional or natural way of life. The introduction of different methods of treatment from a simple drug to brain transplantation, the practice of using plastics and metals in medical treatment are also not natural, but all these are accepted and adopted as the part of human life. Many people of faith consider treatment against diseases as a gift of God. The question of being against the laws of nature has raised objections a number of times in history, especially in the field of health care (Bainbridge, 2003). The introduction of genetic engineering and therapy also has faced many criticisms.[23] However, all these techniques are nowadays seen as great medical breakthroughs and possibly great blessings to human life. Still, a few churches do not support any reproductive treatments based on the reason of being against Natural Law, even within the context of marriage.[24] However, it is noteworthy that these churches are not reluctant to adopt medication, surgery, organ transplantation and pacemakers to extend the life. The Oriental theological anthropology in the light of Biblical truth allows human to "till the earth" (Gen. 2.15). The concept of *epinoia* allows humans to enter into the mystery of nature as a steward, and as a priest with a Eucharistic approach.[25] Mar Gregorios rightly explains this view,

Here a totally fresh attitude is necessary, one which is different from our objectifying-analyzing technique. We shall call it the reverent-receptive attitude. It is the attitude of being open to funda-

mental reality as it manifests itself to us through visible, audible, sensible realities in the creation. This fresh attitude is not to be adopted as an alternative to the scientific technological attitude but as necessary complement. Without this combination the scientific- technological attitude becomes as harmful as the other attitude becomes obscurantist and self-deceiving (Gregorios, 1977, p. 86).

Here Mar Gregorios rejects neither the scientific and technological approach, nor the approach open to the fundamental reality of nature, but he asserts that both need to complement each other. So, human beings, as worldly living beings, may adopt several unnatural ways of life, but we have to be responsible for each and every action of ours.

Seventhly, the metaphor 'playing God' has no positive or negative significance in the Oriental theological outlook towards genetic engineering, because the phrase limits God's qualities and characteristics of God into a single creature's (human) and its single ability (to create or design). God's transcendence is beyond the intellect and abilities of human beings. Humans may be able to tell about God's nature within the boundaries of their knowledge. This explanation is sometimes *apophatic* (negative) and sometimes *kataphatic* (positive).[26] Every attempt to limit God in human intellect is rejected by Oriental theology. Mar Gregorios follows Gregory of Nyssa,

In the first place, as Gregory of Nyssa says, to be infinite is to transcend all boundaries, whether of conception or of time space. The infinite cannot stop at any boundary and must by necessity transcend all – whether the boundaries be intellectual, quantitative or qualitative. And Gregory insists that every finite being must of necessity come to the boundaries of its finitude, whether in concept or being, and the infinite always extends beyond. The definition of the infinite is not that beyond whose boundaries there is nothing, but that beyond every boundary, being is. The transcendence of God is thus not merely conceptual or qualitative or temporal or spatial. It is in transcending every boundary that the infinity of God is manifesting (Gregorios, 1992, p. 21).

In this way, the Oriental understanding of God and human reject the claim of human attempt to play God.

Finally, this metaphor targets only the medical and scientific technologies and interventions. However the advocates (mainly philosophers, politicians and theologians from the Western world) simply ignore the other events, where we could see similar meaning. Their concerns about future human babies, anticipated stigmatization and the embryos including the zygote need to be appreciated. However, if we call such a genetic and medical technology as playing God, then what would we call the thousands of deaths in third world countries due to the lack of provision of basic human needs such as food, water and shelter? It is true that these deaths are the result of an unequal distribution of world resources. In other words, do we ignore the death of our neighbours and repeat the ancient question "Am I the keeper of my brother?" (Gen. 4.9). It is reported,

Though constituting 44 percent of the world's population, the 2,735 million people the world bank counts as living below its more generous $2 per day international poverty line consume only 1.3 percent of the global product and would need just 1 percent more to escape poverty so defined. The high income countries' 955 million citizens, by contrast, have about 81% of the global product. With our average per capita income nearly 180 times greater than that of the poor, we could eradicate severe poverty worldwide if we chose to try-in fact, we could have eradicated it decades ago (Pogge, 2005).

Due to lack of basic needs several human beings including children (the image of God) die every second. This domination of the world's resources by a small number of sophisticated people and the cruel selfishness involved are not being considered as playing God by the so called developed world. If we were to seriously take the notion of playing God, these well-heeled people are playing God, because they keep the resources and reserve to themselves the prerogatives that are properly God's. Is it not unnatural? Is it not the part of their so called 'Original Sin?' They believe that the source of their luxurious life and world resources are their *birthright* or are *inherited* to them and they view the death of poor people at every moment due to lack of food, drink and shelter as their *fate*.[27] In truth, it is the way the wealthy allow them to die; in other words this is a passive killing. No one observes this as "*playing God*."

CONCLUSION

As discussed above, the concept that humans are prohibited to enter the mysteries of the nature and world comes from the two ancient Western philosophical-theological teachings: 1) the Augustinian thought on 'Original Sin,' 2) the Aristotle-Aquinas thought on Natural (Moral) Law. These teachings are not a universal Christian theological and philosophical dogma and hence the metaphor playing God has no relevance in the universal bioethics world especially in the Eastern world, either positively or negatively.

Based on Eastern Christian anthropology, the phrase 'playing God' is absolutely meaningless. As we have seen, the ontological essence of God and humans are different. This *energia* enables human beings to be *co-creators in priestly manner* (*epinoia*). They have the potentiality to discover, invent, sustain or destroy the world since humans are not created as puppets, but they are responsible for their actions. The act of medical professionals or scientists who attempt genetic experiments can never be portrayed as 'playing God.' Thus the absence of the Original Sin and Natural Moral Law in the universal Christian theological context consent to interpret the scientific developments in genomics and genetic engineering as *playing human* or *acting as human* by the use of the creative power from the divine rather than humans are *playing God.* In other words, genetic technology is also an outcome of human potentiality. The scientist could be connoisseur in the required molecular biology or biochemistry, or he could be aware of the glory of God's creation through their astonishing discoveries, but even then it is impossible to ascertain the self-originating and incomprehensible Almighty God with finite human intelligence. Hence at least in the light of Oriental Orthodox theology, it is impossible to use the phrase 'playing God' in appreciation or condemnation of human intervention in genetics. The application of the technology may result to be a friend or a foe, but that does not mean that humans are *playing God* but they are *playing human* or *acting as human.*

The contemporary positive and negative interpretations on the metaphor 'playing God,' as we see above, deliver opposite ethical conclusions. They also underestimate or misconstrue that modern genetic interventions are threats to the God-human concept and to religion as a whole.[28] The critical analysis of this metaphor verifies that it has no relevance in the universal theological perspective towards genetic engineering. According to some bioethicists, the metaphor 'playing God' is an expression of a sense of awe and shows a concern to the issue of responsibility rather than an objection (President's Commission, 1982, p. 53). If the issues are related to the question of responsibility, then why do we need this confusing metaphor as an 'ornament' in bioethics literature? Why can we not analyze the pros, cons and responsibility of each technology? If there is a concern or an awe of emerging technologies, we need to have more discussions and dialogs. In this forum, we could discuss the ethical issues along with the ethical principles such as beneficence, non-maleficence, dignity, justice, autonomy, solidarity, etc. It is also possible to have constructive discussions about risk-benefit assessments and methodological applications such as health technological assessments rather than objections with a banner of 'playing God.' By using the proper scientific approach in the discussion we could make ethical and scientific conclusions and take decisions whether we need the particular technology or not. For instance, genetic therapy is proposed as a treatment for many genetic illnesses. Scientists recommend replacement of a defective gene with a normal and healthy one or providing the natural protein which is absent in an organism, due to genetic defect (Anderson and Friedman, 1995). There are issues for ethical considerations such as informed consent, integrity of the body, therapy versus enhancement, right to know and not to know, stigmatization, trails on patients, etc. Similarly, the proposed stem cells and cloning technology may also contain many ethical problems to be addressed like the moral status of zygotes and embryos, totipotency of cells, parenthood, egg donation and so on. These ethical issues can be discussed and assessed and it might be possible to find ethical elucidation. The proper ethical conclusions may help the policy makers to take the decision either to continue or to stop or to regulate the research and application. Playing God, at least from the Oriental perspective, is a meaningless phrase both in religious and secular context which is not obliging for ethical analysis. It is just beating around the bush rather than getting down to brass tacks.[29]

REFERENCES

Ambrose of Milan. (1995). Explanation of David the prophet. In Jurgens, W. (Ed.), *The Faith of the Early Fathers* (*Vol. II*). Bangalore Theological Publication.

Anderson, W. F., & Friedmann, T. (1995). Strategies for gene therapy. In W. T. Reich (Ed.), *Encyclopedia of Bioethics* (907-914). New York: MacMillan.

Andrews, L. B. (2001). *Future Perfect: Confronting Decisions About Genetics*. New York: Columbia University Press.

Aquinas, T. (1947). *Summa Theologica.* William Benton. Retrieved from http://www.ccel.org/ccel/aquinas/summa.html.

Augustine,. (1887). On Marriage and Concupiscence. In Schaff, P. (Ed.), *The Nicene and post-Nicene Fathers of the Christian Church*. New York: The Christian Literature.

Bainbridge, W. S. (2003). Religious opposition to cloning. *Journal of Evolution and Technology*, *13*(2). Retrieved from http://jetpress.org/volume13/bainbridge.html accessed on 1 May 2008.

Basil. (1978). Letter to Amphilochius. In Philip Schaff & Henry Wace (Eds.), *The Nicene and Post-Nicene Fathers*. London: W M. B. Eerdmans.

Borman, S. (2008a). Bacterial genome made from scratch. *Chemical and Engineering News*, *86*(4), 16.

Borman, S. (2008b). Synthetic genome paves the way to artificial life. *Nature*, *451*(7178), 511. doi:10.1038/451511c

Breck, J., & Breck, L. (2005). *Stages on Life's Way*. New York: St. Vladimir's Seminary Press.

Breitowitz, Y. (2002). What's so bad about human cloning? *Kennedy Institute of Ethics Journal*, *12*(4), 325–341. doi:10.1353/ken.2002.0023

Chapman, A. R. (1999). *Unprecedented Choices: Religious Ethics at the Frontiers of Genetic Science*. Minneapolis: Fortress Press.

Chapman, A. R. (2002). Genetic engineering and theology: exploring the interconnections. *Theology Today (Princeton, N.J.)*, *59*(1), 21–86.

Cho, M. K. (1999). Ethical considerations in synthesizing a minimal genome. *Science*, *286*(5447), 2087–2090. doi:10.1126/science.286.5447.2087

Cohen, H. Floris. (1994). *The Scientific Revolution: A Historiographical Inquiry.* Chicago: University of Chicago Press.

Cole-Turner, R. (1993). *The New Genesis: Theology and the Genetic Revolution.* Louisville: Westminster John Knox Press.

(1998). *Collins English Dictionary*. Glasgow: Harper Collins Publishers.

Congregation for the Doctrine of Faith Vatican. (1992). Instruction on respect for human life in its origin and on the dignity of procreation. In Alpern, D. K. (Ed.), *The Ethics of Reproductive Technology*. New York: Oxford University Press.

Cree, L. M. (2008). A reduction of mitochondrial DNA molecules during embryogenesis explains the rapid segregation of genotypes. *Nature Genetics*, *40*(2), 249–254. doi:10.1038/ng.2007.63

Dahl, E. (2004). Attitudes towards preconception sex selection: a representative survey from Germany. *Reproductive Biomedicine Online*, *9*(6), 600–603. doi:10.1016/S1472-6483(10)61767-1

Deussen, P. (1908). *Philosophy of Upanishads. (A. S. Geden Trans.)*. Edinburgh: T. & T. Clark.

Dorff, E. N. (2001). Stem cell research: Jewish perspective. In Suzanne Holland, Karen Lebacqz, & Laurie Zoloth (Eds.), *The Human Embryonic Stem Cell Debate: Science, Ethics, and Public Policy* (89-94). Cambridge, MA: MIT Press.

Dragnea, B. (2008). Unnatural life. *Nature Materials*, *8*, 102–104. doi:10.1038/nmat2108

Durbin, P. T. (1984). Thomism and technology: natural law theory and problems of a technological society. In Mitcham, C., & Grote, J. (Eds.), *Theology and Technology: Essays in Christian Analysis and Exegesis. Lanham: University Press.*

Dutney, A. (2001). *Playing God.* Melbourne: Harper Collins.

Ebbesen, M., & Jensen, T. G. (2006). Nanomedicine: techniques, potentials, and ethical implications. *Journal of Biomedicine & Biotechnology*, *51516*, 1–11. doi:10.1155/JBB/2006/51516

Fletcher, J. (1970). Technological devices in medical care. In Kenneth L Vaux (Ed.), *Who Shall Live?: Medicine, Technology, Ethics*. Philadelphia: Fortress Press.

Frith, L. (2001). Reproductive technologies. In Chadwick, R. (Ed.), *Ethics of New Technologies*. San Diego: Academic Press.

Gabriel, K. J. (2004). *Gurumukathuninnu.* Kottayam: Sophia Books.

Gareth Jones, D. (2005). Genetic prospects: finding a balance between choice and acceptance. *Perspectives on Science and Christian Faith*, *57*(3), 202–210.

Gareth Jones, D. (2009). *Playing God: scientific, ethical and technological challenges*. Retrieved from http://www.st-edmunds.cam.ac.uk/faraday/CIS/st-edmunds/jones/pdf/jones_lecture.pdf

Graham, R. (1971). Foreword. In Frazier, C. A. (Ed.), *Should Doctors Play God?*Nashville: Broadman.

Gregorios, P. (1980). *Cosmic Man.* New Delhi: Sophia Pub.

Gregorios, P. (2007). *An Eastern Orthodox perspective of God, Man, Nature.* Retrieved from http://www.paulosmargregorios.info. Accessed on 15 November 2007.

Gregorios, P. M. (1977). *The Human Presence: An Orthodox View of Nature*. Geneva: World Council of Churches.

Gregorios, Paulos Mar. (1992). *Human God.* Kottayam: Mar Gregorios Fondation.

Gregorios, P. M. (2007). *Cosmic Man*. New Delhi: Sophia.

Hansen, B., & Schotsman, P. (2001). Cloning: the human as created co-creator. *Ethical Perspectives*, *8*(2), 84. doi:10.2143/EP.8.2.503828

Harakas, S. (1999). *Wholeness of Faith and Life: Orthodox Christian Ethics* (*Vol. III*). Massachusetts: Holy Cross Orthodox Press.

Hefner, P. (1993). *The Human Factor: Evolution, Culture and Religion*. Philadelphia: Fortress Press.

Kalfoglou, A. L. (2005). Opinions about new reproductive genetic technologies: hopes and fears for our genetic future. *Fertility and Sterility*, *83*(6), 1612–1621. doi:10.1016/j.fertnstert.2005.01.090

Kass, L. (2003). Ageless bodies, happy souls: biotechnology and the pursuit of perfection. *New Atlantis (Washington, D.C.)*, *1*, 9–28.

Kilner, J. F. (2000). Human cloning. In Kilner, F., Cunningham, P., & David Hager, W. (Eds.), *The Reproduction Revolution: A Christian Appraisal of Sexuality, Reproductive Technologies, and the Family*. Cambridge: W. B. Eerdmans.

Krishnan, K. J. (2008). What causes mitochondrial DNA deletions in human cells? *Nature Genetics*, *40*(3), 275–279. doi:10.1038/ng.f.94

Long, A. A. (1974). *Hellenistic Philosophy*. London: Duckworth.

Maienschein, J. (2003). *Whose View of Life?: Embryos, Cloning, and Stem Cells*. Cambridge, MA: Harvard University Press.Mallia, P., & ten Have, H. (2005). Pragmatic approaches to genetic screening. *Medicine, Health Care, and Philosophy*, *8*(1), 69–77. doi:10.1007/s11019-004-6752-1

McInerney. Peter K. and Rainbolt, George W. Rainbolt. (1994). *Ethics*. New York: Harper Perennial.

(1997). *National Bioethics Advisory Commission*. Rockville: Cloning Human Beings.

(2003). *National Bioethics Advisory Commission*. Rockville: Beyond Therapy.

O'Keefe, M. (2004). Gender choice: is it playing God? *Christian Century (Chicago, Ill.)*, *121*(8), 12–13.

Pennisi, E. (2008). *Scientist synthesize a genome from scratch*. Retrieved from http://Sciencenow.sciencemag.org/cgi/content/short/2008/124/3.

Peters, T. (1995). Playing God and germline intervention. *The Journal of Medicine and Philosophy*, *20*, 365–385.

Peters, T. (2003). *Science, Theology, and Ethics*. Burlington, UK: Ashgate.

Pogge, T. W. (2005). World poverty and human rights. *Ethics & International Affairs*, *19*(1), 1–7. doi:10.1111/j.1747-7093.2005.tb00484.x

(1982). *President's Commission for the Study of Ethical Problems in Medicine and Biomedical and Behavioral Research*. Washington, DC: Splicing Life.

Ramsay, P. (1970). *Fabricated Man: The Ethics of Genetic Control*. New York: Yale University Press.

Resnik, D. B., & Langer, P. J. (2001). Human germline gene therapy reconsidered. *Human Gene Therapy*, *12*(11), 1449–1458. doi:10.1089/104303401750298607

Rifkin, J., & Perlas, N. (1983). *Algeny*. New York: Viking.

Robert, J. S., & Baylis, F. (2005). Crossing species boundaries. In Thomas A. Shanno, (Ed.), *Genetics: Science, Ethics, and Public Policy* (11-32). Lanham, MD: Rowman & Littlefield Publishers.

Schummer, J. (2001). Aristotle on technology and nature. *Philosophia Naturalis*, *38*, 105.

Siegel, S., Dittrich, R., & Vollmann, J. (2008). Ethical opinions and personal attitudes of young adults conceived by in vitro fertilisation. *Journal of Medical Ethics*, *34*(4), 236–240. doi:10.1136/jme.2007.020487

Smith, D. H., & Cohen, C. B. (2003). *A Christian Response to the New Genetics*. Lanham, MD: Rowman & Littlefield Publishers.

Steiner, H. (2003). Persons of lesser value: Moral argument and the 'final solution. In Garrard, E., & Scarre, G. (Eds.), *Moral Philosophy and the Holocaust*. Burlington, UK: Ashgate.

Sullivan, S. D., & Salladay, S. A. (2007). Gene therapy: restoring health or playing God? *Journal of Christian Nursing*, *24*(4), 199–205.

Takala, T. (2004). The (im)morality of (un)naturalness. *Cambridge Quarterly of Healthcare Ethics*, *13*(1), 17. doi:10.1017/S0963180104131046

(1998). *The New Oxford English Dictionary*. Oxford, UK: Oxford University Press.

Timmons, M. (2006). *Moral Theory*. Oxford, UK: Rowman & Littlefield.

Varghese, P. (1974). *Freedom and Authority*. Madras: CLS-ISPCK-LPH.

Verhey, A. (1995). Playing God and invoking a perspective. *The Journal of Medicine and Philosophy*, *20*, 347–364.

Visudha Kurbana Thaksa. (2003). *Kottayam: M.O.* C Publications.

von Harnack, A. (2009). *History of Dogma – Volume V*. Retrieved from http://www.ccel.org/ccel/harnack/dogma5.ii.ii.i.iv.ii.html.

Webster's Third New International Dictionary. (1976). G. & C. Merriam Company.

Wilmut, (1997). Viable offspring derived from foetal and adult mammalian cells. *Nature*, *385*(6619), 810–813. doi:10.1038/385810a0

Wilmut, I., Campbell, K., & Tudge, C. (2000). *The Second Creation: Dolly and the Age of Biotechnology*. London: Headline.

Winston, R. (2003). Playing God? *Nature*, *426*(6967), 603. doi:10.1038/426603a

Young, J. (2003). *The Death of God and the Meaning of Life*. London: Routledge.

Zoloth, L. (2001). Jordan's Banks: A view from the first years of human embryonic stem cell research. In Holland, S., Lebacqz, K., & Zoloth, L. (Eds.), *The Human Embryonic Stem Cell Debate: Science, Ethics, and Public Policy*. Cambridge, MA: MIT Press.

ENDNOTES

1. It is worthy to note that the word "inappropriate" deserves further clarifications to identify what is really playing God (Chapman, 2002; Chapman, 1999, pp. 52-56).
2. Created co-creator is not a new idea; it has been discussed before in bioethical context by many scholars like Philip Hefner (1993) and Ronald Cole-Turner (1993).
3. The Galilean and Copernicus scientific revolution is the best example for this statement. Galileo was executed by the Catholic Inquisition, because his discovery challenged what the Catholic Church had always taught. See Julian Young (2003, p. 21) and Hillel Steiner (2003, p. 77).
4. Retrieved from http://www.ccel.org/a/aquinas/summa/FS/FS094.html#FSQ94OUTP1
5. In Western Christian world, Ambrose of Milan wrote: "No conception is with out iniquity, since there are no parents who have not fallen. And if there is no infant who is even one day with out sin, much less can the conceptions of a mother's womb be with out sin. We are conceived, therefore, in the sin of our parents, and it is in their sins that we are born" (from "Explanation of David the prophet" I.11.56 (Ambrose, 1995, p. 159)). Augustine also wrote: "Wherefore the devil holds infants guilty who are born, not of the good by which marriage is good, but of the evil of concupiscence, which, indeed, mar-

riage uses aright, but at which even marriage has occasion to feel shame.... Now from this concupiscence whatever comes into being by natural birth is bound by original sin, unless, indeed, it be born again in Him whom the Virgin conceived without this concupiscence. Wherefore, when He vouchsafed to be born in the flesh, He alone was born without sin" (from "On Marriage and Concupiscence" I. 27.24 (Augustine, 1887)).

6 In the scientific world, natural laws are generally refers to interpret the eternal truth behind the cosmic system. However in ethics natural law theory refers the moral principles, which is universally applicable for human behavior. Mark Timmons (2006, p. 66) points out thus: "Moral laws (or principles) according to this theory, are natural in the sense that they are grounded in human nature and thus represent universally valued norms for the behaviour of all human beings;" McInerney and Rainbolt (1994, p. 22) also have described in the same way before: "Being moral is living in agreement with natural laws for human behaviour, naturally good ways of living."

7 Apart from Christianity, Yitzchok Breitowitz, a Jewish philosopher also keeps the same perspective (2002).

8 'Advaita' means oneness, or more precisely not-two-ness. It is a philosophy of non duality, which was founded by Adi Sankara (8th C AD).

9 This Hindu philosophy of non-duality equates human beings as identical with *Brahman* (God). *Atman* (Human self) itself is Brahman. In this philosophical context playing God is meaningful, because *atman's* (self) actions, which leads to truth, is considered as *Brahman's* activities. *Avidya* (ignorance) causes man to move further away from the Self and obscure his/her knowledge of the truth. Hence human actions in connection with *Avidiya* is not considered as true actions *(Dharma)* of Atman.

10 Paul T. Durbin. 1984. Thomism and technology: natural law theory and problems of a technological society. In *Theology and Technology: Essays in Christian Analysis and Exegesis,* Carl Mitcham and Jim Grote, eds. Lanham: University Press, pp. 212-213.

11 Paulos Mar Gregorios (d. 1996) was a world famous Orthodox theologian and philosopher, who served as the Metropolitian of Delhi of Indian Orthodox Church.

12 The theses that human is rational soul and that human is an ensouled body are the corner stones of Augustinian – Aquinas theology, which is rejected by Mar Gregorios; see Gabriel (2004, pp. 52-53).

13 In liturgy, all faithful (as the part of royal priesthood) are called to offer bread and wine which is the symbol of human effort. It is available in the form of wheat and grapes, but the human faculty (*epinoia*) makes them bread and wine and offer as symbol of their thanks giving (*eucharistia*).

14 An Eastern liturgical prayer illustrates that no human can understand God and His mighty deeds fully. See *Visudha Kurbana Thaksa* (2003, p. 154).

15 Mrs. Billy Graham wrote: "If I were an actress who was going to play, let's say Joan of Arc, I would learn all there is to learn about Joan of Arc. If I were a doctor or anyone else trying to play God, I would learn all I could learn about God." (Graham, 1971, cited in Verhey, 1995).

16 There are two approaches in Orthodox theology to understand about God. One is the Apophatic or negative theology that God is best described by what He is not. The Kataphatic or positive theology states that God is in constant touch with His entire creation through His energies, and maintaining its existence. Even though these are called positive and negative theology, they are not

contradictory to each other but complementary. See Harakas (1999, pp. 90-91).

17 In the Western perspective, the image of God in human is the part of his/her soul alone, but eastern Orthodoxy denies this theory and argues that God's image reflects in human, who is composed by body and soul. We believe both are the dwelling place of divinity. (See Breck and Breck, 2005, p. 191.)

18 This denies the thesis of 'natural law that all those things to which humans have natural inclinations.' (Aquinas, *Summa Theologica* I-II, q 94, a 3). Rather the vocation as a human enables to act according to the necessities. The act could be interpreted as unnatural, but that does not mean that it is immoral or unethical.

19 Aquinas' natural law permits human participation in eternal law. However he says "natural law dates from the creation of the rational creature. It does not vary according to time, but remains unchangeable." (Aquinas, *Summa Theologica* I-II, q. 94, a. 5). The human participation in the *energia* explores new knowledge about the created world. This could be new to the understanding of human beings regarding the natural. Unfortunately this new knowledge is interpreted as unnatural and artificial, or something irrational. In the end, human is described as the new law giver and the act is considered as playing God. Nevertheless in the Oriental world, the natural law theory has no significance and hence the new knowledge (positive or negative) cannot be considered as unnatural. The act of creation by God has not yet been completed fully and does not hold an unchangeable law in the deontological way (positive or negative) cannot be considered as unnatural. The act of creation of God yet not completed fully and does not hold an unchangeable law in the deontological way.

20 He wrote against to the contemporary practice of abortion. See Basil, 1978.

21 Some Western authors also acknowledge this biblical truth in the context of bioethics (Cf. Cole-Turner, 1993, p. 99.; Kilner, 2000, p. 138; Hansen and Schotsman, 2003, p. 222.

22 Clones are created from living cells. For example, Dolly, the first cloned mammal, has been created from a cell of a six-year-old pregnant sheep (Fin Dorsel sheep). Then an unfertilized enucleated egg was taken from a Scottish Blackface sheep. After the fusion of the two, the cell started multiplying (cell division) resulting in the formation of an embryo. After one week's growth the embryo was implanted in the uterus of a surrogate sheep. The foetus developed in the uterus and this resulted in the birth of Dolly (Wilmut et al., 1997).

23 For instance, John Paul II has criticized genetic engineering in the beginning stage. However later he has found as blessing and supported for gene splicing when its aims to "ameliorate the conditions of those who are affected by chromosomic disease" (President's Commission, 1982, p. 56.)

24 For example, the introduction of AI, IVF, medicalization of the processes of pregnancy and delivery of child, administering anaesthesia during the time of delivery, introduction of caesarian method of surgical removing of the child from mother's womb, general surgery, organ transplantation like that of heart, and ultrasound scanning are also questioned by the natural law (Congratgation for the Doctrine of Faith, 1992; Frith, 2001). Babies through In Vitro Fertilization are still called as test-tube babies. However, researches show that the ethical attitude dramatically has been changed and the artificiality and unnaturalness of the infertile couples' conception seemed irrelevant for their ethical opinion (Siegel, Dittrich and Vollmann, 2008).

25 In the Eucharist we offer nature's gift as symbol of thanksgiving for the blessing of God.

26 One is the Apophatic or Negative Theology that God is best described by what He is not. The Positive or Kataphatic Theology states that God is in constant touch with His entire creation through His energies, and maintaining its existence. Even though these are called positive and negative theology, these are not contradictory to each other but complementary. See Harakas, 1999, pp. 90-91.

27 Stoic idea of fate and Gnostic dogma of pre-destination has also influenced in the Western philosophical and theological thought (Gregorios, 1980, p. 141).

28 For example, Smith and Cohen (2003, p. 50) remind us that genetic technologies are posing serious challenges to Christian theology.

29 This paper is the result of the research done at the Katholic University, Belgium, Radbond University, Netherlands and University of Padova, Italy. I acknowledge the valuable discussions with Prof. Evert van Leeuwen, Dept. of History of Medicine, Philosophy and Ethics, Radbond University, Netherlands

Chapter 9
A Philosophical Exploration of the Concept of 'Property' in Genetics and Databanking:
Challenges for Bioethics in Asia and Europe

Ole Döring
HGI-Charité Berlin, Germany

ABSTRACT

The chapter criticizes arguments purporting to show that the human body could be made available in the market as property and those arguing that the concept of property could be applicable to the human bodily parts or human DNA. The author argues that the genetic information contained in matter such as DNA cannot be taken for granted as classifiable as property. There are three reasons: DNA is too personal to be commodified; DNA is of familial nature; and commercialization of DNA runs the risk of exploitation of the disadvantaged. Moreover, ethics should venture to clarify interests and stakes in the debate, with sympathy for the vulnerable rather than executing the rationales of powerful groups in economy and society.

WHAT IS AT STAKE

Modern biomedicine and the development of genetic technology raise a number of concerns about the tendency to see a person's body as an accumulation of objects that may and can be separated and commercially transferred. In fact, the view, that sees a human body as an accumulation of objects, is convenient because it suggests and encourages practices of utilization that would be difficult to defend morally without such a view. The primary consequential danger of commodification is that it can lead to exploitation and dehumanization, particularly of vulnerable populations, such as people at the margins of society, thus eventually contributing to de-humanizing societies at large. This danger is most apparent in the pharmaceutical and biotechnology companies' quest to patent and market products derived from human tissues, e.g., the widespread 'biopiracy' in the developing world (Rural Advancement Foundation International (RAFI), 1995).

DOI: 10.4018/978-1-61692-883-4.ch009

Moreover, the metaphysical challenge lies in reducing the human being, in part or in total, to an instrument, be it by other people or by an individual regarding itself. This concern is also expressed in efforts to anathematize and contain any activity or doctrine that might support dehumanization, for example in terms of taboo or ban. Both consequentialist and metaphysical concerns demarcate the conceptual horizons of the cultural philosophical debate about the ethics of databanking.

THE CONCEPTUAL DEBATE

The Human Body Property

Canadian ethicist Williams-Jones has linked an apparent social (or, rather, legal) tendency to accept commercialization of human tissues to three principal factors (William-Jones, 1999):

1. *Cartesian dualism*, which reduces a human being to a person. It conceives of the person as separable into mind and body and emphasizes the primacy of the mind over the body. Historically, such dualism helped demythologize the body and made it a secular object. Scientists gained the freedom to explore the functioning of the body, thereby supporting the development of modern scientific medicine. But this separation, without proper self-critical reflection, also supports the objectification and commodification of the body, for it may be taken to suggest that the body and its parts may be treated like possessions or instruments, alienated from the grounds of human value.
2. A *materialist conception of the person*, which treats the person as a wholly material being, has mostly replaced the dualistic conception or combines with it. As materialism treats the mind as a function of the body, materialism, in a reductionist form (most apparent in popular views of human genetics, such as the 'blueprint' metaphor for the genome), can lead to the body being objectified. Many people now expect all behaviors to be explained in terms of genetic causation. Despite this seeming fatalism, there is nonetheless a desire to gain control over genetic material, for the decoding of the blueprint is seen as the key to controlling a person's life and destiny. At the same time, a reductionist materialist perspective can also lead people to treat their own or other people's body parts as commercially transferable, as seen, for example, in the fact that commercial sperm banks exist for the express purpose of selling 'genetically superior' material to prospective mothers.
3. *The principle of self-determination* (or, misleadingly, "autonomy") in the dominant international bioethics discourse requires conceiving of individuals as fully self-responsible, private persons with strong controlling interests in their bodies. This emphasis on one aspect of autonomy can, however, go so far as to diminish awareness of duties towards one's own body. For example, one argument in favor of a commercial trade in organs is that "autonomous" individuals should be able to do as they please with their bodies. Williams-Jones argues that in combination with dualistic or materialistic (reductionist) concepts of the person, such a conception "may result in the body being objectified, manipulated, dismantled and commodified" (Williams-Jones, 1999, p. 12).

One way of exploring the body as property is to speak of it in terms of *quasi-property rights*. According to utilitarian consequentialists such as Radin, the means of weighing the significance of a person's relationship with an object is the type and level of pain suffered by the loss of the object. If the object is loosely held (e.g., one's car), then it can be freely exchanged or replaced; that is, it is fungible. The relationship to the object is close and personal if the pain of its loss cannot be alleviated by its replacement, e.g., the loss of a family heirloom (Radin, 1982; 1987). This leads to a hierarchy of entitlements, so that the closer the connection with personhood, the stronger the entitlement. Some items (e.g., bodily organs, according to Radin and others) cannot be classified as property at all. To make bodily parts fully alienable would, as Michelle Bourianoff Bray observes, "encourage a perception of body parts as interchangeable commodities and undermine the recognition of the human body as the physical embodiment of the personality" (Bray, 1990, p. 241).

The DNA Property

Obviously, physical DNA and DNA-based information is significantly less closely attached to our self-apperception than the body or its parts. Does commodification of DNA necessarily raise the same kind of concerns? For example, why should anyone be worried about genetic sampling? There are three arguments to suggest that commodification of human genetic material is harmful and unjustified:

1. *DNA is* too personal *to be commodified.* A person's genetic code is a unique identifier of the individual. DNA is fundamentally private and is an important element in determining certain aspects of physical and psychological existence. A person is determined by their DNA in conjunction with social, environmental, and cultural factors, making a person unique. An individual may attach the symbolic meaning of "blood" to his or her DNA, ruling out that it is commodifiable or in the discretion of an individual to commodify it, even when there are practices that support donation or offering.

It is not the matter of or information about the DNA that is problematic (beyond possible issues of benefit sharing), but the ways in which the related information is being used and how it, (including the chain of events that make up such practices in anthropologic terms), impacts on individual's self-perception, especially if they are sensitive and bearing a potentially adverse effect on the individual's health or social situation. At least, it requires special protection. The very least and most fundamental condition for any institutionally organized dealing with DNA is an embedding culture of law, with transparent procedures, separation of powers and effective protections of the vulnerable. This could provide the basis upon which fair understanding could guide he limits and structures of what can and cannot be done.

2. *The familial nature of genetic material* requires us to consider the ramifications of intergenerational and horizontal social relationships and responsibilities. Individual family members cannot profit from the sale of shared genetic material. DNA is not simply a shared resource to claim in good Lockean fashion, as if, for example, one were to harvest trees to make into furniture. Even if we allow the sale of discrete individual body parts and tissue, it does not follow that DNA can be similarly treated.

Perhaps DNA could rather be seen as a shared property (in the double sense of *Eigen*tum and *Eigen*art) of a group of persons. Depending on whether one takes this "property" in a materialistic sense or as an immaterial quality, group members can claim the material as being private and personal. Genetic material is part of a person's genetic heritage and thus at best is a collective property. We are obliged to safeguard rooms and options for those who might turn out to be affected after the facts and entitlements have been (provisionally) fixed.

3. *The commercialization of DNA* is similar in many ways to the sometimes proposed commercial trade in organs, because it puts marginalized peoples at *risk for exploitation*. For example, the Human Genome Diversity Project with its focus on disease resistance and susceptibility in target populations has led many to doubt the HGDP's altruism. Opponents maintain that the primary objective of the HGDP is the patenting of human DNA and exploitation of indigenous peoples, resulting in exploitation similar to that due to the black market commercial trade in organs (Bereano, 1995) or the looting of imperialist powers. These concerns have led to calls for a worldwide ban on patenting of genetic material (e.g., RAFI, 1995; IPCB, 2009; CRG, 2009). The commodification of the human genetic material is thus considered by many people (e.g., indigenous, civil and human rights groups) to be unacceptable because it can so readily lead to exploitation of vulnerable individuals and communities. This is notably a critique based upon consequentialistic or utilitarian considerations, which do not necessarily conflict with metaphysical arguments.

REVIS(IT)ING THE FRAMEWORK

Williams-Jones proposes that widely spread acceptance of genetic reductionist views of the person, in conjunction with an increased emphasis on personal sovereignty and control, can lead to a situation where individuals no longer understand or care about their personhood or physical integrity, or the societal dimensions of their own, body-related actions. At the societal level, this might contribute to the development of a huge market for body parts and genetic material, and lead to exploitation of people and erosion of moral standards. If she is right, here is a strong reason for the responsibility of bioethicists to defend a culturally appropriate and philosophically sound conceptual framework of human body related property.

Integrating the Academic State of the Art into Bioethics

I would like to respond along these lines, albeit with the intention to further amplify, according to the mentioned concerns. Bioethics needs not only to withstand the rationales and powers that reduce human beings to objects of technology or trade. We cannot simply criticize and reject biomaterialism, but ought to begin to be systematically constructive, in order to re-consider and re-ascertain the language and conceptual grounds of humanity, under specific conditions of post-genomics, legally, culturally, politically and ethically.

Perhaps, the renewed urgency of the attention to property issues related to the human body suggests that we should try a fresh start, commencing prior to the onset of the Lockean heritage, and begin with a critique of the reductionistic and rationalistic outlook, as has been undertaken notably by philosophical schools of phenomenology and feminist anthropology. Today, however, we do not only search for the corresponding attitudes towards the body as an irreducible, socially and biologically interconnected whole (*Leiblichkeit*), but have to enlarge the debates' perspectives so as to develop appropriate cross-cultural and transdisciplinary approaches.

Related bridge-building projects have been proposed already, such as by Hartmann: "Drawing on the philosophies of John Locke that individuals have property rights in the body that include identity and individuality to the exclusion of others and Immanuel Kant that a person's intrinsic moral worth is bound up with both body and self, this dignitary property interest is best explained by reference to the human body as an embodiment of self and identity, rather than a commercial or 'quasi-commercial' entity" (Hartmann, 2005).

Transforming the perception of the body to an *embodiment* of dignity appeals to me as a convincing advancement in the conceptual language of bioethics. However, is this also a robust and sufficiently wide construed bridge? Will it hold to discourage relativistic and subtly biological investigations, for example, the gradualistic kind, "how much or little body does it take to become fully dignified?", just as we find it in precedence, in arguments regarding the beginning of a human beings' dignified life or worthiness of protection, in the human cloning and embryonic stem cell research debates? Or will it be durable enough to withstand convenience-driven solutions to organ procurement and transplantation, with medical-reductionistic and technically modulized anthropologies, as we encounter them in debates about brain death versus cardiac death?

It seems to me that the conceptual horizons need to be enlarged, allowing us to deal with perceptions of the body that do not even begin to sympathize with conceiving it as a property in any economic sense. The true challenge for bioethics is not to organize the dissent that exists among those who already accept the foregoing *cultural grammar* of a qualified dualism of the human being that renders one part, quality or property of the human disposable (*verfügbar*) in principle. Rather, ethics should seek to provide strong grounds to explain and argue for the unavailability (*Unverfügbarkeit*) of the human, for any reason or purpose.

Some established opinions in international bioethics merit closer examination. For example, "Whether and how property law applies to the body and its parts depends on a number of variables, especially whether the body is alive or dead, and whether it is a question of the whole body or body parts" (Everett, 2003).

Why would this, ("Whether and how property law applies to the body?"), be the proper question? If we take it for granted that no person has the right to interfere with the integrity or the legitimate self-determination, physically, socially or spiritually, may we even pose such a question and thereby begin to persuade someone to subscribe to the underlying mindset, namely the one that is expressed in and compatible with the given property law?

Reducing the scope of interest to the *mode* rather than the quality of a property-relation to the human being pre-establishes a bias in the discourse, which is more often than not left unexplained. Alternatively, cultivated bioethics, with a global view, includes concern about social concepts and practices that do not depend upon (but might tolerate) standards of property or ownership, and/or certain affiliated body concepts, not only in order to protect vulnerable people, but also because there is no strong reason why they should not have a valid point to be considered in it own right, even when it seems to contradict the powers of the day.

In fact, most major philosophies and religions alert us that we shall not address practice in a language of possession, claiming that ownership or materialistic attachment to worldly things, for example, could be tantamount to hubris or delusion. It is a challenge, how to identify, in the ways we organize biomedicine, the turning point from acts of practical freedom to dehumanized acts of technical rationality.

Notably, this raises a notorious type of problem of political philosophy, in all societies with some degree of moral diversity, and the response to it certainly depends on the respective political order. Countries have found diverse theoretical and practical ways to manage in terms of standardization, of due procedures, legal codes and institutions. For example, Germany has established legislation on the hotly debated and intractable problem of abortion, according to which such an act would be unlawful but not punishable under certain conditions.

However, in countries that do not support civic diversity and notably in those global areas of international conduct that do not enjoy a legal state of civility (*Bürgerlichkeit*), but act under conditions of a natural state (*Naturzustand*, Hobbes on human nature, Kant on perpetual peace) we may not take such protection for granted. Here, namely in view of globalization, bioethics is called to seek ways to pro-actively protect the interests of the disadvantaged, discriminated and vulnerable.

Towards an Empirically Confirmed and Culturally Enriched Framework

At the present stage, we cannot provide positive answers to such questions. We can try to clarify the questions and the stakes.

Beyond the narrow utilitarian scope that can be found in most arguments in the debate, as it is shared by civic stake-holders, business lobbyists and politics, it appears that there are conceptual alternatives. For example, considering ownership of genetic material in general, Safrin speaks of "the corrosive interplay between the patent-based and the *sovereign*-based systems of ownership of genetic material. In *patent*-based systems, genetic material is increasingly 'owned' by corporations or research institutions, which obtain patents over such material. In sovereign-based systems, the national government owns or extensively controls such material. As more patents issue for synthesized genes in developed countries through the patent system, more raw genetic material is legally enclosed by the governments of developing nations, which house most of the world's wild or raw genetic material. This interactive spiral of *increased enclosure* results in the sub-optimal utilization, conservation and improvement of vital genetic material" (Safrin, 2004, p. 642).

So, even within the ownership model, there are qualitatively different conceptualizations of the kind of entitlement connected to property. Generally speaking, *sovereignty* is an open concept that allows us to define property in a plurality of ways, within procedural and principled confinements. It can encourage civic virtues, such as tolerance and attitudes such as precaution and respect. The fact that national sovereigns use it in accordance with the rationales of patenting is not a necessary or logically required but a pragmatic choice. A sovereignty model does have the capacity to support ethical horizons greater than the global market.

Patenting, on the other hand, is an original act that binds property strictly to pre-determined legal entities such as persons or companies, and reconstructs value in reduced positive terms of law, economy or science. It is a materialistic concept with an imperialistic inclination in its normativity that makes sense only under specific conditions of certain economic models.

Ethics, as it subscribes to humanity rather than market rationality, cannot support a property model that submits itself to the confined rationales of patenting. The sovereignty model tolerates a greater variety of options, but needs to be governed in the proper fashion in order to account legitimacy.

INTEGRATING DIVERSE NORMATIVE CULTURES TOWARDS AN ADVANCED FRAMEWORK

Ethics can develop a conceptual framework to safeguard and defend a realm of bodily integrity beyond the reach of any power or sovereign (*Unverfügbarkeit*). Such a framework could not be intended to prohibit any exchange of property but to sustain the tension that makes us continue to criticize, argue, and performatively improve practices of dealing with the body in biomedicine. The chief aspiration thus is to allow ethics to do the job properly on sound theoretical and methodological grounds. In the following, I would like to submit some consideration in order to stimulate the debate.

The semantics of the English word 'property' is interesting because of its ambiguity. There are *intrinsic* and *accidental* qualities of property. The intrinsic meaning refers to what is essentially characteristic of something. For example, humans have a moral property; it is humanly proper to act morally. However, in the accidental sense, we cannot say that morality is our property in that we can dispose of it at will. If we confuse both meanings, we not only commit a logical fallacy, but we also confuse fundamentally different categories, by conceptualizing morality in accidental terms or accidental things or affairs as if they were not possible expressions of morality but morality proper.

If we look into the conceptual impact of normative languages, it appears that, notably within what is called the legal traditions of the West, there are incongruent languages of property, such as in the case of German and English, each normative code carrying a peculiar system of ethics and procedural rationales, with a strong influence of connotative semantics. For example, the fundamental distinction between dignified persons and material objects in the German law makes it most difficult to tell if and how human body materials qualify as belonging to an intrinsically protected person or can be treated under the rules of objects with a certain value. This situation suggests that under conditions of renewed urgency and global relevance of the normative grammar in bioethics, there is a need to discuss the legitimacy of a language of property or ownership in relation to the body (and living things, or nature in general) that includes all significant contributions in human cultures.

We should ask whether 'property' can constitute a bodily self-relationship. What kind of relationship would this make and what kind of "body" can there be in terms of property? I propose that we should not inquire among European or Northern American philosophers only, (although it may certainly do good to revisit their proper interpretation), but invite all cultures to contribute directly or comment on the established answers. At the same time, the established opinions should be revisited. In sum, we need to engage a *global discourse*, (in the appropriate fashion), in order to identify the diversity of views and arguments, before we rush to lend the factual powers ethical legitimacy.

There are conceptual alternatives to defining property in terms of ownership of, legitimate access to, discretion or sovereignty to determine the use of the human body; and there are concepts of the human that rule out any reduced materialistic or objectivistic assessment of the body, because the underlying idea of a human entity that can be separated into body, soul, mind, etc., is not accepted. And there are alternative ways to organize the normative system of human interaction that, for example, transport a different assessment of dealing with the body and its matter.

I would like to randomly pick four examples from the debate that carry alternative concepts and merit closer inspection and thorough systematic analysis in the light of our theme.

Hedging claims. The Chinese philosopher Yu Kam Por (Hong Kong) has started to develop a concept of legitimate entitlements to access the spheres of a human's integrity, based on a classical Confucian term of human social order, *fen*, and an analysis of contemporary uses of this term in common language. This idea grounds the debate in a language of natural entitlements, which need to be individually negotiated regarding the margins but reserve a substance of inalienable integrity.

Being. German social philosopher Erich Fromm has suggested an ethics that is based on a world view of "Being defining Having." Such a conceptual language cannot support the idea of possessing the human body in any materialistic sense and thus rejects all attempts to commodify it.

Propriety instead of Property! Many traditional societies and contemporary schools of thought emphasize a reciprocal sense of entitlements and obligations, which are regulated *in situ* according to a robust set of normative customs. For example, some ethnic groups and religions assess genetics in terms of the 'human blood lines' or as a matter of the sanctity of the universe. Related views on and their practices regarding the human body and its matter are often framed in terms of social relationship and obligations, which have a potential to protect the individual from commodification by embedding the body in social and inter-generational networks of belonging. Such moral systems can be tested for their responsiveness to current conditions.

Equity is a set of legal practices and principles following the English Common Law tradition and supplementing rules of law, introducing a dimension of 'natural justice' into the legal discourse. It allows one to deal with some flexibility with subjective and practical dimensions of judgement of value. For example, a plaintiff whose neighbor will not return his only milk cow, which wandered onto the neighbor's property, may want that particular cow back and not just its monetary value. Or the value of a donated probe of body material to the donor can be assessed in terms different from those of science or medical benefit, or from their market value. In equity, with its emphasis on fairness and flexibility, only general guides apply, known as the maxims of equity. The material force of equity grounds in the moral authority of an accomplished person's conscience, such as, historically, the Lord Chancellor, or, exploratively speaking, in the Confucian Junzi.

These examples for diversity indicate a rich reservoir of contributions to a cross-cultural discourse on a proper normative framework for regulating the practices of dealing with the human body and the questions raised by regarding it in terms of ownership, as property, commodity.

Ethics beyond Governance

Opening the debate to considerations of a culture of humanity will obviously encourage a fundamental critique of capitalism, materialism and alienation. The cross-cultural exercise of comparison of ethics and critique of practice might involve re-assessment of the common sense of non-commercialization, and apply it to the claims of patenting (of life), and the hubris and moral absurdity of bio-markets.

The juridical problem is how to define the human body, the legal person, object and commodity, and their interrelations. What kind of legal category do we use when we assess human body-related property, and whether property can be applied here in the first place? Within a narrow juridical scope, conceptual ambiguity and juridical uncertainty towards the definition of the body and body material, in their distinction from objects might even support the confusion of body and person, and lead to restoring a Cartesian anthropology while at the same time contradicting moral common sense. Thus, depending on a legal or narrow regulatory conceptual framework without a reflected cultural ethical background would make it more difficult to legitimize ethical guidelines that deal with utilizing a person's body or body material, or to see how such an entitlement could be contained.

WHAT CAN PHILOSOPHY DO TO CONTRIBUTE TO ETHICS OF GENETICS?

It is the task of philosophy to clarify the fundamental questions and issues, especially to contribute to the conceptual language of an adequate ethical framework.

A basic question on the table is: Can the debate be framed properly in the language of law, or, is the entire undertaking of utilizing the body and its parts taboo? Does it categorically cover concerns about dealing with information? Should it be assessed according to an approach of equity, rather than positive law? How can the related questions of justice, benefit sharing and donor protection be organized?

In preparation, bioethics should sort out the conceptual correlation between diverse interpretations of property, ownership, possession, entitlement or claim to use, claim to access, control, dispose, etc. related to the human body and the human being. Not only does our understanding of the human body need to be refined in terms that reflect the best of our knowledge and responds to challenges of globalized biomedical modernity, but the semantic field of property needs also to be redefined and re-established in a sense that makes it meaningful and clear.

Some provisional practical conclusions I would like to draw for the time being, that is, while grounds will be prepared for a more adequate framework, are

1. A qualified informed consent with an option for renewable consent is a minimal condition in cases that affect sufficiently developed places and potentially benefiting subjects,
2. Individuals and countries with no apparent self interest or corresponding framing conditions cannot be used as sources (cf. The Belmont Report),
3. Patenting of biological resources with a legal status outside sciences (especially in economics) is extremely problematic because it imposes the rationales of particular cultures on others without an option to opt-out.

These are not new suggestions. The point I am trying to make here is that the basis for such guidelines needs to be strengthened in an adequate and unambiguously ethical manner. In cross-cultural and international, transdisciplinary bioethics, an attitude of open-mindedness with primary concern for the quality of an argument rather than convenience of the doctrine or pragmatic compromises should prevail. Regarding 'property' in genetics, there is a long way to go, if we want to leave the jumbled state of current bioethics and begin to systematically develop a sustainable, cultivated framework.

REFERENCES

Arendt, H. (1999). *The Human Condition*. Sydney, Australia: B&T Publisher.

Bereano, P. L. (1995) Patent pending: the race to own DNA. *The Seattle Times* (Aug. 27). Retrieved from http://weber.u.washington.edu/~radin/guaymi.htm.

Bray, M. B. (1990). Personalizing personalty: toward a property right in human bodies. *Texas Law Review*, *69*, 209–244.

Council for Responsible Genetics (CRG). (2009). Retrieved from http://www.essential.org/crg/crg.html

Döring. Ole. (2007). Grenzen der kommerziellen und medizinischenVerfügbarkeit des menschlichen Körpers. Ein Fallbeispiel aus der chinesischen Medizinethik. In J. Taupitz (Ed.), *Kommerzialisierung des menschlichen Körpers* (283-290). Heidelberg: Springer.

Everett, M. (2003, Spring). The gene business: the body as property in the biotech century. *Bulletin of General Anthropology Division*, *9*(2), 1–5.

Hartmann, R. G. (2005). Face value: challenges of transplant technology. *American Journal of Law & Medicine*, *31*(1), 7.

Human Genome Diversity Project, North American Regional Committee. (1995). *Model Ethical Protocol for Collecting DNA Samples,* in HGDP. Retrieved from http://www-leland.stanford.edu/group/morrinst/Protocol.html#Q0.

Indigenous Peoples Coalition Against Biopiracy (IPCB). (2009). Retrieved from http://www.niec.net/ipcb/

Kevorkian, J. (1992). A controlled auction market is a practical solution to the shortage of transplantable organs. *Medicine and Law,* 11.1-2, 47-55.

Leder, D. (1984). Medicine and paradigms of embodiment. *The Journal of Medicine and Philosophy*, *9*(1), 29–43.

Lysaught, M. T. (1995). Body: social theories. In T. W. Reich (Ed.), *Encyclopedia of Bioethics, 1,* 300-305. New York: Simon & Schuster Macmillan.

Nelkin, D., & Lindee, M. S. (1995). *The DNA Mystique: The Gene as a Cultural Icon*. New York: Freeman.

Radcliffe-Richards, J., Daar, A. S., Guttmann, R. D., Hoffenberg, R., Kennedy, I., & Lock, M. (1998). The case for allowing kidney sales. *Lancet*, *352*, 1950–1952. doi:10.1016/S0140-6736(97)08211-1

Radin, M. J. (1982). Property and personhood. *Stanford Law Review*, *34*, 957–1015. doi:10.2307/1228541

Radin, M. J. (1987). Market-inalienability. *Harvard Law Review*, *100*(8), 1849–1937. doi:10.2307/1341192

Rural Advancement Foundation International (RAFI). (1995). Gene hunters in search of "disease genes" collect human DNA from remote island populations. In *RAFI Communiqué* (May/June). Retrieved from http://www.rafi.ca/communique/19953.html.

Rural Advancement Foundation International (RAFI). (2009). Retrieved from http://www.rafi.ca/

Safrin, S. (2004). Hyperownership in a time of biotechnological promise: the international conflict to control the building blocks of life. *Rutgers Law School (Newark) Faculty Paper 15.* Retrieved from http://law.bepress.com/rutgersnewarklwps/fp/art15. Accessed Dec 12, 2006.

Sheets-Johnstone, M. (Ed.). (1992). The materialization of the body: a history of Western medicine, a history in progress. In *Giving the Body Its Due* (132-158). Albany, NY: State University of New York Press.

Williams-Jones, B. (1999). Concepts of personhood and the commodification of the body. *Health Law Review*, *7*(3), 11–13.

Williams-Jones, B. (2006, October). Bioethics and patent law: the cases of Moore and the Hagahai people. *Wipo Magazine*, *6*, 17–18.

Section 2
Country Experiences

Chapter 10
Biotechnological Patents and Morality:
A Critical View from a Developing Country

Jakkrit Kuanpoth
University of Wollongong, Australia

ABSTRACT

The chapter deals with ethical aspects of patent law and how the global patent regime helps or hinders the development of a developing country such as Thailand. More specifically, it discusses Article 27.3 of the World Trade Organization (WTO) Agreement on Trade-Related Aspects of Intellectual Property Rights (TRIPS), which states that countries may exclude methods of medical treatment, plants and animals (but not micro-organisms) from patent protection. It also provides legal analysis on the issue of whether developing countries can maximize benefits from the TRIPS morality exception (Article 27.2) in dealing with biotechnological patenting.

INTRODUCTION

Biotechnology has now become an important factor for the increasing rate of scientific and technological development, particularly in the fields of medicine and agriculture. A great deal of modern pharmaceuticals have been directly and indirectly developed from biotech inventions. For example, it is estimated that nearly 30 per cent of the present world drug market is accounted for by biotechnological products such as antibiotics, steroids, vitamins, and vaccines. The significance of biotechnology may stem from the fact that some modern drugs developed from biotechnology are able to target traditional diseases more precisely than conventional drugs, and biotech medicines provide more precise cures with fewer side effects (Business Week, 1992, pp. 52-53). In agriculture, plant breeding was once restricted to sexually compatible plants, and generations of offspring were selectively bred to create unique varieties. Currently, genetically modified (GM) crops have the fastest adoption rate of any new technology in global agriculture simply because of their claimed benefits from higher yields and lowered production costs. It is estimated that over 9,000 permits have been issued to field-test GM crops by United States Animal and Plant Health

DOI: 10.4018/978-1-61692-883-4.ch010

Inspection Service (APHIS) since 1987 (Gewin, 2003, p. e8).

Because of the growth of inventions in this area, the concept of patent protection has been changed gradually, and has become the subject of controversial discussions among scholars as to whether or not living organisms could be protected under the patent system. Since living organisms possess special features different from other inventions, particularly their self-replicating feature, this raises many substantial problems in any attempt to protect biotech inventions under the existing patent system. In addition, there are moral concerns surrounding inventions in this area. Before modern biotechnology, morality played a very limited role in patent protection. Since the advent of genetic engineering in the 1970s, however, moral concerns regarding biotechnology patents have significantly arisen, particularly due to the vigorous opposition of patents in the European Patent Office (EPO) by various pressure groups (Schatz, 1998).

A core issue which deserves closer examination is the moral aspect of biotech patents and its role in patent law of developing countries. While this paper attempts to focus on a number of the moral issues in the patent system, it does not aim to provide a theological discussion of morality and patents. In fact it is a rather focused discussion of specific legal provisions, mainly Article 27.3(a) of the World Trade Organization (WTO) Agreement on Trade-Related Aspects of Intellectual Property Rights (TRIPS). Some comments will be given in respect of the issue of whether developing countries can maximize benefits from the TRIPS morality exception in dealing with biotechnological patenting.

PROBLEMS OF BIOTECH PATENTING

The World Intellectual Property Organization (WIPO) provision on the protection of biotechnological inventions of 1984, defines 'biotechnology' thus (WIPO, 1984, p. 4):

Biotechnology concerns the creation of new varieties of plant, new animal breeds and new microorganisms, either by traditional selection methods or by new methods..., genetic engineering.

Biotechnology involves the application of scientific and technological knowledge to the processing of materials by biological agents such as enzymes or cells, in order to provide goods and services for the benefit of mankind (Bull, Holt and Lilly, 1982, p. 21; Llewellyn, 1987). It may involve the production of plants, animals, and micro-organisms, or involve new methods of medical treatment. In creation of new varieties, modern biotechnology basically applies modern techniques such as embryo transfer, or genetic engineering, which is different from conventional methods of selective or cross breeding. The major difference between biotechnology and other inventions is that, the former concerns a modification of existing complexity in living organisms, while the latter involves the creation of complexity by shaping and altering the simple constituents of inanimate material into structures of increasing complexity (Bent, 1987, pp. 6-7; Cooper, 1985).

In the past, the concept of patentable invention was focused only in the field of inanimate matters, such as chemistry and physics. Although biological materials have been used in some industries for a long time, biotechnology was normally excluded from patentability by the Patent Offices and the courts (Beier et al., 1985, p. 25). In 1969, the German Supreme Court accepted in the *Red Dove* (International Review of Industrial Property and Copyright Law, 1970, p. 136) case that an animal-breeding technique could be patentable. It apparently held, for the first time in a European court, that a biotech method was capable of patent protection. However, the Court finally denied the grant of patent because the claimed invention did not fulfill the particular condition of a written description by lack of the repeatable feature.

Subsequently, in 1980, the United States Supreme Court in a landmark case–*Diamond* v *Chakrabarty*–(206 USPQ 193 (1980)) ruled that a bacteria, in which a plasmid from another strain had been inserted, was patentable subject-matter. The Court went further by holding that the statute did not distinguish living matters from inanimate things, but only between the products of nature and man-made inventions. In other words, an alleged subject-matter must not be denied patentability merely because it was alive. It was held also that the patentable subject-matter included "anything under the sun made by man." The problem of patentable subject-matter under United States patent law did not depend on whether or not the claimed invention was a living matter, but on whether it was a result of nature or made by human.

In Europe, the patenting of particular types of biotechnology is prohibited by the European Patent Convention (EPC). Article 53(b) of the EPC states, *inter alia*, that:

European patents shall not be granted in respect of:

(b) plant or animal varieties or essentially biological processes for the production of plants or animals; this provision does not apply to microbiological processes or the products thereof.

Under the WTO TRIPS Agreement, signatories committed themselves to provide intellectual property protection on several subjects. The TRIPS Agreement stipulates in Article 27.1 that "... patents shall be available for any inventions, whether products or processes, in all fields of technology." However, Article 27.3 (b) allows member states to exclude plants and animals (but not micro-organisms) from patent protection. Essentially biological processes for the production of plants or animals other than non-biological and microbiological processes can also be denied protection. But countries are required to protect plant varieties. Since 1999, this specific provision has been up for renegotiations at WTO Council for TRIPS. It could be reworded, deleted or modified in any other way.

The term 'varieties' is crucial. A narrow interpretation of the term means that a bar to patents on varieties does not extend to higher categories - for example, species. As mentioned below, the EPO Technical Board of Appeal opted for this narrow interpretation of plants and animal varieties and awarded a patent on the seeds and the transgenic animals.

In 1988, the United States Patent and Trademark Office issued the world's first patent on a mammal, a transgenic mouse known as the oncomouse. The mouse was particularly susceptible to cancer and thus a valuable research tool. An application for the same invention was filed with the EPO in 1989. The EPO Examining Division at first rejected such claims on the ground that the claimed invention was related to a new animal variety, which is prohibited by the EPC. The Examining Division also stated that the Convention had no contemplation of granting patents on any type of animal. On appeal, the EPO Board of Technical Appeals overturned this decision sending the application back to the Examining Division for re-examination, including consideration as to the meaning of the term 'animal variety' and whether or not the 'invention' is contrary to public order or morality (T19/90, Onco-Mouse/HARVARD, [1990] E.P.O.R. 501).

After such reconsideration, the Examining Division recognized the distinction between 'animal' and 'animal varieties' and held that the Convention excluded only animal varieties from patentability. According to the Division, the term 'animal varieties' should be closely restricted to certain categories of animals, not to animals in general. Since the claims covered all GM mice into which the appropriate genes had been introduced, the invention could not be considered to be an animal 'variety'. It may be noted that the Division did not clearly explain the term 'variety', and this has stimulated controversial debate among those in the patent circles (Merges, 1988).

PUBLIC ORDER AND MORALITY UNDER PATENT LAW

The EPC prohibits the patenting of inventions whose publication or exploitation of which would collide with *ordre public* or with morality. Article 53(a) provides:

European patents shall not be granted in respect of:

(a) inventions the publication or exploitation of which would be contrary to 'ordre public' or morality, provided that the exploitation shall not be deemed to be so contrary merely because it is prohibited by law or regulation in some or all of the Contracting States.

The term 'public order' was defined in EPO terms in Case T19/90 Onco-Mouse/HARVARD as "the protection of public security and the physical integrity of individuals in society, which includes the protection of environment." The term 'morality' was explained by the EPO Technical Board of Appeal in the Plant Genetic Systems case[1] as follows: morality "is related to the belief that some behaviour is right and acceptable whereas other behavior is wrong, this belief being founded on the totality of accepted norms which are deeply rooted in a particular culture." From this, public policy mainly focuses on the state policy of a country, and morality refers to social conducts judged by the norms of a particular society.

In June 1998, the revamped version of the draft Biotech Directive was adopted by the European Council of Ministers. Just like the corresponding provisions in the harmonized patent laws of the EPC states, the Directive 98/44/EC on Biotechnological Inventions contains a moral safeguard clause in Article 6, which states:

1. Inventions shall be considered unpatentable where their commercial exploitation would be contrary to *ordre publique* or morality; however, exploitation shall not be deemed to be so contrary merely because it is prohibited by law or regulation.
2. On the basis of paragraph 1, the following, in particular, shall be considered unpatentable:
   a. Processes for cloning human beings;
   b. Processes for modifying the germ line genetic identity of human beings;
   c. Uses of human embryos for industrial or commercial purposes;
   d. Processes for modifying the genetic identify of animals which are likely to cause them suffering without any substantial medical benefit to man or animal, and also animals resulting from such processes.

The EPC is not the only patent system to incorporate a morality exception. Patent law of many countries sets out the morality exception to patentability.[2] The exception is incorporated into countries' patent law because it is provided under the TRIPS Agreement. Article 27.2 of TRIPS states that:

Members may exclude from patentability inventions, the prevention within their territory of the commercial exploitation of which is necessary to protect ordre public or morality, including to protect human, animal or plant life or health or to avoid serious prejudice to the environment, provided that such exclusion is not made merely because the exploitation is prohibited by their law.

Article 27.2 of TRIPS does not aim at a uniform and universal substantive definition of '*ordre public*' and 'morality.' The interpretation of the terms thus is left to the decision of each member country. It may be noted that like Article 53(a) EPC, the TRIPS provision does not deal with the morality of patenting an invention, but rather with the morality of publicizing or exploiting it. A patent application may be rejected if the invention can be used only in manner which would violate public order or morality. However, TRIPS clearly states that a country cannot use this clause to reject a patent application simply because the invention is not in line with general policy. In other words, approval or disapproval of the exploitation of the invention by national law does not constitute per se a sufficient criterion for the grant or denial of patents under Article 27.2.

We now consider whether or not the morality exception can and should be used by developing countries as an objection to particular categories of biotech inventions

1. Human beings

The European Biotech Directive explicitly prohibits patents on the human body at various stages of its formation and development. It also prohibits certain inventions concerning human beings such as human cloning, germ-line genetic modification of human beings, and the use of human embryos for industrial or commercial purposes (Biotechnology Directive, Art.6(2). A human gene is patentable only when it is isolated or produced by means of a technical process. The sequence or partial sequence of human genes can also be the subject of a patent if the applicant provides an explanation of industrial application of the gene sequence (Biotechnology Directive, Art. 5).

The patent law of many developing countries is silent on whether inventions relating to human beings are patentable. Unlike plants and animals, there is no provision in the TRIPS Agreement that allows patients to be refused for inventions relating to human beings. However, it may be correct to say that developing countries can deny patent protection for this sort of inventions on moral grounds.

Patent applications for inventions involving humans must be subject to stricter moral criteria than any other invention (Hoffmaster, 1991; Stercks, 1998). The standard may be that the human body is a vehicle for human dignity which is universally recognized as a fundamental human right. Allowing patent rights over human body or body parts is tantamount to making humans an instrument and therefore humiliating human dignity. Based on moral considerations, the human body and organs, human-derived products, including cell lines, genes and nucleic or amino acid sequences, as well as the knowledge of all or part of the structure of a human gene should not be eligible to be patented.

2. Animal varieties

The EPO Technical Board of Appeal considered the rule on morality in Onco-Mouse/HARVARD. The patent application concerns the technique of inserting oncogenes into the genome of two founder mice. This markedly increased their susceptibility to develop cancerous tumors such as breast cancer or lymphoma, which would in turn cause considerable suffering to the animals. Onco-mice are claimed to be useful in research, for example for the assessment of anti-cancer drugs, but the ethical issue surrounding this application was whether the invention was against morality. The Technical Board of Appeal held as thus:

Whether or not order public/morality were a bar to the patenting of an experimental transgenic mouse depended mainly on a careful weighing up of the suffering of the animals and possible risks to the environment on the one hand, and the invention's usefulness to mankind on the other.

On the question of public order or morality, the EPO Examining Division to which the Onco-Mouse case was referred back finally granted the patent by comparing the suffering inflicted on the animal and possible risks to the environment with the potential interests to mankind. The Examining Division considered the invention not contrary to morality, because the "onco-mouse purpose, of facilitating cancer research and treatment, was of paramount importance for the welfare of mankind." However, the Division implied in its decision that the question of morality was to be decided on a case-by-case basis. This led to the EPO's rejection of another patent application (the Upjohn application) concerning a mouse destined to be used not for cancer research experiments but for the testing of cosmetics (European patent application no. 89913146.0).

In this case, the EPO construed the scope of 'public order and morality' rather widely by applying it to preclude patenting of inventions which involved animal suffering but also genetically manipulated animals that might entail unforeseeable and irreversible adverse effects to the environment (T19/90, Onco-Mouse/HARVARD, [1990] E.P.O.R. 501). This interpretation is in line with the TRIPS Agreement which requires WTO members to look into morality of exploiting the invention, rather than morality of patenting it.

The EPO practices provide general guidelines for interpretation of the exclusion on grounds of public order or morality. Efforts however have been made to comply literally with this prohibition while at the same time interpreting the terms 'public order' and 'morality' flexibly so as to permit patent protection for certain aspects of living organisms. However, developing countries who are WTO members are not obliged to adopt this line of practices, and are able to exclude all aspects of animal and plant subject matters from their patent law as explicitly stipulated in Article 27.3(b) of TRIPS, and can also allows patents relating to animals to be refused on the basis of a moral exception under Article 27.3(a). For example, a country may reject the patent application involving techniques that cause animal suffering even if the applicant can show substantial medical benefit from exploitation of the invention. It may also deny patent protection for genetic animal inventions that raise environmental concerns on the same ground.

3. Plant varieties

One important aspect of modern technologies is the genetic improvement of crop plants. Genetic engineering leads to the development of new plants with desirable properties. It is claimed by those who engage in modern plant breeding techniques that the development of science in this area involves considerable research and requires a multitude of inventions and discoveries. Patent law, which provides partial monopoly protection, helps by encouraging investment in inventive activity. However, granting intellectual property rights on plant materials has so far proven to be a highly contentious issue at the various international forums.

Article 27.3(b) of TRIPS authorizes WTO members to deny patenting of certain kinds of biotech inventions including plants. But members are required to protect plant varieties. This implies that a 'plant' is distinguishable from 'plant varieties'. What is 'plant varieties' is explained by the EPO Board of Appeal in LUBRIZOL/Hybrid plants as follows (T320/87, LUBRIZOL/Hybrid plants, OJ EPO (1990), 71):

'Plant variety' means a multiplicity of plants which are largely the same in their characteristics (i.e. homogeneity) and remain the same within specific tolerances after every propagation or after every propagation cycle (i.e. stability).... Only possession of both these criteria, homogeneity and stability, would be a prerequisite for a 'plant variety.'

Article 27.3(b) was included within TRIPS at the insistence of Europe. In Europe plant varieties are excluded from patent protection both under the EPC and the Biotech Directive. The exclusion of plant varieties from patentability may be traced back to 1961 when the International Convention for the Protection of New Varieties of Plants (UPOV) was adopted. The UPOV Convention was signed in 1961 establishing a plant breeders' rights system as an alternative to patents for plant varieties. The exclusion of plant varieties from patentability under the EPC, was to avoid double protection because plant varieties can be protected by plant breeders' rights under the UPOV Convention.

With regard to the protection of plant varieties, WTO members have three options. They may protect plant varieties by (i) patents, (ii) an effective *sui generis* system, or (iii) a combination of both systems (Overwalle, 1999). Unlike the protection of intellectual property rights in other areas, TRIPS does not list other international treaties regarding plant varieties to which governments must adhere (*i.e.* it does not mention UPOV). This implies that the negotiators deliberately kept the term '*sui generis*' broader than the UPOV requirements for plant breeders' rights.

Many developing countries are reluctant to embrace plant patents unequivocally when determining compliance with their obligations under the TRIPS Agreement. Thailand, for example, has responded by looking to *sui generis* rights and not the patent system. The Plant Variety Protection (PVP) Act B.E.2542 was enacted in Thailand to comply with TRIPS Article 27.3(b). It is a *sui generis* form of protection tailored for the purpose of the protection of new and traditional plant varieties. To protect products of plant breeding activities, the law has certain features in common with patents for industrial inventions, but contains safeguards and social welfare mechanisms including the breeders' and farmers' exemptions. It also recognizes the important role played by farmers and local communities as custodians of traditional crop cultivars. The PVP Act acknowledges and compensates farming communities for their contributions in plant conservation and breeding activities by allowing them to register traditional plant varieties. It vests ownership over these plant genetic resources not in individuals but in the community in which the plants are endemic. Use of plant genetic materials for any purposes requires authorization from a competent official and a profit-sharing agreement must be signed. The law also enables the establishment of a Plant Variety Protection Fund, which would tap a portion of the royalties paid by those who use plant genetic resources for commercial purposes.

The adoption of the PVP Act reflects the objective of Thailand to recognize the importance of plant varieties. The exclusion stemmed from political and social resistance to the granting of an absolute monopoly to private enterprises. There were two reasons underlying the establishment of the Thai '*sui generis*' PVP system. First, patent is considered inappropriate to protect plant varieties. The patentability requirements (*i.e.* novelty, inventive step, and industrial application, as well as a sufficient written disclosure) are too difficult for plant varieties to fulfill. Secondly, plants are vitally important for agriculture which is the backbone of the Thai economy. Absolute monopoly rights under patents might have hindered free access to germplasm and seeds which are a basic material for the production of food crops.

It may be noted that the UPOV system clearly ignores issues of morality and equity in protecting plant varieties. This has proved problematic and led to 'bio-piracy.' Often, plant genetic resources, together with customary and informal knowledge, have been made publicly available and exploited without adequate compensation, and occasionally the resources have been claimed as proprietary rights of researchers or companies. Such illicit and uncompensated appropriation of genetic resources is exploitative and is cause for concern among developing country governments as well as among indigenous and local communities.

Traditional community and indigenous people are left without recourse to challenge the system and the financial implications of intellectual property litigation remain beyond their reach (Correa, 1999). The system for plant variety protection should be designed to take account of the needs of custodians, breeders and developers of plant varieties. Since developing countries have been fighting for the recognition of their traditional knowledge, as well as genetic resources, as new intellectual property components, moral standards should be introduced into the system of plant variety protection (Dutfield, 2000). The morality concept should be extensively defined to adequately deal with bio-piracy practices. A plant variety would be considered against morality, and thus denied registration, if it is developed based on the use of genetic resources or traditional knowledge without any prior informed consent and equitable sharing of benefit.

4. Methods of medical treatment

The TRIPS Agreement provides for the exclusion of methods of treatment from patentability. Article 27.3(a) states:

Members may also exclude from patentability:

(a) diagnostic, therapeutic and surgical methods for the treatment of humans or animals;

This provision permits WTO members to deny patentability to medical practices, such as surgery, therapy or diagnosis on human or animal body. Patents for the methods of treatment are denied on moral grounds in order to protect medical practitioners from the restrictions of monopoly privileges, and at the same time allowing for healthy competition to enhance the well being of the public.[3] Patent protection would "operate as a restraint rather than a stimulus to research, investigation and the creation of ultimately useful, practical and valuable innovation" (Black, 1989, p. 13; Grubb, 1986, p. 147). Phillips and Firth quite rightly state that, "the protection of life and health are universally recognized objectives which transcend the sordid realm of proprietary rights" (1990, p. 51).

Under the TRIPS provision, the medical practices which are excluded from patentability fall into two categories: (a) methods of treatment by surgery or therapy, and (b) diagnostic methods. The exclusion is now considered.

a. Methods of treatment by surgery or therapy

Methods of treatment by surgery may be defined as treatment of disease or injury of the human and animal body by operation or manipulation. Therapeutic methods refer to the medical treatment of disease generally (Panchen, 1991). The methods of treatment must be directed towards the curative treatment of disease or illness on human and animal bodies.[4]

As an exception, this provision must be construed very narrowly. Methods in relation to contraception or pregnancy, for instance, could be patentable as they are not concerned with curative treatment. This would be the same as the treatment of removed body tissues and fluids which are not returned to the body, a method of improving bodily appearance, a method of cosmetic treatment, and a method of getting rid of lice on human beings.

On the contrary, when a method of treatment has its therapeutic aim to deal directly with disease or illness, it is not a subject of patentability. The excluded subject-matters may include processes of gene therapy, the use of biological agents that indirectly cause genetic modifications, an *in vitro* method for preparing altered cells (for somatic cell gene therapy), a method for relief of pain, discomfort or incapacity, a method of dental surgery, a method of removing plaque and cleaning teeth, a method of treating and filing dental cavities, etc.

Methods of treatment which could be excluded from patentability include not only curative treatments, but also those treatments for prophylactic purposes (Panchen, 1991, p. 885). Therefore, methods of prevention or immunization such as a surgery for removal of tonsils or a method of vaccination are not patentable inventions.

b. Methods of treatment by diagnostic method

So far as diagnostic methods practiced on human and animal bodies are concerned, this sort of medical practice may not also be patentable. The diagnosis basically involves the identification of malfunction or disorder for the purpose of medical treatment (Panchen, 1991, p. 885). The EPO Technical Board of Appeal ruled that the method of treatment must provide an absolute result, which enables medical practitioners to decide on a particular course of medical treatment, for example blood test to identify disease. If such a method only produces an intermediate result which, in order to identify the medical disorder, has to be used with the results of other tests, it could be patentable. The instances of this method are those of X-ray investigations and blood pressure measurements.[5]

The exclusion is permitted in order to make medical treatment methods freely available to all. But it may be noted that the law does not prohibit patenting of a substance or composition used in the treatment of human or animals. This means in effect that the legal status of medical products and methods from the point of view of patentability is unequal. Regarding the rationale for this difference, Bainbridge argues that a medical practitioner is in a position where he is expected to pass on his knowledge to others and does not expect financial recompense for his ideas. In contrast, a pharmaceutical company operates in a competitive industry and needs a patent to recoup its R&D investment (Bainbridge, 1992, p. 271).

The above opinion seems to suggest that medical practice is always a social service which has to be conducted for the benefit of society. True as this may sound, it does not consider the fact that the monopoly position of the pharmaceutical industry might lead to high prices for pharmaceutical and other medical products considered to be basic health care requirements. The view also ignores the fact that public order and morality must play an important role in every sector of medicine, not only among health professionals, but also among manufacturers and distributors of life-saving drugs.

Although most countries contain a compulsory licensing (*i.e.* the system which allows the use of patented invention without committing a patent infringement), in their patent law, it is doubtful if the system can safeguard the public interest in case of a real emergency situation. Article 31 of the TRIPS Agreement requires WTO members to follow procedures which will delay the grant of a compulsory license or make the compulsory licensing scheme unworkable. In view of the possible consequences for the costs of health care, developing countries should not rely on the compulsory licensing to tackle patenting of medical treatment methods, but may consider excluding this type of inventions from patentability on the basis of moral considerations.

CONCLUSION

The patent system was not designed to protect animate material and its use in respect of living material has met with criticisms around the world. The most public concern of developing countries in regards to biotechnology is not just about the ethically unacceptable technologies like reproductive human cloning, or modifying genetic identity of human beings. The introduction of patents for biotech inventions may also have wider implications for developing countries, particularly on their agriculture and food sectors. Patents over medical treatment may generate adverse effects on health care, in the same way as pharmaceutical patents currently restricting access to medicines in the poor countries.

Apart from its socio-economic impact, another problem of extending patent protection to biotech inventions concerns legal and technical difficulties in the administration of micro-organism and biological material patents. How will the developing country governments search the prior art of living-organism and biological method claims? How will the inventive step of the biotech inventions be examined? What regulations should developing countries adopt to deal with possible hazards resulted from the release of micro-organisms into the environment? These practical problems are to be taken into consideration if the poor countries are required to protect biotech inventions in their patent system.

The EU Biotech Directive has extended patent protection to the whole or parts of living organisms, which will have worldwide implications. One of the major implications of this event beyond Europe is that the member states of the European Union are now likely to stop their support to the developing countries in fighting for the worldwide right to exclude the patenting of plants and animals under the WTO-TRIPS Agreement. Article 27.3(b) of TRIPS may be deleted or revised obligating member states to provide patent protection for living organisms and higher life-forms. Of course, without the active resistance of Europe, the right of developing countries to exclude biotechnological inventions is likely to be lost in the TRIPS revision.

Allowing protection for all types of biotech inventions under patent law would obscure the nation's vision on a fundamental area of technology that should not have been protected or should be protected at the minimum. It is therefore imperative that the moral aspect of patent legislation be critically reexamined with a view to expunging the biotech inventions from the law. It is equally reasonable to suggest that developing countries should only grant minimum protection to high technology areas, otherwise they stand to gain very little from patenting such technology.

REFERENCES

Bainbridge, D. L. (1992). *Intellectual Property*. London: Pitman Publishing.

Beier, F. K. (1985). *Biotechnology and Patent Protection*. Paris: OECD.

Bent, S. A. (1987). *Intellectual Property Rights in Biotechnology Worldwide*. New York: Stockton Press.

Black, T. (1989). *Intellectual Property in Industry*. London, Edinburgh: Butterworths.

Business Week. (1992). March 2.

Cooper, I. P. (1985). *Biotechnology and the Law*. New York: Clark Boardman.

Correa, C. M. (1999). Access to plant genetic resources and intellectual property rights. Commission on Genetic Resources for Food and Agriculture, Background Study Paper No. 8. FAO. Retrieved from http://www.fao.org.

Dutfield, G. (2000). *Intellectual Property Rights, Trade and Biodiversity*. London: Earthscan.

Gewin, V. (2003). Genetically modified corn - environmental benefits and risks. *PLoS Biology*, *1*(1), e8. doi:10.1371/journal.pbio.0000008

Grubb, P. W. (1986). *Patents in Chemistry and Biotechnology*. Oxford: Clarendon Press.

Hoffmaster, B. (1991). Between the sacred and the profane: bodies, property, and patents in Moore case. *Intellectual Property Journal*, *7*, 115–148.

International Review of Industrial Property and Copyright Law. (1970).

Llewelyn, M. (1997). The legal protection of biotechnological inventions: an alternative approach. *European Intellectual Property Review*, *19*(3), 115–127.

Merges, R. P. (1988). Intellectual property in higher life forms: the patent system and controversial technologies. *Maryland Law Review (Baltimore, Md.)*, *47*, 1051–1075.

Overwalle, G. v. (1999). Patent protection for plants: a comparison of American and European approaches. *IDEA the Journal of Law and Technology*, *39*, 143–194.

Panchen, K. E. (1991). Patentability in the field of therapy and diagnosis. *International Review of Industrial Property and Copyright Law*, *22*, 879–880.

Phillips, J., & Firth, A. (1990). *Introduction to Intellectual Property Law*. London: Butterworths.

Schatz, U. (1998). Patentability of genetic engineering inventions in European patent office practice. *International Review of Industrial Property and Copyright Law,* 29, 2-16.

Sterckx, S. (1998). Some ethically problematic aspects of the proposal for a directive on the legal protection of biotechnological Inventions. *European Intellectual Property Review*, *20*(4), 123–128.

WIPO. (1984). *Industrial Property Protection of Biotechnological Inventions. WIPO Pub. Biot/CE/I/2. Bull, A. T., Holt, G. & Lilly, M. D. (1982). Biotechnology: International Trends and Perspectives*.Paris: OECD.

ENDNOTES

1. T356/93, Plant Genetic Systems/Plant cells (opponent: Greenpeace), [1995] E.P.O.R. 357.
2. For example, Thailand's Patent Act B.E. 2522 states in Section 9 that: "The following inventions are not protected under the Act: … (5) inventions contrary to public order, morality, health or welfare."
3. Decision of EPO Technical Board of Appeal, T116/85, WELLCOME/Pigs I, OJ EPO (1989), 13.
4. Decision of EPO Technical Board of Appeal, T144/83, DUPONT/Appetite suppressant, OJ EPO (1986), 301.
5. T385/86, BRUKER/Non-invasive measurement, OJ EPO (1988), 398.

Chapter 11
Social Issues Related to Gene Patenting in Latin America: A Bioethical Reflection

Eduardo Rodriguez
University of Chile, Chile

Fernando Lolas
Pan American Health Organization / World Health Organization, Chile

ABSTRACT

The chapter reports on the experiences of both experts and lay people on the level of knowledge and social representations of genomic research and its applications in a number of Latin American developing countries. Issues discussed include access to prevention and therapeutic methods related to genomic medicine in Latin America, risks associated to genetic modifications in humans, lack of equity in the access to health benefits, control by biotechnological companies, commercialization of gene sequences through patents which leads to commercial exploitation of underdeveloped countries, among others

INTRODUCTION

With the advent of genetic knowledge and the possibility of gene patenting a commercial mentality has been dominating the genetic research area. Issues related to gene patenting such as globalization of regulations, health policies, relation between developed and developing countries, equity, economic power of biotechnological companies and accepted uses of biotechnology deserve bioethical reflection.

The patenting of human genes has been possible through the completion of the Human Genome Project, sequencing the entire human genome and characterizing the function of many genes (International Human Genome Sequencing Consortium, 2001; Craig Ventor et al., 2001) and upon the acceptance of patent offices of developed countries. There are great expectations regarding the social applications on health of genomic information as a field where patents of human genes can have industrial use. Many genes are related to hereditary diseases; 1,112 genes have been identified related to Mendelian inheritance diseases (Catalog of Mendelian Inheritance Diseases, 2010) and many others act on multifactorial diseases, such as cancer or diabetes. This has created interest in

DOI: 10.4018/978-1-61692-883-4.ch011

patenting sequences with use in health industry for diagnosis, prognosis and treatment.

Furthermore, the biotechnological industry is growing enormously due to the creation of genetically modified organisms to improve crops and cattle with profits by patenting.

Data in this chapter are drawn from Latin American journal articles and from our own investigation on social representations of genomics in four Latin American countries (Argentina, Chile, Mexico, and Peru).1

ADVANCES IN GENOMIC RESEARCH IN LATIN AMERICA

In general, there is an impression that Latin American countries are not prepared to respond to the explosive development of genomics and genetic engineering which has taken place in developed countries. There is little interest in most Latin American governments for research in this area since they believe that other priorities are more important. As a result, Latin American countries act mostly as consumers with the added problem of the little information that lay civilians possess. There is also lack of legal norms to regulate this field in general. This contributes to the generation of certain anxiety since there are fears that certain issues raise by the expansion of genomic research could be manipulated and used for the interest of a few.

In general, developing countries are left behind in biotechnology and genomic medicine development, but there are some exceptions at Latin America such as Brazil, Cuba, Argentina and Mexico in the development of genomics. Cuba has linked biotechnology to its health care sector. Brazil represents the biggest market at Latin America. Argentina is moving towards being a force at Latin America in transgenic market. Mexico has recognized the potential for genomics in addressing public health issues.

Brazil has 71 biotechnological companies, most working on transgenics and some have developed genetically engineered health care products, such as insulin, vaccines, kits for diagnosis and immunization; since 1998 the Human Genome Project for Cancer has been functioning under the sponsorship of Ludwig Institute of the United States and the Foundation for Research Protection at Sao Paulo (Fundación de Amparo a la Investigación del Estado de Sao Paulo—FAPESP) (Bisang et al., 2009, p. 73-74). Brazil has created the Genomic Institution ONSA in 1997 uniting several laboratories. The first development was sequencing the first Latin American microbial genome, the bacteria *Xylella fastidios* (Simpson et al., 2000). Furthermore, several projects have been initiated which link universities and research institutions with regional agriculture and health problems. Examples are: the sequencing of *Chromobacterium violaceum,* a human pathogen; the sequencing of *Herbaspirillum seropedicae* of farm value; the sequencing of RNA transcripts of human cancers; and the sequencing of RNA transcripts of the disease caused by *Leishmania chagasi* (Simpson, 2001).

An example in Cuba has been the development of the first human vaccine with a synthetic antigen for meningitis B. The vaccine against Haemophilus influenza type B (Hib) infection made in Cuba is much cheaper and safer product than other existing vaccines. This was possible due to the commitment and involvement of government bodies, public research institutions, universities and the health system (Thorsteinsdóttir et al., 2009).

Argentina has used its Intellectual Property Rights laws to develop the pharmaceutical sector for competition in the global marketplace and constitutes an example of a developing country moving forward in creating national guidelines, approval procedures and research institutes to evaluate the risks of genetically modified organisms.

Genomic medicine has become a priority to the Mexican government as a means of finding new strategies to tackle common diseases. Thus, in 2003, Mexico launched a plan to develop a genomic medicine program which led to the establishment of the Institute of Genomics Medicine and a National Platform for Genomics Medicine which has modified the health care system (Sanchez, 2003). As a result of this program, Mexico has genotyped over 1,200 people from different regions of the country and study possible relationships between genetic make-up and health problems with greater burden, such as macular degeneration, hypertension, obesity, infectious diseases, cancer, diabetes and cardiovascular diseases, with the potential to reduce health care costs.

Chile has also recently favored the development of genomic research, which is coordinated by the National Commission for Science and Technology. Several universities as well as biotechnology companies participate in research, producing genetically modified organisms in agriculture. Examples are: potatoes genetically modified for resistance to viral and bacterial diseases, such as *Erwinia caratovora;* of a bacterial strain *Corynebacterium glutamicum* overproduces trehalosa. Chile is also working in the genetic characterization of current and antique indigenous populations, studying mommies, and also specific mutations and genetic markers for particular diseases, such as cystic fibrosis, diabetes and cancer (Paola et al., 2002; Francisco et al., 2003).

SOCIAL ISSUES RELATED TO HUMAN GENE PATENTING

Review of the scientific and legal literature shows that the application of genomic research in patenting can have social and legal consequences in need of bioethical reflection.

Patenting Criteria

Regarding human genes patenting, its legal and intellectual legitimacy in relation with ethics are discussed (Madrid, 1999). While the objective criteria contained in patent law are novelty, inventiveness and industrial use, some authors argue that the current tendency to patent human gene sequences has derived in a progressive loss of limits between invention and discovery, and is opposed to the principle of non commercialization of the body and its parts (Bergel, 2000). In this regard, the practice is controversial since the 1997 UNESCO Universal Declaration on the Human Genome and Human Rights (1997) declared that the human genome in its natural state cannot give rise to pecuniary benefit. Furthermore, countries which possess the technology adopt a comprehensive patent system control and dominate the market (appropriating information shared by all human beings) (Marco et al., 2001). Patent offices from developed countries have circumvented the problem accepting as inventions the isolation and modification of gene sequences for industrial purposes (for example, European Directive 98/44/CE, June 6, 1998).

The initial plan of the Human Genome Project Consortium was to generate genetic data with free access avoiding competition which could restrict access to information and would make human genes patentable and genetic tests a big business. The intrusion of Celera Genomics generating the whole human genome sequence and the interest of biotechnological companies in patenting has prevailed and the gap between developed and developing countries is increasing because of the technology. Bergel considers that this practice damages human dignity since it manipulates human genetic information with commercial interests (Bergel, 2002).

APPLICATIONS OF HUMAN GENE PATENTING

Patenting of human genes has applications in medicine. There are many possible advances in genomic medicine. It is expected that knowledge of the human genome will offer new ways for prevention, diagnosis and treatment of diseases with a hereditary component. Many developments are recognized: new vaccines with moderate immunogenicity for hepatitis and malaria; drugs obtained by genetic manipulation, such as insulin, growth hormone and interferon; development of molecular neurobiology for treatment of psychiatric diseases; production of tissue activators such as t-PA; production of monoclonal antibodies; molecular diagnosis using micro-arrays; pharmacogenetics or therapy based on the individual genetic characteristics of the person with respect to respond to drugs; molecular epidemiology to know risk factors, geographic distribution of diseases and prevention; new therapeutical interventions using genetic engineering; gene therapy, introduction of genes which activate drugs for destruction of cancer cells, stimulation of the immune response, inactivation of mutant oncogenes, activation of tumor suppressor genes, rybozimes or RNA with catalytic activity for destroying specific proteins, antisense therapy; and paternity tests (Lolas et al., 2004).

Controversial Applications

Even though genomic research and its applications are positively viewed, it is recognized that there are manipulations contrary to human dignity and fundamental human rights, demanding regulation by legal norms and sanctions. It is feared that the idea of genetic determinism and the power of genetic technology may engender the myth of perfect health leading to abuses in developing countries with poor resources and low understanding of the real power of genetics (Claude, 2004). There are also concerns related to the resurgence of eugenics, like the practice of eugenic abortion and embryo selection, not for medical purposes (Santos, 1997; de Constantino et al., 2006).

The Human Genome Diversity (HGD) Project attempts to understand DNA human variability and the evolutionary history of human populations. Blood and tissue samples have been obtained from indigenous peoples of 722 communities (10% of the different linguistic populations) (Patents, Indigenous Peoples, 1993). This raises troubling questions and has been resisted by indigenous populations. Several problems arise: how informed consent has been obtained since the concept and purpose is difficult to understand by many indigenous populations, who will own the genetic samples, how intellectual property rights of indigenous people will be protected since there are no international policies or regulations governing the trade of human genetic materials, and who will profit from the commercialization of products derived from the samples. This Project may be opening the doorway for widespread commercialization and potential misuse of samples and data, since there are no safeguards to prevent genetic manipulations with the samples collected. There have been complaints that samples of Brazilian ethnic groups are sold at U.S. Coriell Institute of Medical research (http://www.coriell.org/).

The Genographic Project, led by Spencer Wells of National Geographic, a team of renowned international scientists, and IBM researchers, uses cutting-edge genetic and computational technologies to analyze historical patterns in DNA to better understand human genetic roots, including indigenous populations (https://genographic.nationalgeographic.com/genographic/lan/en/index.html). The use of indigenous populations remains controversial (Machiola, 2009).

SPECIFIC CONCERNS

Data gathered through interviews and questionnaires to biomedical researchers, lawyers, students and lay civilians in Latin American countries shows worries about (Rodriguez et al., 2005): little access to prevention and therapeutic methods related to genomic medicine, the risk associated to genetic modifications in humans, lack of equity in the access to health benefits, commercialization of gene sequences through patents which leads to commercial exploitation, the possibility of physical or psychological damage through of stigmatization or genetic discrimination, the possibility of genetic modifications or abortion for eugenic reasons, the necessity of confidentiality, the little participation of indigenous communities in the studies done on their DNA, sometimes without proper informed consent, the necessity of legal regulations to prevent the pathway towards enhancement genetic modifications or reproductive human cloning and of regulating access to genetic information. Biomedical researches and legislators worry about the commercial mentality associated with genetic research and its applications which may lead to increase the excessive gap that exists between developed and underdeveloped countries. Biomedical researchers appreciate more the benefits of the human genome project, but experience the lack of governmental support for research. University students worry about the lack of equity in access to genomic medicine in the population and the instrumentation of human beings. Lawyers and legislators are worried about the regulation of genetic information because of its possible manipulation by power interests, of possible eugenic selection of embryos and of possible genetic discrimination by health insurers and employers (Rodriguez et al., 2005).

SOCIAL ISSUES REGARDING PATENTING OF GENETICALLY MODIFIED ORGANISMS

Intellectual Property Rights

Patenting organisms and their DNA promotes the concept that life is a commodity and the view that living beings may be considered as "gene machines" to be exploited for profit. Patents derive from concepts of individual innovation and ownership, which may be foreign to cultures which emphasize the sharing of community resources and the free exchange of goods and knowledge. Genomic technologies by their very nature represent a challenge to existing values and systems, and induce changes in traditional concepts of nature.

Current global situation regarding intellectual property is characterized by a dense regulatory framework linked to international commerce. There is a framework imposed to national states by multilateral treaties. The new system for intellectual property rights is characterized by: increases in protection levels, widening of matters included in patents, globalization, close link with regulations on investments; a multi scale and simultaneous logic for negotiations and regulations (Llancaqueo, 2006). The multi scale system is expressed at different levels with international agreements and treaties: Agreements on Intellectual Property Rights related of the World Trade Organization which link intellectual property rights to commerce (Trade-Related Aspects of Intellectual Property Rights TRIPs, 1994), treaties of the World Intellectual Property Organization (WIPO) with a mandate from the United Nations for promoting intellectual property rights, General Agreement on Tariffs and Trade (GATT) and International Union for the Protection of New Plant Varieties (UPOV). National regulations must adhere to the system in order to survive economically. Nowadays there are financial pressures to patent pharmaceutical products and living beings and the national faculties to regulate intellectual property rights are restricted. National regulations must be adapted to international norms and a unique system for patenting is being imposed (GRAIN, 2003), with coercive economical sanctions to countries which do not fulfill international norms (Roffe, 2006).

In parallel with this system of protection of intellectual property, there is another framework system related to the protection of the environment in which the access to genetic resources and traditional knowledge is regulated. This system is also linked to commerce and is expressed mainly in the Convention on Biological Diversity (Rio de Janeiro, 1992) and the Kyoto Protocol (United

Nations Framework Convention on Climate Change, in force 2005) documents, which has been criticized for emphasizing the commercialization of biodiversity and payment for environmental services by indigenous populations (Llancaqueo, 2006). Among these services, ecotourism and ethno-tourism, bioprospecting and access to genetic resources and traditional knowledge are included. This service market is regulated by the General Agreement on Commercial Services of the World Trade Organization. Functions of nature, such as water, oxygen and biodiversity are considered as goods and environmental services. When communities and farmers sell their resources as environmental services, they lose control over them. As a result there is a change of focus, from a territorial right focus to a focus on market of transitional services which generate utilities (Instituto de Estudios Ecologistas del Tercer Mundo, 2006).

The Problem of Biopiracy

The Convention on Biological Diversity (CBD) has stated the commitment of signing countries to "the conservation of biological diversity, the sustainable use of its components and the fair and equitable sharing of the benefits arising out of the utilization of genetic resources" (Convention on Biological Diversity, 1992). One of the methods to reach these goals is by ensuring "appropriate access to genetic resources and by appropriate transfer of relevant technologies, taking into account all rights over those resources and to technologies, and by appropriate funding." However, biotechnological companies are interested in keeping at a minimum the regulation over access to genetic resources and in that already existing germplasm collections be excluded from any new access regulation or benefit-sharing schemes, for which they are working in negotiations. Parties seeking access to genetic resources or traditional knowledge enter into a contract with the sovereign entity that grants that access. There have been many reports of biopiracy activities where germplasm collectors have entered countries as tourists. Although the Convention on Biological Diversity allows for mechanisms to frame laws to prevent biopiracy and promotes to respect, preserve and maintain knowledge, innovations and practices of indigenous and local communities embodying traditional lifestyles relevant for the conservation and sustainable use of biological diversity with equitable sharing of benefits arising from the utilization of such knowledge, innovations and practices (article 8), the difficulty is how to put it into practice.

NGOs such as RAFI, GRAIN and the Third World Network have been networking to raise general awareness of the phenomenon of biopiracy. Patent claims over genetic resources are based on traditional uses, and are not novelties, for which they are considered as biopiracy-based claims. They are acts of plagiarism of indigenous knowledge about plants and animals bred and used by local communities for millennia. Some Latin American countries have contested patents granted by the U.S. Patent and Trademarks Office (PTO) for biological materials, especially plants, taken from indigenous people, such as Bolivia, which successfully defeated Colorado State University's application for a U.S. patent on *Chenopodium quinoa* of biopiracy (RAFI, 1998). Quinoa is an important cereal in the diet of indigenous people at the Andean Region of Latin America. Since ancient times local communities have cultured varieties adapted to the severe conditions of Andean mountains. In 1994, researchers from the University of Colorado obtained patent 5.304.718, monopolizing market control over plants with masculine sterility of "Apelawa," a traditional variety of quinoa from Bolivia. Later, researchers recognized that they only had collected the plant without any innovation.

On the other hand, Article 27.3(b) of TRIPS legitimizes private property rights in the form of intellectual property over life and processes entailed in modifying life forms. But these are rights for individuals, companies, and states, not for indigenous peoples and local communities. There seems to be a contradiction between the work of United Nations towards promoting international standards for the protection of indigenous peoples and the protection of traditional knowledge through the CBD and the international agreements of FAO over access to genetic resources and farmers rights and the way in which the World Trade Organization Agreements handles these issues.

Furthermore, biotechnological companies are establishing as well private ownership rights over plants and their components by UPO.V (1991 Act of UPOV convention). But the UPOV Convention of 1991 is derived from the need to protect the interests of plant breeders in developed countries; it does not derive from the needs of users in developing countries. UPOV was established in 1961, gathering 37 countries under a common norm which protects the interests of plant breeders. Although initially UPOV was accepted only by developed countries, the situation is changing currently due to pressures to developing countries to establish intellectual property rights on biodiversity in order to fulfill the norms of free commerce. Governments are told that patents are the key to attract investments in biotechnology in order to improve economy and food safety, but most Latin American countries lack an effective patent infrastructure. UPOV grants to owners of plant varieties able to transmit its characteristics to new generations the right to absolute commercial control. Farmers cannot exchange or sell seeds and they must pay royalties to the owner of the seed variety, so that they are allowed to keep seeds for the following year. Furthermore, a license is required for cultivating these seeds. But the 1991 convention restricted the possibility of keeping seeds for next cultivation and the possibility of free innovation of new varieties under the rationale to compensate conventional plant breeders against biotechnological companies which obtain ample patent rights over genes and plant species with the creation of transgenics.

As a result, the UPOV system is criticized in Latin America since there are no provisions for benefit sharing with farmers, it is not possible to reuse seeds, requires expertise which most farmers lack and genetic resources are privatized affecting biodiversity (Grain Report, 1998).

The United Nations Declaration on the Rights of Indigenous Populations (2007) recognizes their right to take decisions in matters affecting them (Article 18), to participate with informed consent (Article 19) and to keep, control, protect and develop genetic resources, seeds, medicines, knowledge about living beings properties and develop the intellectual property of their cultural patrimony, traditional knowledge and cultural expressions (Article 31). These rights need to be protected by national legislations.

Recent commercial treaties between developed and developing countries incorporate traditional knowledge into the norms for protection of intellectual property in order to limit biopiracy. However, imposing a uniform international model of protection, the right to self determination by indigenous populations has been limited (Rodriguez, 2006). These treaties regulate explicitly or implicitly the access and patentability of resources, biodiversity and cultural and intellectual indigenous patrimony (Llancaqueo, 2006).

There have been efforts to regulate the transfer of transgenics. The Cartagena Biosafety Protocol is the first internationally binding legal instrument that regulates the handling, use and movement between frontiers of genetically modified organisms. The Protocol is significant for establishing the basis of international law to regulate the trade of genetically modified organisms, which it recognizes as inherently different and carrying special risks and hazards.

There have been also efforts to protect access to genetic resources. The Andean Community (Bolivia, Colombia, Peru, Ecuador and Venezuela) has attempted to legislate jointly on access to genetic resources and plant breeders rights under the Cartagena Agreement (1996) (Decisions 391 and 345). A policy to prohibit patenting living beings, except microorganisms has been enacted. But there are pressures to introduce the patenting of transgenics in these countries. The Andean Pact (Decision 486, Article 15) establishes what should not been considered inventions: All or part of living beings encountered in nature, natural biological processes, biological material existing in Nature or isolated from living beings, included genome or germplasm. In line with participation in benefits related to genetic resources, Costa Rica made an agreement with Merck and Co. ensuring a realistic share of royalties from marketable products obtained from bioprospecting (Chadwick, 2003).

Peru acknowledges ancestral knowledge with a policy to prevent foreign biotechnological companies from patenting products known to have been developed by traditional knowledge of indigenous Peruvians. But matters are changing; the recent law 29316 accepts U. S. patent criteria and does not recognize a special treatment for patents that use genetic resources and traditional knowledge as recognized by the Andean Community. An agreement for access to the genetic resources, a certification of source of origin and an authorization for use of traditional knowledge is not required, thus surpassing the requirements of the Andean Community (Decission 486 Andean Community), leaving this declaration as only a document of good intentions (Vitrano, 2009).

It is objectionable that intellectual property rights are respected only when there are industrial applications and profits involved. No consideration is given to the innovation that took place in local communities over the centuries, gives countries the option of formulating their own *sui generis* for plants as an alternative to patent protection, a way has not been found for doing so due to financial pressures from developed countries. Most patents based on indigenous knowledge appropriation violate the criteria of novelty and/or non-obviousness since organisms are altered by already known genetic engineering techniques. the other hand, and innovation at the community level is often of a collective nature. In relation to the protection of biodiversity, farmers' rights and indigenous peoples' rights should be recognized as collective rights and used to protect their intellectual property and resources. Intellectual property rights represent Western notions of "ownership and property right(s)" extended to all facets of nature, but this concept is foreign to indigenous cosmologies which revere the sanctity of all life, and mandate human responsibility to serve as stewards, not owners, of the natural world (Harry, 1995). Informal knowledge is developed over a period of time in a cumulative manner, codified in texts or oral traditions and produced in particular biological and cultural contexts. For indigenous populations the territory is the material expression of the net of relations that build collective knowledge. Furthermore, collective intellectual rights are derived from territorial rights since territory and knowledge conform an indivisible unit (Llancaqueo, 2006, p. 15). Currently, indigenous populations are pressed towards two possible positions: commercialization of ethnic knowledge and indigenous patrimony using the intellectual property rights patenting system without respecting their own idiosyncrasy or the denial of any use of *sui generis* adaptation for not complying with the international system. In general, traditional knowledge is not recognized as "previous knowledge" in patenting system.

According to the World Intellectual Property Organization commission the topics to be considered in order to elaborate systems and norms for the protection of indigenous traditional knowledge are the following (OMPI, 2001, p. 23):

- Agreement over principles and objectives for protection
- Respect for knowledge systems and their preservation

Equity and fair distribution of benefits derived from the use of knowledge

- Rise in the use of knowledge
- Understanding the relations between the official system of intellectual property and the juridical system customary of indigenous communities

Development of methods for solving juridical and administrative problems

Benefit Sharing

The problem is that under TRIPS, there is no obligation to share benefits with the state or communities in countries of origin. There is little that a country of origin can do to enforce in benefit-sharing rights which have been recognized in the convention on biodiversity. Legal challenges are prohibitively expensive and there is no guarantee of success. If the courts of a particular country operate in a context favorable to granting patents, there is little that can be done by a country of origin to ensure that biopiracy does not take place and to obtain sharing of benefits. There is need for agreements between patent authorities of foreign countries and national authorities from countries of origin of genetic resources for establishing benefit sharing arrangements derived from patents.

Intellectual property regulations seem to increase profits of biotechnological companies restricting the protection of genetic resources, the flow of germplasm and benefit sharing with communities; it is difficult for breeders to keep their products in the market. Under the rules of the General Agreement on Tariffs and Trade (GATT) (Final Act, 1994), all member countries must bring their national Intellectual Property Rights laws into conformity with provisions of the new agreement on TRIPs. This agreement obliges member governments to provide for "the protection of plant varieties either by patents or by an effective *SUI GENERIS* system ("of their own kind") or by any combination thereof." Simultaneously, governments are given the option to exclude from patentability "plants and animals other than micro-organisms and the essentially biological process for the production of plants or animals other than non-biological and microbiological processes." TRIPs incorporates a chapter on intellectual rights and industrial ownership, globalizing basic concepts. It establishes that patents can be obtained for all inventions of products or procedures if they are novel, an invention is incorporated and they are susceptible for industrial use (Art. 27. I). The idea is to have a universal patent system, while respecting specificities of national legislations. However, TRIPS may deprive nations from their right to determine the balance of private and public benefits in order to meet national development goals. Nevertheless, some of the norms are based on ethical considerations. For instance Article 27b of GATT agreement authorizes members to exclude the patentability of inventions whose commercial exploitation in their territory must be prohibited in order to protect public order or morality, including the protection of health or the life of persons or animals, preserves plants or avoids serious damage to the environment.

The international trend shows strengthening and widening of patent protection. Some Latin American countries, such as Argentina (Law 24.481 on invention patents and utility models), have modified the legislation on patents according to the standards adopted in the Uruguay Round, adopting the agreement on TRIPs.

There is also concern about accepting broad patents for securing the market to patent holder preventing competition, but stifling research. An example is the European patent on transgenic soybean, awarded to the company Agracetus, claiming "a soybean seed which will yield upon cultivation a soybean plant comprising in its genome a foreign gene effective to cause the expression of a foreign gene product in the cells of the soybean plant." This means that the patent covers all transgenic soybeans. In fact, the grant of this patent was initially challenged by Monsanto as being too wide, and therefore unfair, until Monsanto later acquired Agracetus and thus, the ownership of the patent. Later Argentina has undergone pressures from Monsanto to pay royalties for the use of transgenic soybean, claiming to be the only holder of patents for the soybean, even though this patent has not been recognized in Argentina. Monsanto has threatened Argentina to pay a fine of 15 dollars for each ton of soy bean exported to Europe. In Argentina, farmers have the right to keep and reusing seeds, established by the law for seeds. Since 1999 Monsanto has a norm expressed by its distributors to charge a percentage for reusing seeds ("extended royalties"). Due to the pressure by Monsanto, the Minister of Agriculture of Argentina presented a law proposal for "global royalties," called Funds for Technological Compensation. By this mechanism the government applies taxes to farmers to safeguard the interests of multinational international companies. Due to the protest of farmers this law has not passed, but Monsanto threatens to charge greater fees in the ports of exportation entry. Another issue is that transgenic soybeans have passed through smuggling from Argentina to Paraguay and Brazil. Monsanto has achieved that in these countries the transgenic soybeans have been legalized and royalties paid (Ribeiro, 2009).

Specific Concerns

Data gathered through interviews and questionnaires to biomedical researchers, lawyers, students and lay civilians at Latin American countries show (Rodriguez et al., 2005) that commercialization of genetic products for agriculture and farming is seen as negative since it will mostly benefit the international biotechnological companies and not local farmers. Although biomedical researchers see benefits in the generation of transgenics for increasing productivity and improving properties, such as vigor, duration, nutrition and plague resistance, lay civilians and university students tend to consider genetically modified organisms as dangerous for health, especially of inducing cancer and as artificial which is equated to lack of trustworthiness for consuming them.

There are objections to the commercialization of genetically modified organisms, since the tendency is to use the genetic richness of Latin American natural products for patenting new organisms in developed countries. This can have negative effects on Latin America since those patents can be used to restrict access and change farming practices. It is argued that big companies use genetic material from Latin American countries to create genetically modified organisms and commercialize them without financial gain for these countries. In spite of promises of "technology transfer" and "feeding the world" the tendency so far is that biotechnological companies care only for making profits. The importance of biosecurity to avoid risks to health and the environment is acknowledged.

Risks

Although the patenting of life forms should have an industrial use, there is an exigency based on the "precautionary principle" to evaluate and minimize risks before getting into the market. Biotechnology has come under sharp criticism and media focus for its potential negative impacts and contingent risks due to its potentiality to change the course of nature and life. There are questions raised by the use of the new biotechnology applications such as whether it will assure sustainable development, affect biodiversity and environmental ecology, create new diseases, or that the genetically modified organisms are made to suit market needs rather than to help the poor. Many biotechnological companies from developed countries may use developing countries to test crops designed for needs of developed countries, not necessarily those of the developing countries. This is more serious considering the lack of controls to protect the environment in many developing countries. Although there is interest in Latin American countries for developing better crop yields, achieve nutrition needs and for reducing the need to spray pesticides, on the other hand, the biotechnological companies tend to play down the difficulties that countries may have in managing the environmental risks posed by genetically modified organisms. Engineered microorganisms resistance could result in the evolution of new and harmful pests and new ones may be developed for bioterrorism purposes (Jackson et al., 2001; Taubenberger et al., 2005).

Although there are fears in the press, there are in fact very few studies which show possible toxic risks or adverse effects on health because of consuming transgenic food (Roig and Gómez, 2000), and those found are withdrawn from market. Some of the potential adverse effects identified are: allergy, resistance to antibiotics, lack or modification of the nutritious value of food, presence of toxic components, emergence of new non treatable diseases and possible damage to wild species (Reyes and Rozowski, 2003; Paparini and Romano-Soica, 2004). There is, therefore, an ethical exigency for guaranteeing test trials for every transgenic introduced in the market. Some transgenics may lead to impoverished soil. For example, a group of scientists at Oregon State University engineered a variety of *Klebsiella planticola*, bacteria able to decompose plant material, with the idea of converting agricultural wastes to ethanol fuel. Although this goal was achieved, later on was discovered that the new product also destroyed much of a beneficial mycorrhizal fungus essential to the recycling of nitrogen through plant roots and for soil maintenance (Hill, 1994).

Genetically-engineered fish and shellfish pose additional ecological risks, since aquatic organisms have the capacity to survive in nature, moving easily from the controlled environment to the wild. In Chile, for example, there is concern over the possible industrial use of transgenic salmon with growth hormone, which grows much faster than wild salmon, and may compete with native species, supplanting them and deploying ecosystems by high grade of food consumption (Gonzalez, 2009).

There are worries about the possible effect of a decrease in biodiversity because of the use of genetically modified organisms in agriculture. The causes of this decrease are many, but in relation to the current system are related to the substitution of traditional varieties with high genetic diversity by transgenic seeds with a high degree of uniformity (FAO, 1999).

An example is what is happening in Argentina, where the agricultural production system has become dominated by one crop: the transgenic Roundup Ready soybean developed by Monsanto. This crop is resistant to the herbicide glyphosate and relies on repeated herbicide applications to control weeds. But nature finds ways to evolve around it. Already, strains of Roundup-resistant weeds have appeared in Argentina, requiring ever-heavier doses of the herbicide, killing off microbes and degrading soil quality. Heavy herbicide applications and widespread planting of Roundup Ready soybeans has also led to increases in pest and disease severity. The rate at which forests in Northern Argentina are being turned into soy plantations is 3-6 times higher than the world average. This massive destruction of forests has sparked violence and protests by agrarian families and is changing the local climate to a semi-desert causing both droughts and flooding (http://apis.ufl.edu, accessed 5 August 2009). Monoculture makes the country very vulnerable to environmental changes. Furthermore, the emphasis on finding and isolating plants with the most marketable traits leads inevitable to the decline of other plant species and varieties, as only the new techno-varieties tend to be cultivated based on market prices and incentives. This is already happening in developed countries such as the US, where the focus on commercial varieties has led to the loss of many varieties of plants in seed bank storage and some non-commercial crops such as chufas, martynia and rampion have been lost entirely

(Fowler and Mooney, 1990). Furthermore, the patents on transgenic seeds are preventing farmers from saving seeds for replanting unless they pay royalties to the companies. Patents promote unsustainable and inequitable agricultural policies if procedures continue to erode genetic diversity.

Some technologies are viewed as not respecting nature. An example is the terminator genetic engineering technology that causes plants to release a fatal toxin in the second generation of seed, so they cannot reproduce themselves, introducing infertility (SIMAS, 2007).

The arguments for and against transgenics are based on visions of the new technology from widely different ethical perspectives, involving scientists, philosophers, commerce, politicians, journalists, religious and lobby groups and the public. These arguments involve values and the controversies have polarized society into the proponents and opponents, with once seemingly trustworthy and ethically sound scientists being viewed with suspicion by many. In the polemics intervene beliefs, questioning topics such as the role of God and the sacrality of nature; and also power relationships, such as the immense power of biotechnological companies and the ownership of life forms through patents. The moral and ethical concerns are important factors in influencing the aversion of public over risks related to transgenics (Callahan, 1996;.Wadman, 1996; Geary, 1996; Newton et al., 1999).

MEDIA ROLE

New technological advancements have modified the role of information in society giving it extra power and responsibility. Although there is some public awareness of genomic issues, frequently the public is misled due to lack of formal education on genomics and of information based on real facts. The media may treat genomic issues in a sensational way, provoking an imbalance between media agenda and public needs. This leads to a lack of awareness of some possible threats that genetic material poses, as well as the health risks that it involves. Furthermore, the role of the media is also such that other threats are overly exaggerated. Thus, some authors suggest that some important aspects of genetics such as genotoxic effects on the environment which affects humans are not considered public health needs due to lack of media coverage (Gonzalez, 2007). The media may portray an additional charge of imagination beyond scientific facts. For example movies presenting human clones as aggressive, the movie Jurassic park presenting the resurgence of extinct species. Transgenics have been presented emotively, with talk of *Cashing in on Hunger, Demon Seeds, Terminator Technology and Frankenstein Foods* (Jonathan, 1999). Carelessly presented accounts of genetic engineering developments by the media have aroused concerns that human health will be adversely affected by consumption of transgenic crops and products derived from them (The Economist, 1999, pp. 93-95).

On the other hand, other press reports seem to favor industry interests. Transgenic use appears without risks and enhancement characteristics are viewed positively, while scientific reports on possible negative effects on environment are not presented.

The media provides information in a horizontal or vertical way, horizontally provoking a dialog on the internet in which all may intervene equally but with knowledge gaps; and vertically expressing the views of experts and the voices of those with a special interest which may manipulate the direction of thinking. Internet is used by non-governmental organizations to raise conscience on abuses and injustices committed by the use of transgenics which may be useful in some instances, but in others it may provoke negative attitudes toward transgenic food consumption without critical thinking.

Since there are myths appearing in the media which may distort the public analysis, it is critical to focus the discussion in a wide scope and integrative perspective, engaging the scientific community, together with other public sectors of the society, for a debate associated with the positive and negative aspects of new discoveries around human genome and the use of transgenics. This is imperative, since the scientific advances do not by themselves cause social problems; rather they are caused by the decisions that society takes in terms of how technological advances is used (Maccioni, 2004).

An important point is that part of the advocacy done by environmentalist groups and the fears expressed by the civil society are focused on arguments that are not always scientifically sound. The idea of rejecting genetically modified organisms because they are artificial or because human beings are acting against sacred laws of nature through introducing such modifications cannot be scientifically sustained. In a poll carried out by the FAO in Latin America, it was found that civilians had a negative perception of transgenic food due to lack of knowledge on how they are generated, lack of confidence toward biotechnological international companies, fear of unknown innovations, lack of confidence toward regulatory measures, and awareness of the threat to biodiversity (http://www.ric.fao.org/redes/redbio/default.htm, accessed 5 August 2009).

Role of Bioethics

Bioethics has a role in clarifying the complex topics of social consequences derived from problems arising from the use of genomic research products and information. The bioethical debate over present and prospective applications of genetic research and the social consequences of access to genetic information must rely on rational background based on scientific data and sound philosophical reflection. Clearly, lack of knowledge and reflection creates anxiety. Therefore it will be helpful to improve information about these issues in a critical way at all levels of society. Globalization of information and the sensationalization by media about genetic issues make public participation on bioethical debates over these issues all the more necessary. Nonetheless, public debate is limited by lack of knowledge, sensationalism of media portrayal and fundamentalism of religious beliefs. There are also powerful interest groups which manipulate the debate pressing over their own ideologies.

Therefore, we recommend the need of educating the people on genomic issues and social implications, particularly clarifying which are the real and available benefits and risks of biotechnology in order to avoid or clarify conflicting issues such as the natural/artificial split, since some popular press tends to equate what is artificial to evil.

Bioethics represents an adequate instrument for the critical analysis of the scientific activity. We consider that bioethics should not be a discourse only for experts, but the ideal medium to achieve exchange between lay persons and experts. Bioethics is dialogical venue *par excellence,* enabling exchange of opinions among those affected as equals.

Law, as a normative system has the function to safeguard order and propriety in the society, which points towards its justification by principles and arguments based on morals in order to be accepted. In the field of patenting there are also ethical considerations taking into account with bioethics having a role in reflecting about limits to patenting, scope of protection to safeguard public interests and rights of specific groups.

Bioethical reflection has a role in pointing out areas in need of regulation. For example, there is a need of legal regulation in order to control the possibility of arbitrary genetic manipulations contrary to human dignity. There is also little regulation for the introduction of genetically modified organisms; some are still under evaluation for safety and nevertheless they are introduced as if their safety were already established. While there are no reports of these organisms producing cancer or causing physical damage on human beings, there are questions to be answered, such as the impact on biodiversity or the possible transfer of genes from one species to others with properties not wanted in wild plants such as resistance to herbicides or to insects or microorganisms.

Bioethics has the role not only of applying the four core principles to clinical cases under an interpersonal perspective in genomic medicine, but to see the issue of genomics as a social issue. Topics such as ownership and patents, genetic manipulation, allocation of resources, public policies for regulating genetic manipulations in humans and living beings require the involvement and reflection of the whole society (Kottow, 2002). The globalization of genetic banks is creating new concerns over group interests, since genetic information affects beyond individuals and may modify the meanings associated to the terms "individual" and "community." Concepts such as solidarity, benefit sharing, equity, public participation, and collective identity need to enter into the bioethics debate (Chadwick, 2003).

The issue of a responsibility towards future generations is a new challenge in genomics derived from the genetic technological progress which makes possible to introduce genetic variations affecting future generations, which calls for bioethical reflection. Hans Jonas (1995) has reflected on this principle, which affect not only humans but also to all living beings. The increased power of biotechnology has created concern about responsibility, in the sense that with greater power there is greater responsibility for avoiding evil and channeling technology towards good goals (Gonzalez, 1999). Responsibility means reflecting on the actions to be taken, balancing risks and benefits for social development. Human beings have a great power over nature thanks to biotechnology; life can be manipulated and altered deeply. The principle of responsibility of Hans Jonas implies at least two duties: that future generations will count with an environment and biodiversity comparable to present one; and that the genetic identity of human beings will not be altered (Jonas, 1979). Rules for decision making must consider the ignorance we have for the consequences may pose to ecology the uncontrolled introduction of genetically modified organisms. On the basis of the principle of non-maleficence, this issue creates obligations in terms of environmental and health care policies in order to avoid, as far as possible, the worsening of genetic endowment (Neri, 2009). Care must be taken to limit the slippery-slope tendency whereby acceptance of one controversial technology inevitably leads to acceptance of more and more, the benefits of which become progressively less certain, which calls for a definition on acceptable and unacceptable activities and the setting of limits (Weil, 1996).

Recommendations

Bioethically speaking, the following guidelines should be reflected and considered on practical uses when dealing to accept or deny the use of patented genetically modified organisms:

- Ensure prior notification, field testing results, comprehensive impact assessment, full public disclosure, mandatory labeling to be used and fully informed advanced agreement of affected nations and communities for every experimental and commercial release and transfer of a transgenic or its products.
- Ensure comprehensive methodology for a case-by-case evaluation of each transgenic or its product, alternatives, socio-economic study, human health, and ecological impacts of its full life cycle including disposal on environments other than that in which it was developed.
- Adhere to the precautionary principle performing a risk assessment for each transgenic.
- Give all nations and communities the right to determine what is an acceptable level of risk and the right to refuse any release or transfer of a transgenic or its product.

- Ensure investment of commercial developers and/or the commercial exporters of transgenics or its products in the implementation of the protocol attached to each transgenic, risk management and for safeguarding strict liability.
- Monitor transgenic performance, impacts, and range of habitat after any release or transfer.
- Establish national reports, compliance reviews, mechanisms for enforcement and possible dispute resolutions.
- Provide capacity building of use of transgenics in farming and agriculture.
- Ensure full participation of all interested persons and organizations with transparency of information.
- Guarantee the prohibition of varieties which lead to genetic erosion
- Make provisions for respecting community and farmer's rights should be maintained.
- Establish benefit sharing arrangements from genetic resources that should be made with country and community of origin.
- Measure the extent to which traditional knowledge of indigenous peoples and local communities has been incorporated into development and resource-management decision-making processes with prior informed consent, fair and equitable sharing of benefits and in situ conservation of lands and territories used by indigenous peoples and local communities.
- Include in national laws the right of farmers to continue with their traditional practice to save, use, exchange and sell farm-saved seed and propagating material, which may use *sui generis* protection of plant varieties.
- Protect the right of countries to exclude plants, animals, micro-organisms and any parts thereof and microbiological processes for plant and animal production.
- Avoid broad-scope patents that limit access to a wide segment of germplasm.
- Protect rights claimed by plant breeders over materials deposited in international gene banks
- Preserve phytogenetic resources and germplasm biobanks of wild varieties. They may have possible future uses and will never be recovered if lost.
- Disclose in patent applications the country of origin of the biological resource and associated knowledge, and obtain consent of the country providing the resource and knowledge, to ensure equitable sharing of benefits.
- Develop a system for documentation and protection of traditional knowledge to avoid bio-piracy.
- Achieve access to technology and transfer of technology to facilitate better conditions for agreements.
- Protect natural varieties and organic ways of seed production, avoiding transgenic contamination by norms and monitoring, regardless of transgenic production.
- Regulate nationally and globally for the protection of biodiversity.

REFERENCES

Bergel, S. D. (2000). Aspectos Éticos y Jurídicos del Proyecto Genoma Humano: Patentamiento de Genes y Secuencias. *Medicina*, *60*, 729–730.

Bergel, S. D. (2002). Los Derechos Humanos entre la Bioética y la Genética. *Acta Bioética*, *VIII*, 315–329.

Bisang, R., Campi, M., & Cesa, V. (2009). *Biotecnología y Desarrollo." Comisión Económica para América Latina y el Caribe*. CEPAL.

Callahan, D. (1996). Biotechnology and ethics: a blueprint for the future. Keynote: setting and communicating the limits in biotechnology. Retrieved from http://www.biotech.nwu.edu/nsf/callahan.html.

Cartagena Agreement. (1996). Retrieved from http://www.comunidadandina.org.

Catalogue of Mendelian Inheritance Diseases OMINM. http://www.ncbi.nlm.nih.gov/sites/entrez?db=omim

Chadwick, R. (2003). Genomics, publich health and identity. *Acta Bioethica*, *9*(2), 209–218. Retrieved from http://www.scielo.cl/scielo.php?script=sci_arttext&pid=S1726-569X2003000200007&lng=es Accessed 5 August 2009. doi:10.4067/S1726-569X2003000200007

Claude, V. (2004). Genética y Bioética en América Latina. *Acta Bioética*, *10*(2), 155–166. Retrieved from http://www.scielo.cl/scielo.php?script=sci_arttext&pid=S1726-569X2004000200004&lng=es Accessed 5 August 2009.

Convention on Biological Diversity. (1992). United Nations. June 1992.

Craig Ventor, J. (2001). The sequence of the human genome. *Science*, *291*, 1304–1351..doi:10.1126/science.1058040

De Constantino, B. F. M. G., de Miranda, G. Y., & Coutinho, A. F. G. (2006). Saúde pública e ética na era da medicina genômica: rastreamentos genéticos. *Revista Brasileira de Saú de Materno Infantil*, *6*(1), 141–146. Retrieved from http://www.scielo.br/scielo.php?script=sci_arttext&pid=S1519-38292006000100017&lng=en Accessed 5 August 2009. doi:.doi:10.1590/S1519-38292006000100017

FAO. (1999). *The State of the World's Plant Genetic Resources.*

Final Act Embodying the Results of the Uruguay Round of Multilateral Trade Negotiations. (1994). Marrakesh, 15 April 1994.

Fowler, C., & Mooney, P. (1990). *Shattering: Food, Politics and the Loss of Genetic Diversity.* University of Arizona Press.

Francisco, R., et al. (2003). Análisis de ADNmt de restos esqueletales del sitio arqueológico de Tiwanaku y su relación con el origen de sus constructores. *Chungará (Arica),* 35(2). Retrieved from http://www.scielo.cl/scielo.php?script=sci_arttext&pid=S0717-73562003000200006&lng=es&nrm=iso. Accessed 31 August 2009. DOI: 10.4067/S0717-73562003000200006.

Geary, J. (1996). Battle of the bean genes. *Time*, *28*(October), 46–47.

González. Elio A. Prieto. (2007). Deterioro genómico y manipulación genética: Dessequilibrio en la prioridad de las agendas públicas. *Acta Bioethica*, 13(2), 223-231. Retrieved from http://www.scielo.cl/scielo.php?script=sci_arttext&pid=S1726-569X2007000200010&lng=es. Accessed 5 August 2009.

González, G. (1999). *Derechos humanos. La condición humana en la sociedad tecnológica.* Madrid: Tecnos.

Gonzalez, G. (2009). Alerta en Chile ante el supersalmón. http://www.tierramerica.net/2001/0812/articulo.shtml. Accessed 5 August 2009.

GRAIN. (2003). ¿Un sistema mundial de patentes? El Tratado sobre el Derecho Sustantivo de Patentes de la OMPI. Documento de Análisis. http://www.grain.org

Harry, D. (1995). Patenting of life and its implications for indigenous peoples. Kellogg Foundation. http://user.uni-frankfurt.de/~ecstein/gen/iatp/iprinfo7.html. Accessed 17 August 2009.

Hill, Richard L. (1994). OSU study finds genetic altering of bacterium upsets natural order. *The Oregonian,* August 8.

Instituto de Estudios Ecologistas del Tercer Mundo. (2006). *Encuentro taller internacional 'servicios ambientales': la naturaleza como mercancía.* Quito: Mayo.

International Human Genome Sequencing Consortium et al. (2001). Initial sequencing and analysis of human genome. *Nature, 409,* 860–921.. doi:10.1038/35057062

Jackson, R. (2001). Expression of mouse interleukin-4 by a recombinant ectromelia virus suppresses cytolytic lymphocyte responses and overcomes genetic resistance to mousepox. *Journal of Virology, 75,* 1205–1210. doi:10.1128/JVI.75.3.1205-1210.2001

Jonas, H. (1995). *El Principio Responsabilidad.* Barcelona: Círculo de Lectores/Herder.

Jonathan, R. (1999). Ethics and transgenic crops: a review. *Electronic Journal of Biotechnology, 2,* 5–6. Retrieved from http://www.scielo.cl/scielo.php?script=sci_arttext&pid=S0717-34581999000200003&lng=es Accessed 5 August 2009.

Kottow, M. H. (2002). Public health, genetics and ethics. *Revista de Saude Publica, 36*(5), 537–544. Retrieved from http://www.scielo.br/scielo.php?script=sci_arttext&pid=S0034-89102002000600001&lng=en Accessed 5 August 2009. doi:.doi:10.1590/S0034-89102002000600001

Llancaqueo, V. T. (2006). El nuevo regimen internacional de derechos de propiedad intelectual y los ederechos de los pueblos indigenas. In Berraondo, M. (Ed.), *Pueblos Indígenas y Derechos Humanos.* Centro de Políticas Públicas y Derechos Indígenas, Instituto de Derechos Humanos, Universidad de Deusto.

Lolas, F., Rodríguez, E., & Valdebenito, C. (2004). El Proyecto del Genoma Humano en la Literatura Biomédica en cuatro países Latinoamericanos. *Acta Bioética, X,* 167–180.

Maccioni, R. B., Muñoz, J. P., & Maccioni, C. (2004). Dimensiones bioéticas de la investigación sobre el genoma humano. [revista en la Internet]. *Acta Bioethica, 10*(1), 75–80. Retrieved from http://www.scielo.cl/scielo.php?script=sci_arttext&pid=S1726-569X2004000100010&lng=es Accessed 5 August 2009. doi:10.4067/S1726-569X2004000100010

Machiola, J. I. (2009). Banco de genes indígenas alimenta sospechas en Argentina y el Mundo. Retrieved from http://www.biodiversidadla.org/content/view/full/16097. Accessed 17 August 2009.

Madrid, R. (1999). Cuestiones Jurídicas en el Proyecto del Genoma Humano: Presente y Perspectivas Futuras. *Humanitas, 15,* 20–24.

Marco, S., & Miazato, I. E. S. (2001). Bioethics, Intellectual Property and Genomics. *Rev. Hosp. Clin. Fac. Med. S. Paulo,* 56(4), 97-102. Accessed 5 August 2009. Retrieved from http://www.scielo.br/scielo.php?script=sci_arttext&pid=S0041-87812001000400001&lng=en. DOI: 10.1590/S0041-87812001000400001.

Neri, D. (2009). On the concept of eugenics: preliminaries to a critical appraisal. *Cadernos de Saude Publica.* Retrieved from http://www.scielo.br/scielo.php?script=sci_arttext&pid=S0102-311X1999000500004&lng=en Accessed 5 August 2009. doi:.doi:10.1590/S0102-311X1999000500004

Newton, P., Brown, D., & Clover, C. (1999). Alarm over "Frankenstein" foods. *Electronic Telegraph,* Issue 1358, 12 February 1999. http://www.telegraph.co.uk.

OMPI. (2001). Panorama general sobre las cuestiones relativas a la propiedad intelectual y los recursos genéticos, los conocimientos tradicionales y el folclore. Documento preparado por la Secretaría, Comité Intergubernamental sobre Propiedad Intelectual y Recursos Genéticos, Conocimientos Tradicionales y Folclore, Primera sesión, Ginebra 30 de abril a 3 de mayo de 2001.

Paola, R. P., et al. (2002). Composición genética de la población chilena: Distribución de polimorfismos de DNA mitocondrial en grupos originarios y en la población mixta de Santiago. *Rev. méd. Chile* [revista en la Internet], Feb, 130(2), 125-131. Accessed 31 August 2009. Retrieved from: http://www.scielo.cl/scielo.php?script=sci_arttext&pid=S0034-98872002000200001&lng=es. DOI: 10.4067/S0034-98872002000200001.

Paparini, A., & Romano-Soica, V. (2004). Public health issues related with the consumption of food obtained from genetically modified organisms. *Biotechnology Annual Review*, *10*(1), 85–122. doi:10.1016/S1387-2656(04)10004-5

Patents, I. P., & Diversity, H. G. (1993). RAFI Communique, Rural Advancement Foundation International, Ottawa, Canada. (May 1993).

Report, G. (1998). Diez razones por las que la UPOV es un mal negocio. Conflictos entre comercio global y biodiversidad. Retrieved from http://www.grain.org/briefings/?id=76 Accessed 19 August 2009.

Reyes, M. S., & Rozowski, J. N. (2003). Alimentos transgénicos. *Revista Chilena de Nutricion*, *30*(1), 21–26.

Ribeiro, S. (2009). Monsanto y la soya Argentina. http://www.etcgroup.org/es/materiales/publicaciones.html?pub_id=64. Accessed 5 August 2009.

Rodríguez. Silvia. (2006). TLCs: el conocimiento tradicional en venta. *GRAIN, Abril 2006.* http://www.grain.org/briefings/?id=198. Accessed 19 August 2009.

Rodríguez, E., et al. (2005). Social, ethical and legal attitudes towards genomic research in four Latin American countries. *Electronic Journal of Biotechnology*, 8(3). Available from http://www.ejbiotechnology.info/content/vol8/issue3/full/9/index.html. ISSN 0717-3458.

Rodríguez, E., Valdebenito, C., Kanner, E., & Lolas, F. (2004). El Proyecto del Genoma Humano y las Regulaciones jurídicas en cuatro países latinoamericanos. *Jurisprudencia Argentina*, *IV*(5), 42–50.

Rodríguez, E., Valdebenito, C., Misseroni, A., Fernández, L., Outomuro, D., Schiattino, I., & Lolas, F. (2004). Percepciones Sociales sobre Genómica en Cuatro Países Latinoamericanos. Implicaciones Ético Legales. *Derecho y Genoma*, *21*, 141–164.

Roffe, P. (2006). América Latina y la nueva arquitectura internacional de la propiedad intelectual: de los ADPIC-TRIPS a los nuevos tratados de libre comercio. In *Diálogo Regional sobre Propiedad Intelectual, Innovación y Desarrollo Sostenible.* Costa Rica, 10-12 Mayo: UNCTAD/ICTSD.

Roig, J. L. D. &Gómez, Mercedes. (2000). Riesgos sobre la Salud de los Alimentos Modificados Genéticamente: Una revisión bibliográfica. *Revista Espanola de Salud Publica*, *74*(3), 255–261. doi:10.1590/S1135-57272000000300003

Rural Advancement Foundation Internacional (RAFI). (1998). Quinoa patent dropped: Andean farmers defeat U.S. University. *RAFI Genotype* – May 22, 1998. http://www.rafi.org. Accessed 5 August 2009.

Sanchez, G. J. (2003). Developing a platform for genomic medicine in Mexico. *Science*, *300*, 295–296. doi:10.1126/science.1084059

Santos, M. A. (1997). Aspectos científicos de los principales avances de la genética humana. El Impacto Social de la Manipulación Genética. *Humanistas, 9*, 16–27.

Schiattino, I., Silva, C., Lolas, F., Valdebenito, C., Rodríguez, E. (2005). Descripción de las percepciones sobre el proyecto genoma humana en Chile, Perú, Argentina y México, *Revista Quirón,* 36(1/3).

Schiattino, I., Silva, C., Lolas, F., Valdebenito, C., & Rodríguez, E. (2005). Percepciones y estados emocionales sobre el proyecto genoma humano en actores sociales seleccionados en la Región Metropolitana, Chile. *Revista Chilena de Salud Pública, 9*, 154–161.

Seeds of discontent. (1999). *The Economist,* 20 February 1999, 93-95.

SIMAS Comunicación para el desarrollo rural. (2007). Terminator: Las semillas suicidas son semillas homicidas. Retrieved from http://www.simas.org.ni/simasnoticia/359. Accessed 5 August 2009.

Simpson, A. J. (2000). The genome sequence of the plant pathogen *Xylella* fastdiosa: the *Xylella* fastidiosa consortium of the organization for nucleotide sequencing and analysis. *Nature, 6792*, 151–157.

Simpson, A. J. G. (2001). Genomics in Brazil. *Research Coordination, Pan American Health Organization* (online). Retrieved from http://www.paho.org/english/hdp/HDR/ACHR-02-Simpson.PDF. Accessed 31 August, 2009.

Taubenberger, J. K. (2005). Characterization of the 1918 influenza virus polymerase genes. *Nature, 437*, 889–893. doi:10.1038/nature04230

Thorsteinsdóttir, H., Daar, A. S., Sáenz, T. W., & Singer, P. A. (2009). Building a biopharmaceutical innovation system in Cuba: growth through linkages. In Mytelka (Ed.), *Pharmaceutical Innovation (Bio)Building K. L.: Pharmaceutical Innovation Systems in Developing Countries*. Maastricht: Institute for New Technologies (INTECH), United Nations University.

Vitrano, G. (2009). Perœ al servicio de transnacionales y biopiratas. www.selvas.org. Accessed 17 August 2009.

Wadman, M. (1996). Genetic resistance spreads to consumers. *Nature, 383*, 564. doi:10.1038/383564a0

Weil, V. (1996). Biotechnology and ethics: a blueprint for the future. Biotechnology: social impact and quandaries. http://www.biotech.nwu.edu/nsf/weil.html.

ENDNOTE

1 Grant Department of Energy award DE-FG02-02R 63435. Representation of Genomics Research among Latin American Laymen and Bioethics: An inquiry into the migration of knowledge and its impact on underdeveloped communities. Bibliography: Lolas, Rodriguez and Valdebenito, 2004; Rodriguez et al., 2004; Rodriguez et al., 2004; Rodriguez et al., 2005; Schiattino et al., 2005; Schiattino et al., 2005.

Chapter 12
Indonesian Legal Perspectives on Biotechnology and Intellectual Property Rights

Theofransus Litaay
Satya Wacana Christian University, Indonesia

Dyah Hapsari Prananingrum
Satya Wacana Christian University, Indonesia

Yakub Adi Krisanto
Satya Wacana Christian University, Indonesia

ABSTRACT

The authors focus their attention on the structure of Indonesian law and policies on biotechnology issues; they also address some issues related to bioethics and research activities and economic activities, such as the issue of bioprospecting and biopiracy on Indonesian biodiversity, and how the legal and governance structure within Indonesia are designed to cope with this issue. An issue that looms large is about intellectual property rights.

INTRODUCTION

This chapter will discuss the existing laws, regulations, arrangement, and policies on the field of biotechnology in Indonesia.

In Indonesia, those areas are developing not at a rapid pace. The main concern, for the time being, is more on the economic and poverty reduction issues. In fact, the development of laws and policies in this area in Indonesia did not intend to address the biotechnology issues exclusively; they serve various kinds of issues. Nevertheless, they play an important role when it comes to the issue of biotechnology.

By writing this chapter, the writers are trying to describe the structure of Indonesian law and policies on biotechnology issues. The writing will also address some issues related to bioethics and

DOI: 10.4018/978-1-61692-883-4.ch012

research activities and economic activities, such as the issue of bioprospecting and biopiracy.

The areas of discussion in this chapter will embrace several fields as follows:

- Law and policies on biodiversity, biosafety, and biotechnology. This section will explain the issues and connections between those fields.
- Intellectual property rights (IPR issues). This part will discuss and explain the IPR legal frameworks on patent, trade secret, and plant variety protection.
- Bioprospecting and biopiracy issues. This section will raise the attention to the emerging but neglected issues in protecting biodiversity in Indonesia.

INDONESIAN LAW AND POLICIES ON BIODIVERSITY, BIOSAFETY, AND BIOTECHNOLOGY

Laws and policies on biodiversity, biosafety, and biotechnology are interdependent and interconnected. This section will explain the relationship between policies in those fields. Besides, the section will also discuss the challenges faced by the government and the stakeholders.

Implementation of UN CBD and Challenges

Indonesia is a member United Nation's Convention on Biological Diversity of CBD. The country also ratified the Convention in 1994 (Law No. 5/1994).

At the national level, Indonesia had implemented CBD member's obligations through strategy-making, policymaking and its implementation. However, according to the Indonesian third national report, the country is still struggling to achieve some targets in the implementation part.

At the policy making level, there are more than 15 biodiversity-related-laws that are related to biological diversity issues. These are, among others:

- Law No. 5/1990 on Conservation of Bio-Nature Resources and Ecosystem.
- Law No. 16/1992 on Plant and Animal Quarantine.
- Law No. 12/1992 on Plant Breeding System.
- Law No. 5/1994 on Ratification of United Nations Convention on Biological Diversity
- Law No. 6/1994 on Ratification of United Nations Framework Convention on Climate Change
- Law No. 7/1996 on Food.
- Law No. 23/1997 on Environmental Management.
- Law No. 41/1999 on Forestry.
- Law No. 29/2000 on Plant Variety Protection.
- Law No. 18/2002 on National System of Research, Development, and Application of Science and Technology.
- Law No. 7/2004 on Water Resource.
- Law No. 32/2004 on Regional Development.
- Law No. 21/2004 on Ratification of Cartagena Protocol on Biosafety to the Convention on Biological Diversity.
- Law No. 4/2006 on Ratification of International Treaty on Plant Genetic Resources for Food and Agriculture
- Law No. 26/2007 on Spatial Management.
- Law No. 27/2007 on Coastal Areas and Small Islands Management.

Concerning UN-CBD, the Environmental Management Ministry is the national focal point and the leading sector in Indonesia. The role is to coordinate the national agenda to achieve the targets of CBD. This is a huge task for the ministry, which has no vertical office at the provincial and district government level. No wonder, some members of parliament demand for the strengthening of the position of the ministry.

CBD needs to be implemented in an integrated manner. In reality, at least there are around five ministries that handle the implementation of UN CBD. The Environmental Management Ministry will face difficulties when it has to reach a wide range of separate government bodies to address an issue. There is a tendency in Indonesia to slice a piece of an international policy framework into different fields and share it proportionally to various government bodies. Therefore, we are rather diverging in our policy-making rather than converging with the resources. This is in contrary to the country economic reality.

Biosafety and Biotechnology

A legal instrument to release biotechnology products (whether it is a plant, fish, animal, or animal feed) to market or consumer is the Government Regulation on Biological Safety of Genetically Engineered Product (Government Regulation No. 21/2005). This regulation was formulated based on the Cartagena Protocol on Biosafety to the Convention on Biological Diversity that has been ratified in 2004 (Law No. 21/2004).

The Government Regulation above is a specially designed provision to regulate biotechnology product in Indonesia. From the legal-hierarchy perspective, the regulation is an improvement from the older regulation–in 1999–which was a mere joint-Ministerial decree.

Those regulation also established the Biological Safety Commission (BSC/KKH) and Biological Safety Technical Team (BSTT/TTKH). BSC and BSTT are assigned to the function of licensing and evaluating a particular biotechnology product-safety issue prior to its research and development.

According to the Biosafety regulation, any decision regarding food safety and/or feed safety of biotechnology product will engage several government agencies. Those agencies are among others the National Agency of Drug and Food Control (NA-DFC), Agriculture Ministry, Forestry Ministry, and Environmental Management Ministry. The involvement of the Environmental Management Ministry was also an improvement. In the previous joint-Ministerial Decree, the Environmental Management Ministry was unlisted. According to the author's resource there was a anti-biotechnology trend inside the Ministry at the time that made them reluctant to join the decree with other ministries. Nevertheless, the new administration brought new ideas to the ministry that changed the policy atmosphere.

Another effort aimed at developing the issue of biosafety and biotechnology was made by providing the channel for public consultation. For that purpose–and as mandated by the CBD–an Indonesian Biosafety Clearing House (Indonesia BCH) has been established. Indonesia BCH could be accessed online at http://www.indonesiabch.org. This link to CBD's Biosafety Clearing House Portal in http://bch.cbd.int.

In order to control biotechnology research safety, the Agriculture Ministry issued the Procedure of Research and Development on Genetically Modified Product. Based on this procedure, any research on genetic engineering needs a prior license from the Head of Research and Development Agency of Agriculture Ministry. Based on Biosafety regulation, the focal point is Enviromental Management Ministry. This is in line with CBD arrangement.

Regarding the food safety issue, the government provided Government Regulation on Food Safety, Quality, and Nutrition (Government Regulation No. 28/2004). For the food safety and feed safety issue, there exist more than one body in Agriculture Ministry itself that handle the issue. Among others are Agriculture Quarantine Agency of Indonesia, Food Security Body, Directorate of Quality and Standardization, and all the Directorates of the related-commodities. This situation itself created coordination problem. Competing interests among agencies may have resulted in stagnation in policy-making. A breakthrough steps in reorganizing the decision making process is apparently urgent.

INTELLECTUAL PROPERTY RIGHTS ISSUES

Biotechnology is one of the solutions to solve food and health problems in many countries including Indonesia. However, majorities of biotechnology applications controlled by developed countries and multinational companies and in high cost due to the issue of intellectual property protection.

At the international level, global instruments to address intellectual property protection issues refer to WTO's TRIPs Agreement (Trade-Related aspects of Intellectual Property Rights). Based on this agreement, WTO member developed the national standards to address the issue.

Intellectual property right (IPR) protection has both positive and negative impacts. On the one hand, biotechnology products could be protected from piracy; while, on the other hand, one should use certain strategy in order to get the necessary biotechnology products. Discussion in this paper regarding IPR will be based on Indonesian Patent Law (hereinafter referred to as IPL), and completed with additional information that related to Indonesian Trade Secret Law (TSL) and Indonesian Plant Variety Protection Law (PVPL).

Issues on Biotechnology Patent and Patent Law in Indonesia

According to IPL (the latest version of the law is Law No 14/2001), patent is an exclusive right granted by the State to the applicant (the inventor) for the technological invention. The State granted this right for a period time for the patent holder to enjoy the right by producing the patented invention or to give consent to another party for the implementation of the patent (IPL Article 1 verse (1)).

The law defines invention as "an idea that is manifested into a specific problem solving in the field of technology that can be a product or a process, or improvement and development of prior products or processes" (Unofficial translation by the author). In 1992, United Nations Convention on Biological Diversity (CBD) also addressed the issue of biotechnology. According to CBD, "Biological diversity" means the variability among living organisms from all sources including, inter alia, terrestrial, marine and other aquatic ecosystems and the ecological complexes of which they are part; this includes diversity within species, between species and of ecosystems (CBD Article 2 paragraph 1). While, "Biotechnology" means any technological application that uses biological systems, living organisms, or derivatives thereof, to make or modify products or processes for specific use (CBD Article 2 paragraph 3).

Under the light of Indonesian Patent Law, the CBD's Biotechnology definition fits into the field of protection of the law. Meanwhile, based on its Biological diversity definition, there are some aspects of the invention in this field that is exempted by the law and this paper will discuss it later.

By owning the patent, the law will protect the patent holders in implementing their rights legally. The patent holders–based on IPL Article 16 paragraph (1)–enjoy the exclusive right to carry out a patent owned and prohibit other parties from enjoy the same right without the patent holder's consent. This provision also stipulates that:

a. In the case of product-patent, the patent holder empowered for the making, using, selling, importing, leasing, granting, or providing for the sale, rental, or products that were patented.

b. In the case of production-process patent, the patent holder is the sole right holder to use the given production process patent for making goods and other actions referred to in the letter (a) above.

This Article also provides a limit for the exclusivity of this right as stipulated in IPL Article 16 paragraph (2), that in the case production-process patent, the power to prohibit the importation of goods only applies for the products produced solely from the use of the patent-owned process.

In Article 16 paragraph (3), the law provides an exemption that the two previous paragraphs be exempted if the patent utilization is for the interest of education, research, experiment, or analysis without harming the proper interests of the patent holder.

The patent granting authority is the Directorate General of Intellectual Property Rights in the Ministry of Justice and Human Rights. The authority will grant the patent to the invention that met three substantive requirements, which meet the novelty requirement, contain an inventive step, and can be applied in industry (Article 2 Paragraph (1)).

In 1989, Article 7c of the old IPL stipulated that patent right could not be granted to the discovery of new varieties of plants or animals, or about any process that can be used for breeding plants or animals and its results. The Article was deleted in the 1997 IPL, which resulted in opening the possibility to process the patent on living organisms.

The Ministry of Environment and many environmental NGOs opposed that provision. Their struggle managed to influence national parliament during the making of new IPL in 2001. The 2001 IPL re-arranged the provision by granting exclusion to patent inventions of all living things except the microorganism.

The exclusions in this provision are in line with TRIP's Article 27 paragraph (2) and paragraph (3). Those paragraphs provided a basis for Indonesia to exclude a patent for something contrary to morality, to protect human, animal or plant life or health or to avoid serious damage to the environment.

Other exceptions in this provision are diagnostic methods, treatment and surgery for humans and animals, plants and animals other than microorganisms and essentially biological processes for the production of plants and animals other than non-biological or microbiological. Besides that, WTO members must provide protection for plant varieties either through patents or through a system of *sui generis* systems (as needed). In 2000, Indonesia provided this protection through the Plant Variety Protection Law (PVPL).

By observing the above mentioned IPL norms, Article 7 of the new IPL in 2001 excludes granting a patent for invention of:

a. Process or product, which announcements and usage or implementation contrary to the laws and regulations in force, religious morality, public order, or morality.
b. Method of examination, treatment, medication, and/or surgery that applied to humans and/or animals.
c. Theories and methods in the field of science and mathematics; or,
d. All living creatures, except microorganism and essential biological processes for producing plants or animals, except non-biological process or microbiological process.

Based on the above article, several types of biotechnology inventions–especially regarding the common social concerns–cannot enjoy biotechnology patent protection. The law provides that patent right will not be granted to specific food crops or livestock consider the fact that they are consumed by most of Indonesian people.

This provision has never been implemented effectively to lower food price in the era of food industrialization in Indonesia. The fact is that the price is composed by many other components of cost, including royalty-fees on other types of intellectual property rights such as trademark, industrial design, copyright, and even patent right on substance or additive products.

Biotechnology product may enjoy patent protection for twenty (20) years since the Date of Acceptance and that period cannot be extended. Patent gazette will record the Date of Commencement and the end of the patent period of each patent holder.

Patent office–representing the government–holds the power to implement a patent or to appoint a third party to implement it (based on Government Regulation No 27/2004, Article 2 Paragraph (3)). This policy is implemented as far as the specific invention is in the concern of national defense and security (Article 2 Verse (1)), or in the event that there is an urgent need based on public-interest-reason (Article 2 Paragraph (2). Implementation of patent-related biotechnology as referred to Article 2 paragraph (1) Regulation No. 27/2004 also includes the implementation of patents in the field of chemical weapons and biological weapons.

Implementation of Patent based on the public-interest-reason as referred to in Article 2 paragraph (2) also covers the following sectors:

a. pharmaceutical products needed to tackle diseases widespread outbreak;
b. chemical products related to agriculture; or
c. animal drugs needed to cope with pests and diseases infectious animal widely.

Regarding the pharmaceutical products to tackle disease widespread outbreak, during the Bird Flu outbreak in Indonesia last year, a law and political "drama" took place. The Indonesian Minister of Health–Dr. Siti Fadilah Supari–took a surprised decision by refusing deliver the virus sample to WHO. According to dr Supari, WHO is in the cooperation with the pharmaceutical company that produces the Tamiflu. Using the Indonesian virus sample the company will produce the medicine that covered with Patent protection and Indonesia must pay big amount of money to get the medicine. Based on this idea, the Indonesian Ministry of Health decided to directly engage with the Tamiflu producer rather than through any go-between entity.

The issue above was widening into political arena when the Minister of Health accused that the scheme to stole "Indonesian virus" was conducted through the NAMRU 2 laboratory facilities in Indonesia. NAMRU 2 is a long-term join project between the government of Indonesia and United States in order to conduct research on tropical disease. The Minister went to the parliament to gain political support for closing the facilities. The facility is still function to date.

That story continues when President Susilo Bambang Yudhoyono won the election for his second term in office. He appointed Dr. Endang Rahayu Sedyaningsih as the new Minister of Health replacing Supari. Political uproar took centre stage during a television live interview with Dr. Supari in television welcoming the new Minister of Health. Speaking live on television, the outgoing Minister accused her successor as the perpetrator of selling the virus sample to "the foreign parties." It was such a coincidence that Sedyaningsih was one of the Indonesian researcher at the NAMRU 2 project. The accusation was rejected by Endang. Her position in that project is officially appointed by the Health Ministry. She also provide that she always work based on tight rules and procedures.

This case shows how biotechnology and patent issue had been a sensitive political issue in Indonesia. Another related-case to the biopiracy issue will be discussed in another part of this paper.

Associated with patent issues in biotechnology, an interesting case should be observed. A Japanese cosmetics company recently got nine patents on products derived from plants from Indonesia. Some of these plants are *Rapet* wood, *Sambiloto*, *Legi* wood, *Lempuyang*, *Brotowali*,

Beluntas, *Pulowaras*, *Cubeb*, etc. all of which have been used for hundreds of years for herbs, herbal medicine and community medicine in this archipelago (Hira, 2000).

Issues on Biotechnology Commercial Information Issues and the Trade Secret Law

Indonesia regulates undisclosed biotechnology information based on Trade Secret Law (TSL; Law no. 30/2000). According to this law, trade secret is undisclosed information in technology and/or business that contains an economic value due to its utilization in business activity and its secrecy is kept by the owner. This law provides legal basis for confidential agreement among parties in a contract or in a working relationship.

Theoretical basis for this provision is property theory, contract theory, and tort (http://www.indonesialawcenter.com). Those theories provide the underlying opinion for the valuation of trade secret or undisclosed information as the asset of its owner, it could be protected on a contractual basis, and the breach of confidential agreement is a form of the breach of contract.

Article 2 of TLS provides that the scope of confidential information is production method, processing method, selling method, or other information in the field of technology and/or business with economic values and classified.

Protection for this kind of information solely rests in the hand of the information holder. The holder should take necessary, sufficient and proper steps to protect the information. There are two parties here, the information holder and the employee. It is necessary for the information holder to provide a proper and sufficient procedure of access to information, in order to create a legal basis for the owner to assert his/her right.

Breach of a trade secret will take place when somebody by purpose:

- Disclose the undisclosed information.
- Break the trade secret agreement.
- Break the written or unwritten obligation to protect the trade secret.
- Obtain trade secret by breaking the law.

Some exception rules are applied to the act of breaking the trade secret as follows:

- Disclose or use the undisclosed information for the interests of defense and security, health, or public safety.
- Engineering act on the trade secret product belongs to another party for the sake of existing product development.

Issues on Biotechnology and the Plant Variety Protection Law

Plant Variety Protection (PVP) is a special protection given by the State to the applicant. The authority for granting the right is the Plant Variety Protection Office. Based on the Plant Variety Protection Law (PVPL, Law No. 29/2000), this type of IPR is known as PVT that stands for "Perlindungan Varietas Tanaman."

PVP is specifically granted by the PVP registration office to the breeder and/or the PVP right holder. The right holder may use their own varieties of plant breeding or give approval to another person or other legal entity to use for a certain period.

PVPL defines plant varieties as a group of plants of a species or species of plants marked by shaped, plant growth, leaf, flower, fruit, seeds, and expression characteristics of the genotype or combination of genotypes that can be distinguished from the type or the same species by at least one that determines the nature and reproduced if nothing changes (Article 1 paragraph (3)).

The law also defines plant breeding as a series of research and testing activities or activities of the discovery and development of a variety, in accordance with standard methods to produce new varieties and maintaining the purity of seed varieties produced (Article 1 paragraph (4)).

PVP registration office will grant PVP right to the variety that met requirements of novelty, uniqueness, uniformity, stability, and recognition (Article 2 paragraph (1)). Detail explanation of each trait is provided by the law.
PVP right will not be granted to the variety which usage will against the law, disturb public order, morality, religious values, health and environmental preservation (Article 3). This right is granted for a period of 20 years for annual crops and 25 years for perennial crops.
PVPL is similar to IPL, regarding the rights and obligations of the right holder. PVP holder has the right to implement the right and give approval to the person or other legal entity to use a variety of seeds and crops that are used for propagation (Article 6 paragraph (1)). The use of crops for the propagation, which comes from a protected variety, must be approved by the PVP owner.

Article 6 Paragraph (1) also applies to three types of varieties:

- The essential derivative varieties derived from a protected variety or varieties that have been registered and recognized.
- Varieties which cannot be clearly distinguished from protected varieties referred to in Article 2 paragraph (1).
- Varieties produced by always using a protected variety (Article 6 Paragraph (2)).

The right to use the variety as stated in Article 6 Paragraph (1) includes the following activities:

1. Producing or reproducing the seed.
2. Preparation for propagation.
3. Advertising.
4. Offering.
5. Selling or trading.
6. Export.
7. Import.
8. Reservation for the purposes of production, preparing for the propagation, advertising, offer, sales, exports, and imports.

The use of essential derivative varieties in the above-mentioned Article 6 paragraph (2) must be approved by the PVP holder and/or the owner of the original varieties under following conditions:

- Essential derivative varieties derived from varieties that have received PVP right or had been recognized based on law and regulations and not a previous essentially derived variety.
- Varieties are basically maintaining the expression of essential characteristics of its original varieties. However, it could be clearly distinguished from the original varieties.
- Essentially derived varieties referred to above can be obtained from natural mutation or induced mutations, somaclonal variation, selection of individual plants, back cross, and transformation by genetic engineering of original varieties.

The law stipulates that original variety to produce the essential derivative should have been recognized and registered by the Government.

One important discussion in Indonesia–in the area of plant biotechnology–is the transgenic seed and genetically modified organism. For example, in the case government policy to promote cottonseed (provided by Monsanto) in South Sulawesi in 2001. This policy puts Indonesia in the position of the transgenic seed's consumer genetic engineering. Legal protection is obtained from the PVPL. Many protests by local NGOs gained media attention nationwide and in the end government decided to stop its project on transgenic cottonseed. However, the transnational company continues the project in other regions.

Regarding the transgenic cottonseed, one high official at the Ministry of Agriculture explained that the seed is a foreign seed. As a foreign seed, it is possible that the seed contain pests. The natural predators for that pests does not exist in Indonesia. In order to anticipate this situation, a particular bacterial gene had been transferred into the seed to kill the pests. The result was very productive for Indonesia. Nevertheless, due to continuing protests from the communities, Ministry of Agriculture eventually decided to stop the program. That decision was taken even though the Ministry of Agriculture won the court dispute from the first instance court up to the Supreme Court–based on class action lawsuit brought by the NGOs toward the government.

Efforts are still made by the researchers in universities in Indonesia to produce transgenic seed for other commodities. The ongoing researches are on drought-proof and flood-proof rice variety, drought-proof sugarcane, and better quality teakwood.

BIOPROSPECTING AND BIOPIRACY

Biopiracy Issue

Many food products and traditional medicines are the result of traditional knowledge development. For multinational corporations or trans-national corporations, knowledge is highly prized as intellectual capitals that can be further developed as into unique and valuable products.

Such large corporations, many of which are based in developed industrial countries, are operating on a global scale. The corporations are able to capture traditional products from the field, fund research and development for products and derivatives, and devise versions of the product with new trade-dressing. The product will be labeled as "new" patents and inventions protected by intellectual property right mechanisms. This pattern is known as the capitalization of knowledge and intellectuality possessed by the owner of the new knowledge, and which usually becomes the *intellectual capital* of the country that developed the new product.

The "newly-found" medicine's patent will be registered and claimed as the property of the corporation or research institutions. Their researchers are often claiming themselves as the inventor of the medicine. In such situations, the history and origin of the product are completely erased. The role of traditional communities as the original founder or inventor of the product will be also erased, as the intellectual capital becomes the property of the patent right owner. This is clearly an ethical problem.

To prevent such occurrences, existing regulations provide that patent rights cannot be granted to the applicant if an invention is proven to be exist and/or invented by another party prior to the application (the prior art requirement). Hence, the action by other parties claiming ownership to what is actually a community's biodiversity-based product is, in fact, an act of biopiracy.

There are many definitions on biopiracy. The International Development Research Centre (IDRC, Canada) provided three viewpoints on biopiracy, which are legalistic view, critical view, and industry view (Viewpoint, 2010). According to IDRC, legalistic view defines biopiracy as "the appropriation of biological resources without the prior informed consent of the local people and/or of the competent authority of the respective state, for access and benefit sharing, under mutually agreed terms."

Another legalistic view provided that biopiracy could be categorized as "illegal collecting" that "can infringe on the sovereignty of nations, decrease the economic health of indigenous communities, deplete or destroy species"[1] To date, there have been a number of important biopiracy cases, namely: the Rosy Periwinkle case, the Neem Tree case, the Enola Bean case, and the Hodia Cactus case (Commercialization in Traditional Medicine, 2010). Descriptions of the above cases show clearly that biopiracy also related to unbalanced relationships between rich and poor countries. Poor countries own and provide biodiversity resources. However, it is the rich countries that often are the ones to benefit from them.

There also exists a gap between such countries regarding standards used to value the ownership of the resources. The Neem tree case is an example where according to an American pharmaceutical company claimed that their patent was valid because the information regarding traditional medicine of Indian farmers was never published in any scientific journal and so Indians could not claim themselves as the first inventor of the product.

The challenge for the government is clear, how to develop a strong border management (for both national and provincial borders) in Indonesia. Strong border management will help the country to control biopiracy. The capacity of related agencies needs to be improved in order to handle the issue of biopiracy traffic (Litaay and Karetji, 2009, p. 20).

The earlier-mentioned story regarding the H5N1 virus sample case could be also connected to biopiracy issue. When Indonesia decided to stop sending samples of the avian flu virus to the World Health Organization, the country was negotiating a contract to sell the samples to Baxter Healthcare an American vaccine company. For WHO, the strains of the H5N1 virus circulating in Indonesia are crucial to develop up-to-date vaccines and following mutations in the virus (Indonesia May Sell, but not Give, 2009).

The Indonesian reason in this case is justifiably that sometimes the samples are used to produce high price vaccine that could not be afforded by the people from where the samples came. Clearly, Indonesia was looking for compensation. For the Indonesia Health Ministry, Indonesia cannot be share samples free, and they demand for a rule of the game. Using an anti-biopiracy position, the spokesperson of Health Ministry said, "Just imagine, they could research, use and patent the Indonesia strain" (McNeill, 2007).

The policy itself triggers a polemic both internationally and nationally. For many researchers, Indonesia should work closely with WHO in order to be able to access the best laboratories in the world. That kind of cooperation requires information sharing. The fact is until now Indonesia is still working with WHO.

Required Protection

Litaay and Karetji's (2009) study shows that Indonesian government must begin to take steps to prevent biopiracy. This must start with an understanding that the prevention of biopiracy is critical as a valuable resource in order to benefit from the wealth of this biodiversity and the importance of protecting the environment.

Legal foundation for the proposed policy above is provided by the UN *Convention on Biological Diversity* 1993 (UNCBD) where Indonesia is one of its member states. CBD asserts the right of resource-owner-states to control the access of its biodiversity. Article 15 of UNCBD recognizes the right of government to regulate access toward genetic resources inside the territory of a particular country.

One of the goals of the UNCBD is to enable developing countries to enjoy the benefits of their knowledge and resources. According to the agreement, the party who is in need of information (the bioprospector party) is required to gain informed consent in order to access any kind of resources in a particular region and is consequently required to share the benefit with the owner country.

In order for this to proceed, local and national legislation is needed. If the national level is slow in implementing this process, local regulations at provincial and/or sub-provincial level can be set up first, especially on products that are specifically unique to the region.

Another approach is by preventing unilateral patent applications by foreign parties. India's example is one best practice to learn. The government of India developed a database of Indian traditional medicines which was then published. This enables corporations to check related information and the prior-art situation online when they plan to apply for patent applications for new products.

The Indonesian government has never consolidated such information in the past, although steps in this direction is possible by utilizing much of the existing information collected in prior studies by many parties. Several books have previously been published, which provide information on traditional medicines; however, such publications were not developed as intellectual property information but often as publications to support the tourism industry. Such publications could be collated and strengthened to support the protection of traditional knowledge.

Publications on traditional knowledge can also help in initiating further advanced database development to serve as scientific information with sufficient authority as scientific reference. If this kind of information is published online, increased access will allow better collaboration with relevant stakeholders including patent registration offices globally. This will also support government agencies in their efforts to control unlawful parties from registering traditional knowledge without recognition of its origins. The database itself will be protected by copyright law to ensure protection and benefits to indigenous communities. Use of information provided via the database by other parties can also be conducted under contract, providing further legal protection.

By applying these considerants in policy, institutions and communities in Indonesia can and should gain more benefits from the globalization of information, while enhancing Indonesia's capacity as an information producer rather than spectator.

One important note that needs attention is the fact that the existing intellectual property right approach is orienting more towards individualistic subject-centered ownership. Development of intellectual property law on traditional knowledge needs to take on a more communal perspective.

Through the introduction of such specific rights, application of traditional knowledge could be controlled, developed, and protected including genetic and human resources, seeds, derivative products such as traditional medicine and health practices, herbal plants, and knowledge of flora and fauna.

Free and prior informed consent means agreement from all community members to make decisions based on their traditions and practices, free from external manipulation and coercive action. The consent may only be gained after the outsider expresses his/her intention fully and clearly regarding the activities and its scopes, in a manner that is fully understood by the community members.[2]

CONCLUSION

Indonesia has made plenty of efforts in implementation of UN CBD, WTO Agreement and other instruments. Nevertheless, the country needs to tackle some challenges such as reorganization, modernization of bureaucracy, capacity building of its agencies in order to improve its performance in achieving national targets based on UN CDB.

In IPR field, the country had also improved the quality of its legal system. The challenge is to improve the quantity of national product that will enjoy the protection. If the national production is low, the benefit of the legal system will only be enjoyed by foreign producers.

Biodiversity is a source of ecological sustainability and economic development. Meanwhile biopiracy is the threat to ecological interest and economy interest. Protecting biodiversity and utilizing it in an ecological-friendly way will enable the country to create a new economy that sustainable economically and ecologically. The government needs to pay more attention in this area and take necessary and effective measures to tackle the problems.

REFERENCES

Commercialization of traditional medicine. (2010). Retrieved from http://en.wikipedia.org/wiki/Biopiracy. Accessed 4 February 2010.

Gollin, M. A. (2009). *Biopiracy: the legal perspective.* Retrieved from http://www.actionbioscience.org/biodiversity/gollin.html. Accessed 7 January 2009.

Hira, D. G. (2000). Beberapa kasus paten atas kehidupan dan biopiracy. Email discussion, July 9.

IDRC Canada. (2010). *Viewpoint – what is biopiracy?* Retrieved from http://www.idrc.ca/idrcbulletin/ev-64405-201-1-DO_TOPIC.html. Accessed 4 February 2010.

Iwantoro, S. (2006). The role of agriculture quarantine agency of Indonesia (AQAI) as a trade instrument in global trade liberalization. PowerPoint presentation in Nanning, China, November 2, 2006. Director General of Agriculture Quarantine Agency of Indonesia.

Iwantoro, S. (2007). *Issues and concerns affecting biosecurity in Indonesia. PowerPoint presentation.* Director General of Agriculture Quarantine Agency of Indonesia.

Litaay, T. (2007). *Hak Kekayaan Intelektual.* Salatiga: Widya Sari.

Litaay, T. (2007). Kemendesakan Perlindungan Pengetahuan Tradisional di Indonesia. In Louhenapessy, D. (Eds.), *Biosecurity and Indigenous Knowledge – Workshop Proceeding. PSKTI UKSW-CDU-CRCNPB-BaKTI.* Salatiga.

Litaay, T., & Karetji, P. (2008). *Unity in (bio)diversity? Indonesian traditional knowledge protection in research activities.* Salatiga: PSKTI UKSW.

Litaay, T., & Karetji, P. (2009*). Public policy and protection of indigenous knowledge.* Paper presented in Asia Pacific Sociological Association Conference, 2009.

McNeill Jr. Donald G. (2007). Indonesia may sell, not give, bird flu virus to scientists. *The New York Times*, February 7, 2007. Retrieved from http://www.nytimes.com/2007/02/07/world/asia/07birdflu.html. Accessed 21 November 2009.

Muchtadi, Deddy. (N.d.) *Keamanan pangan produk hasil rekayasa genetik.* Powerpoint presentation. Departemen Ilmu dan Teknologi Pangan FATETA. Institut Pertanian Bogor.

PP No. 27 tahun (2004). tentang Tata Cara Paten oleh Pemerintah.

UU No. 28 tahun (2000). tentang Perlindungan Varietas Tanaman.

UU No. 14 tahun (2001). tentang Paten.

Suharto. (2007a). *Tugas dan fungsi pusat informasi dan keamanan hayati dalam pelaksanaan pengawasan keamanan pangan.* PowerPoint presentation 30 October 2007. Pusat Informasi dan Keamanan Hayati / Centre of Information and Biosafety, Agriculture Quarantine Agency of Indonesia.

Suharto. (2007b). *Peran karantina dalam rangka perlindungan sumberdaya genetik dan keanekaragaman hayati.* PowerPoint presentation 11 December 2007 in Pontianak, West Kalimantan. Pusat Informasi dan Keamanan Hayati / Centre of Information and Biosafety, Agriculture Quarantine Agency of Indonesia.

Untung, K. (2008). National policy on biological diversity: community management of biosecurity. *Special Co-Publication between Kritis-Journal of Interdisciplinary Development Studies (Indonesia) and Learning Communities-International Journal of Learning in Social Contexts (Australia)*, 228-238.

ENDNOTES

1. This approach is also called 'take and run.' See Gollin, 2009.

2. Litaay's own view is based on a study conducted on Philippine's IPRA 1997, for other definition and elements compare UNCBD Article 15(5).

Chapter 13
Human Biobanks:
Selected Examples from and beyond Europe

Brigitte Jansen
BioEthicsLaw e.V., Germany & University of Madras, India

ABSTRACT

The chapter presents a careful comparative study on ethical and legal aspects of human biobanks both in Europe and elsewhere. The rapid expansion of human DNA sampling and data collection has taken place in the last few years, but the legal and ethical perception of this situation looks very different in European countries and beyond. The author focuses her attention on the European Union, especially in Estonia, where a population wide gene back has been established; moreover, she also discusses what is happening in Macedonia, a relatively neglected country in Eastern Europe, as well as Australia, India and Israel.

INTRODUCTION

Traditionally, biobanks are defined as a systematic organized collection of cells, tissues or blood samples which are stored to be retrieved for analysis for a more or less long time. This definition includes the idea that universities, researchers etc. have their own working collections which are also increasing in the last time (Cambon-Thomsen et al., 2003, p. 145). Since large population biobanks were set up, this definition is not really broad enough as large population biobanks are a good basis for health surveys, research etc. because this allows monitoring of the health status on a permanent basis.[1] So it seems helpful to make a distinction between[2]:

- **Diagnostic biobanks and treatment biobanks:** Collection of human biological material delivered for medical examination, diagnostics and treatment, and
- **Research biobanks:** Collection of human biological material and information directly deriving from analyses of this material for research purposes.

DOI: 10.4018/978-1-61692-883-4.ch013

The consequence of such working definition could change the understanding of our current biobanks[3] in a more practical way. It is obvious that for example for research in biobanks the physical sample itself may become less important, because more tests and analyses can be run automatically with the collection of samples. In particular genomic data could become more important (Mattick, 2003), and we shall concentrate on this aspect. Biobanks are increasingly viewed as an international resource; thus, standardization and collaboration between biobanks will become common practice (Hirtzlin et al., 2003).

The chapter will describe the situation in several countries where the discussion and development of DNA banks is especially advanced. Furthermore, indications will be given as to how the situation is developing, especially in European countries.

COUNTRIES

DNA Banks in Australia

In Australia there are a number of DNA banks for research and clinical use. Already at the end of the eighties, the "Human Genetics Society of Australasia" (HGSA) (1990), which represents both Australia and New Zealand, foresaw that the development of genetic tests would lead to an enormous growth of bio-banks. Therefore, the "Guidelines for Human DNA-Banking" were developed, to be presented in July 1990 to set standards in medical/legal and ethical areas.

These guidelines take stock of the differences between clinical services and research and require that the researchers have certain duties towards the families whom they are investigating. Part of this is that the responsible State office that is carrying out the clinical genetic support service makes this public, and engages in exact documentation which is to be made available to all health care givers of the family. Future services for the family must be taken into consideration, even after the end of research. This aspect related to the care of the family in the Australian HGSA guidelines is stressed here as less emphasis is placed on this area in the other national guidelines.

According to the guidelines, the cases are carefully delineated as to when it is necessary or not to get consent for a blood probe and the retention thereof. In certain cases, formal consent is absolutely not necessary; in other cases, this depends on the situation. For example, the guidelines assume that the purpose of prenatal tests for recessive genetic disorders such as cystic fibrosis is generally known and that only parents and their children are affected by this (The Human Genetics Society of Australasia, 1990, p. 4). In this case, a formal agreement is not necessary. On the contrary, Huntington's chorea requires samples from extended family members who may not be motivated enough to undergo tests nor even clearly understand the purpose of the tests. In this case, consent seems to be quite necessary.

The secrecy of test results is also held to be a high ideal by the HGSA. As such, no results should be passed on to third parties without written permission and this clearance is only allowed to a small number of exactly specified people (The Human Genetics Society of Australasia, 1990, p. 4).

In the same year, the Australian Health Ethics Committee (AHEC) that was set up by the National Health and Medical Research Council (NHMRC) published the *Guidelines for Genetic Registers and Associated Genetic Material* (1999). These guidelines, which came into power on January 1, 2000, are to take into consideration all aspects related to the establishment and the organization of a genetic register.[4] There are certain key characteristics that are to distinguish these genetic registers from other collections of data. The registers record genetic disorders where each register only should collect data relating to one disorder or a narrowly related group of disorders. At the same time, one should ascertain family members who have a higher health risk because of such abnormalities so that one can engage in preventative measures. The essential purpose of the genetic register is to give relatives of the registered person the possibility to utilize information about genetic disorders for preventive purposes. Therefore, the relatives need to be contacted as well. In this area, one must respect the ethical principle that both the registered person and the relative must first agree that they will be informed at all: the right to ignorance must in any case be respected.

These guidelines also take as a premise that "in general, consent should be obtained before identified information about a person is included in a register" (NHMRC, 1999, p. 17; Crosbie, 2000, p. 11). All parties should be informed that participation in the register is voluntary and that they can change their mind at any time - to the extent of withdrawing identified genetic information and samples that they have made available. Should a future law regulate this, such information could potentially be passed onto other people, courts, or organizations, for example Social Insurance Security or the police (NHMRC, 1999, p. 27).

The Genetic Privacy and Non-discrimination Law (draft) of 1998 that was passed after lengthy consideration, does not expressly name biobanks as something that should be regulated, but includes the collection of DNA samples in the regulation process since this relates to the protection of privacy (Crosbie, 2000, pp. 13ff). In contrast to the guidelines, this is a law which as such has an obligatory character and partly regulates the same areas as the guidelines. Among other aspects, it is worthy of mention that unless the donor has given his/her written consent for further use, the sample must be destroyed upon the termination of the test or investigation. This does not exclude his/her right to demand the sample back at any time.

Similar to the relevant Israeli guidelines, to be discussed later in the chapter, great value is placed upon the situation that the information is clearly comprehensible to the donor.

DNA Banks in India

As India is one of the most populous countries in the world, it engages in extensive genetic research.

Actually this situation and also the ethnic diversities make it interesting for genetic studies and testing for special conditions. India has a lot of experience with smaller screening projects. Actually, for a few years now, two biobanking projects with different protocols have been up and running under the guidelines of the Medical Council of India. These projects are

1. The Indian Genome Variation database (for five years) which was setup in the year 2003 and
2. The Women's health biobank which was established in 2004 with support from the UK.

Indian Genome Variation (IGV) Database: The Project and its Aim

IGV has been set up with the funding of the Indian government as a network between six constituent laboratories of the Council of Scientific and Industrial Research (CSIR)[5] and a private company with the aim to initiate a network program on predictive medicine using repeats and single nucleotide polymorphisms. The base is supposed to be 15,000 individuals drawn from Indian subpopulations. These genes have been selected on the basis of their relevance as functional and positional candidates in many common diseases including genes relevant for pharmacogenomics. This is the first large-scale comprehensive study of the structure of the Indian population with wide-reaching implications.

The research is focused on studies related to asthma, diabetes, neuropsychiatry disorders, cancer, coronary artery disease, clot disorders, high altitude disorders, retinitis pigmentosa, predisposition to malaria as well as other infectious diseases and drug metabolism (Indian Genome Variation Consortium, 2005).

The protocol and the description of the scientific part of the project show that naming the population with a particular set of tag SNPs will allow a better interpretation of the biological significance for use in future studies of association, population history and population relatedness. Nevertheless, it has important ethical and social ramifications. The investigators try to avoid any social backlash that could destabilize the very unique fabric of Indian society, i.e. unity in diversity. They decided against disclosing the identity of the populations. This is because the way a population is labelled in this project and described in publications will have implications for all members of the population, as all of them (and all members of closely related populations) might be affected by the interpretation and use of findings of future studies. However, this study clearly spells out that individual variation in the population. Risk population for a given predisposition cannot be ruled out. As the consortium stated, the samples collected from different populations are bar-coded with each population being given a specific code revealing the linguistic affinity of the population, the geographic zone to which the population belongs as well as the type of population, e.g., large endogamous population, isolated population or special population.

Women Health Biobank

The idea of the women health biobank project in India was published and set up in 2004 (Kennedy, 2004). The reason and the published argument to do this was that in European countries women health problems frequently appear on any list of research priorities because it will be seen as a common problem. This is different in India, especially in South India because of the existing commitments to fertility control research or in other words: the value of woman in the Indian society is strongly linked to her fertility. The other argument was that, in addition, the possibility exists of enriching the cohort by collecting family members in a way that will not be feasible in the UK. Beside these very questionable reasons, several scientific questions are also raised, for example how feasible would it be to collect a cohort that is truly representative of the population at large? The project follows the protocol of a smaller collaborative project investigating the genetic epidemiology of endometriosis which had also involved the Centre for Cellular & Molecular Biology in Hyderabad, the University of Oxford and hospitals in Hyderabad and Kolkata. Similarly, the population based Hereditary Cancer Registry at the Chennai Cancer Institute is collecting data and blood samples from families with hereditary cancers with the help of government hospitals and private nursing homes in the Chennai metropolitan area. It is clear that a large number of women are needed for such a huge project.

The foreseen procedure and description for data collecting and samples, consisting of standardization of data, bar-coded samples, and informed consent and confidentiality, is obtained from all participants. As stated the study is governed by adequate ethical and legal standards regarding the use of personal information and biological samples both now and in the future under the regime of a common legal framework in India.

The aim of this Women Health Biobank[6] is to provide the research community and the pharmaceutical industry in India with a unique resource in the post-genome world and also to study the interaction between genotype and environmental factors for a range of debilitating conditions, provided that sufficient numbers of participants are recruited.

Regulation and Acts Related to Biobanks and Genetic Testing in India

The rules for biobanking in India are quite different and we try to describe the basis first before. Both described projects are following the "Ethical Guidelines of Indian Council of Medical Research

(ICMR)." These guidelines have been enforced since 2000. The Indian Medical Council itself was established under the Indian Medical Council Act of 1956 with several amendments (1964, 1993, 2001 and 2005). The ICMR is the regulatory body for this area.

The last one, the Indian Medical Council (Amendment) Bill, came out in 2005 by the government of India, but it was rejected by the parliament in March 2006 and is still in the phase of revision and, therefore, not enforced yet.[7]

These guidelines have been enforced since 2000, and can be divided in 2 parts. The first part is the general statement and the statement of general principles (Ethical Guidelines, 2000, pp. 8ff). These national "Codes" were drawn from international codes and universal principles, based on the international common guidelines in medical research and a second part, which is more focused on specific principles.

In the actual guidelines, you will find the chapter "Statement of specific principles for human genetic research" (Ethical Guidelines, 2000, p. 39).

In this chapter, it is pointed out that in the field of genetic testing appropriate communication skills are necessary. There is articulated a likelihood of social stigmatization and discrimination in schooling, employment, health and general insurance, which requires much greater care in recruiting subjects in research studies, obtaining informed consent and maintaining confidentiality of research findings, than in any other area of research. As is well known, genetic testing can be misused, in particular for the pre-selection of sex. In order to prevent this special law was enacted in 1994, namely The Prenatal Diagnostic Techniques (Regulation & Prevention of Misuse) Act of 1994.

If one sees the pedigree studies, it is interesting in the light of the constitutional law that special privacy and confidentiality concerns arise in genetic family studies. In the Indian constitution from 1950 itself (http://lawmin.nic.in/coi.htm), "privacy" was not clearly pointed out. It came into play through a High Court decision in the year 1964 recognizing that there is a right of privacy implicit in the Constitution under Article 21, which states, "No person shall be deprived of his life or personal liberty except according to procedure established by law" (http://lawmin.nic.in/coi.htm).

This article states that in families each person is an individual who has the right to keep the information about himself or herself confidential and that "Family members are not entitled to know each other's diagnosis." This statement is very optimistic. If you see the cultural background, it seems very complicated to achieve this goal, because there is no link to penal law to support this right etc. The main problem is that women's rights are not given in every part and class of the Indian society and it also differs from country to country. But there is a lot of progress which you can see in some changes in the law.[8] Nevertheless, in this type of research informed consent has to be given. The guideline has foreseen from the early beginning, besides the individual informed consent, that also community consent could be given (Ethical Guidelines, 2000, p. 8).

To take part in a cohort study like the women health biobank creates also a lot of pressure on the designated families or, if tribes are designated, on the group, because it can be regarded as an unfriendly act. In India, minorities such as tribes etc. are keen to be identified as somebody who belongs to a minority. The reason for this is that minorities are privileged in the taxation system. There is also a quota system to get a place in the University, etc. The position of a physician in terms of trust and other things is higher. The society believes also in the progression of science. Often patients join programs under the idea that they have to do the doctor a favor.

In the guidelines, the question of data confidentiality and protection is described. Codification is necessary and safeguards have to be established. There is a clear distinction between primary and secondary use. Normally, the use of the samples shall be reserved for the purpose for which informed consent is given. For the secondary use, the institutional ethic committee has to approve and to examine that it is in the scope of the original informed consent and that also the anonymity of these samples is granted.

For India, as a country with a lot of ethnicities, the topic human genome diversity is very important and regulated by the Minority act. In the guidelines this topic is covered by referring to a document of the Department of Biotechnology, Government of India, which had set up a clear definition under which conditions a project like the women health biobank could be set up. There it is pointed out clearly that in a project like the IGV, appropriate safeguards and regulations have to be established in order to ensure anonymity of the sample (not traceable) and protection of the rights of the people. It is also pointed out clearly that the analysis of DNA samples shall be carried out by Indian scientists or laboratories. International collaboration, if any, shall be carried out with well-documented MOU, which is approved by the Institutional Ethical Committee. This should include the scope of utilization of exchanged material and related IPR issues, as well as concerns for human rights. A major concern regarding these studies is the possibility that generated information may produce ethnic disharmony. Therefore, great care is necessary for the handling of this data, particularly in reference to the release of news to media and publication of research results.

No legislative measures concerning data protection, however, have been considered to date. This means that there is no direct Data Protection Law established, but there are several other laws and decisions in force to which bio-banking can be linked to. The Information Technology Act was set up in May 2000. The Information Technology Act is a set of laws which intended to provide a comprehensive regulatory environment for electronic commerce (http://www.mit.gov.in/it-bill.asp). The Act addresses computer crime, hacking, damage to computer source code, breach of confidentiality etc. (Section 72) and viewing of pornography.[9] Chapter X of the Act creates a Cyber Appellate Tribunal to oversee adjudication of cyber-crimes such as damage to computer systems (Section 43) and breach of confidentiality (Section 72). This section can be used as a kind of the data protection to strengthen the position of the Medical Council. Breaking confidentiality falls under a penalty. It is sometimes discussed that the Indian courts can apply the corresponding rules of the English Common Law but there exist also high court decisions to which in case of data protection in biobanks could be referred to. So the Supreme court decided in the *Auto Shankar* case, for example, that every citizen has the right to safeguard his or her privacy and that nothing could be published on areas such as the family, marriage and education, "whether truthful or otherwise," without the citizen's consent, but carved an exception to this rule for material based on public records and information about public officials' conduct that is "relevant to the discharge of their duties" (Failure to Define Law on Privacy, 2001).[10] On the other hand, in the IT Act only a pseudo privacy is given because the providers and others have to observe all activities of their clients and to give these data to the authorities.[11] The Right to Information Act[12] regulated only the rights of the citizen to get appropriate information from the authorities. The confidentiality of data including penalties if it will be broken is also pointed out in the Public Financial Institutions Act of 1993.[13]

DNA BANK IN ESTONIA

Development, Design and Content

In the year 2000, the Estonian Parliament passed the Human Genome Research Law[14] by 42 positive and 3 negative votes. This legislation regulated the establishment and the operation of a genetic data bank and the use of genetic information relating to the phenotype and genetic characteristics of the Estonian population. The Estonian Genome

Project (EGP), which is based on this legislation, started with a pilot project in October 2002 that could be successfully concluded in February 2003. In the final phase, it should contain genetic data of one million of the 1.4 million Estonians (Decoding Estonia, 2010). Preparations for this project alone were extremely extensive and took four years. It is estimated that it will take an additional five years to set up the genetic bank.

According to those responsible for the project, the prerequisites are good as the population is large enough to offer an adequate number of samples for sicknesses. In addition, Estonia possesses a developed infrastructure and a computer based national health system. Allegedly the legal and ethical environment is transparent, the employees qualified, the biotechnology industry in full expansion and the project financially feasible. Furthermore, there seems to be an ethnic mixture of the population that would be quite attractive for pharmaceutical enterprises. In contrast to the isolated location of Iceland, there was a plethora of invaders who all "left their genes behind" in Estonia. Therefore, one is in the position of looking for the genes connected to sicknesses that are probably present in most Europeans (Decoding Estonia, 2010).

The goal of the project is to use genetic studies and health examinations to identify genes that cause or influence ethnic sicknesses and develop new medicine against them (Kruuv and Nomper, 2002). Long-term studies about the connection between genes, environmental factors and illness should lead to a better health service and also bring benefits in the short term. In the long term, the use of DNA based diagnosis should lead to the condition that this kind of "informed consent" does not suffice in a purely individual sense to develop the guarantees through "informed consent" (Metspalu, 2002, pp. 42-44). The economic advantages allegedly come from the fact that investments will be made and jobs created: Both a better level of education and the construction of a biotechnology industry will be the result (Metspalu, 2003).

On the basis of an intensive information campaign, already 43% of the Estonians are ready to donate their samples and provide data; 36% want to inform themselves before coming to a decision and only 6% are against the idea. The participation in this project is voluntary and data that has been garnered can be deleted subsequently if the participant so wishes. The data belong to the Estonian state and were administered firstly by a private genome foundation.[15] This non-profit organization was set up in 2001 by the Estonian government and is under the control of the Social Ministry (Metspalu, 2003). It is led by a Supervisory Board whose nine members are appointed by the Parliament, the Government and the Estonian Academy of Sciences. The Supervisory Board has as its purpose the promotion of the development of genetic research, to collect information about the health and genetic status of the Estonian people and to use the results of the genetic research for the benefit of the public health. It should be stressed here that in comparison to Iceland, the Estonian Republic wishes to set up a gene bank itself and thus has the final word in decisions.

The financing is only covered to a small extent by the government; the bulk of the expenses are to be covered by domestic and foreign businesses. The cost of the design of the gene bank is estimated at about 150 million US dollars, of which the foundation need merely provide seed capital of $70,000 (Kruuv and Nomper, 2002, p. 2).

A short period of time after the law was passed, the Estonian Gene Foundation established the public law firm EGeen Ltd. in Estonia. This company initially belonged completely to the foundation but is to be divvied up among potential investors. This company carries out the first analyses and evaluations of genetic data in order to produce the electronic gene cards for the donators. Pharmaceutical companies can pay for information out of the Egeen data files in order to countercheck their own research results (Kruuv and Nomper, 2002, p. 32). Thus, it was hoped that the results of the DNA experiments will show a profit through licensing and that a part of the resulting money will flow back into the Genome Foundation which would then use this for the project itself and the operation of the data banks.[16] The subsequent US based company EGeen International Inc. managed to receive exclusive licensing rights during the first five years of the analyzed gene data. EGeen International is looking for further financial sponsors in order to gain a sufficient financial basis for the Estonian mega-project.

This model was not successful, however. In 2005 the Estonian government fully overtook the organization and the financial support for the biobank. As an result of some mess, the Estonian government had to set up rules for the organization of the biobank, and since 2007, the biobank has been run by the University of Tartu. Therefore, the revision of the Human Gene Research Act came to be enforced.[17]

Legal Funding and Organization

According to the rules of the Human Gene Research Act, the Project Foundation is to advance the development of research in the service of the Government, collect information about the health of the population as well as using this genetic information and the results of genetic research to improve the public health. For this purpose, the foundation is to build up and maintain the genetic bank and to organize the storage of the samples and data and to coordinate the use thereof and the genetic research related to this, etc. According to the HGRA, the "chief processor" can delegate the competencies for the carrying out of the project in individual cases to the so-called authorized processors. However, the responsibility to encode and decode must remain in the control of the chief processor. In the revision of this part of the act, the Chief Processor of Gene Bank is the University of Tartu, which has outlined clearly defined objectives. In §5 it is stated that

1. The Chief Processor of the Gene Bank is the University of Tartu, whose objectives as the Chief Processor are to:
   a. Promote the development of genetic research;
   b. Collect information on the health of the Estonian population and genetic information concerning the Estonian population;
   c. Use the results of genetic research to improve public health.[18]

The instruments to achieve these objectives:

2. ..., the competence of the chief processor is to organize the samples, compose the statements of health and genealogy, to code and decode, preserve, destroy, and release them, to carry out genetic researches and to collect, preserve, destroy and release genetic data.
3. The chief processor has the right to delegate the rights of processing, except for coding and decoding, to an authorized processor on the basis of a contract in the cases and under the conditions prescribed in this Act.

In addition, there is a monitor that supervises the quality level. The inspection of the manner of data acquisition (including the questionnaires) and the coordination of the data collector and the genetic bank is also in his hands. The donor's participation in giving probes and data is voluntary and must take place on the basis of informed agreement for scientific purposes, with no paper trail leading back to the donor. The collection of health data (genetic tests fall outside this category) is quite extensive, for along with the more narrowly interpreted health status (IDC, 2010) one also collects the relevant phenotype (genealogical, lifestyle decisions, environmental factors and tolerance for medicines). Health and genealogical data as well as blood probes will be collected by family doctors and general practitioners in their role of "authorized processors" for the genetic bank. On the basis of this, detailed studies should be carried on a selection of the health data, and if so desired to contact the donor again for a further written agreement for additional studies.

The donor has no title to a report about his or her state of health or for the use of the research results. However, he or she can turn to the supervisory board of the foundation to demand that data which might disclose his/her identity through decoding or illegal disclosure be destroyed. To prevent this, the donor can also demand the destruction of the tissue samples, the description of the DNA and the description of his health status. Above and beyond this, the genetic donor can exercise his right to go back on his agreement.

It is also important, as a goal, that the financing of the projects, as well as further possible use of the data, the ownership of data and samples including the encoding and decoding processes will be made clear for the future. In the agreement form, one item of essential information relates to the basic decision of the gene donor. Additional information can be given in a so-called gene donor information kit.

The gene types are to be discovered on the basis of the phenotype. This process requires a highly effective technology where personal data is carefully encoded and cannot be connected in any way with the donor during its use. In any case, the requirements for data protection related to matters as important as a person's health status is held to be quite sensitive data which must be thus protected in an organized manner.

The protection of the genetic donor is based upon the confidential treatment of his information and his right to the destruction of data betraying his identity,[19] non-discrimination by his work place and insurance company and last but not least the involvement of an ethical committee. Thus, after encoding, the donor can choose to remain anonymous or can publicize his identity as donor; he can have free access to his personal data or decide to ignore this; add additional information or (as far as this is already stored in a data bank) refuse the expansion, renewal and review of such descriptions of his health status. These rights reflect Article 10 of the bioethics convention, where it is stated that the right to privacy and confidentiality of personal medical data must be guaranteed. [20] The purpose of the independent ethics committee is to protect the health, human dignity, identity, privacy and other fundamental rights and freedoms of the donor.

A central aspect as regards to the organization of this agreement is the possibility of individual withdrawal. According to §12 Paragraph 4, 7, a gene donor has the absolute right to take back his approval—at least until the tissue sample or the description of his health status is encoded. In the above case, all information and blood probes will be destroyed. Afterwards, the donor can appeal to the Chief Processor at any time to have the data destroyed—this is done by decoding (Kruuv and Nomper, 2002, p. 9). However, the destruction of data that make the decoding and anonymizing possible may lead neither to the destruction of other data nor of biological material. The donor has the right to this only if his identity is disclosed in an unjustified manner.

The question is whether this application is against Articles 5 and 16 of the Bioethics Convention whereby everyone has the right to withdraw his agreement at any time. Opposing this interpretation, the Estonian legal opinion that it is enough that the donor can withdraw his probes and data until encoding, for Article 16 refers to a physical intervention which is not relevant here. Here it is merely a matter of the research of tissues that are separate from the body and made anonymous. Thus, it is thought that this legal settlement as regards to the withdrawal of consent suffices (Kruuv and Nomper, 2003, p. 8) and is in harmony with the Convention.

All in all, the organization and participatory form of the Estonian biobank are seen to be a good example for the design of legislation in other countries (Redecker, 2003, pp. 63-79).

DNA BANKS IN ISRAEL

The Israeli National Helsinki Committee for Human Genetic Research adopted in 2002 the Guidelines for Review of Research involving DNA Collection, Storage, and Testing. The goal of the guidelines is to support researchers in the preparation of protocols for approval of projects by the above committee. In these protocols, the research project has to be described exactly, where the samples and data come from, how the privacy of the person involved can be protected, especially as far as the "Genetic Information Law" of 2000 is concerned. Biobanks are not specifically mentioned, but collection, storage, and testing of samples are exhaustively handled. Thus, it must be demonstrated what the source of the samples is, where they are located, where they will be tested and by whom, how long the tests will be retained and whether further tests will be carried out. In line with the Genetic Testing Law, it is important whether identifiable or unidentifiable samples are used. Should it concern identifiable samples, the methods of storage must be exactly described. Should it be unidentifiable information, a clear portrait of the method of protection, the method of division from the identifiable information, as well as who has access to the data is a must.

The rights of the people affected are also discussed in detail, among other things, their risks and advantages, their right to withdraw their approval, whether or not they will receive individual results of the identifiable genetic tests—should such be of clinical value—whether they will receive a comprehensible research result, whether their agreement to further use has been given and possibilities of financial involvement in the patent. However, it should be added that generally speaking, there is no assumption of such participation.

Furthermore, the information documents relating to participation in the research and the "informed consent" paper must be submitted.

The guidelines contain at the end of these directions further information for the researchers, where it is emphasized that genetic information has a unique aspect in so far as the health of family members is concerned. Here it is less about direct health risks than the range of dissemination of such a kind of information, which could lead to invasion of the private sphere or repercussions causing discrimination and stigmatization. According to the guidelines, these are risks that not only can affect the individual and her family but also the whole society.

It is again clearly stated in the guidelines as to the special risks of stigmatization towards specific populations that have a high risk of genetic problems, should this information be gathered for the research. Furthermore, there is the danger that the publication of such studies could identify participants of this research or indicate disadvantageous physical or behavior of subgroups (minorities) of this population. Therefore, in many of these research plans, individual agreement does not suffice and the scientist must clearly state whether, according to the specific conditions of the research project, community consent is requisite. However, this community consent may under no condition replace the individual agreement.

Here a detailed proposition for a "Model Informed Consent" is delivered, including among other things a codicil where the sample is to be made available for all future research. This passage must be separately signed, even when this is simply for a limited research plan. It is emphasized that the results of the genetic research will in no case become part of the medical file of the participant. Only those in the research staff who are permitted by law may check the medical information and the results of the genetic diagnosis.

In addition to this Model Informed Consent for Participation in Research involving DNA Collection, Storage or Testing, that must be signed by the participant before the beginning of the research project, an information paper will be handed to him in which the content is described once again in a short and clear manner.

As shown by the previous report, these guidelines are extremely extensive, providing basic preconditions for genetic research and the evaluation of samples and data in biobanks. The guidelines try to preserve an extensive protection of the participant in the research without essentially restricting this research now or in the future.

MACEDONIAN DNA BANK (HDNAMKD)

It is somewhat less known in Europe that Macedonia is also building up a DNA databank. This databank is operated by the Institute of Immunobiology and the Institute for Human Genetics at the University of Skopje. Different goals are followed with this data bank. On the one hand, the genetic variety of the Macedonian population and the minorities is to be measured; on the other hand the connection between genetic diversity and genetic sicknesses is to be researched. For this purpose, a genealogical data bank will be built up. This biobank is financed by the Macedonian government.

The data bank is divided up into three different research and project areas as Table 1 illustrates:

Defining samples in the field of anthropological research is not without explosive connotations as the samples are structured and defined according to the following criteria: population group, language and religion. Considering things from an ethical point of view, it seems extremely dangerous to utilize these "soft" criteria as the results are prone to potential misuse towards ethnic groups. Table 2 integrates the content of this biobank once again.

Table 1. Areas of the project, number of DNA samples present, December 2002 (Source: http://www.hdnamkd.org.mk)

Project	DNA: stored samples			
	2002	2001	2002	2000-2002
Anthropology (Illustration 2)	46	279	652	978
Patients not relevant to project	111	249	75	435
Patients relevant to project	19	159	103	281
Total	176	687	831	1694

"Nonrelevant" patients relate to data and material from diverse medical projects outside of both institutes where no family relationship is established. No family data was acquired. At the moment, the bank contains about 409 samples. Essentially this concentrates upon ten different kinds of sicknesses: rheumatism, heart disease, Diabetes mellitus 1 + 2, etc.

"Relevant patients" refers to projects where DNA samples are connected up with family and patient data. This bank emphasizes spinal marrow transplantation, related renal transplantation and autism. At the moment, about 232 samples from 232 people can be found in the bank.[24]

The Macedonian Constitution protects individual rights. [25]

Each doctor has the right to apply for participation in the project. He has to give the necessary information to the patients and the patients have to give informed consent. Informed consent form has to be signed by the doctor and the patient. The owner of the anthropology data is Institute of Immunobiology and Human Genetics (IIBHG), Faculty of Medicine, Skopje, Republic of Macedonia. Banked DNA from the patients is property of the depositors (IIBHG and MD) who sent the material for DNA banking.[26] Nevertheless whatever the reasons are, doctors and patients have the right to request destroying of the samples.

The samples itself can be used for scientific purposes from research institutions all over the world.

CONCLUSION

The reflections of the social, ethical perception and the legal aspects of genetic data banking

Table 2. Form of the Macedonian population and the samples

Population	Nationality	Other Languages	Mother tongue	2nd language	Religion	DNA-Samples
Macedonian	Macedonian	Macedonian	Macedonian	Macedonian	Orthodox Catholics Protestants Muslim	353 0 0 4
MKDAlbaner	Albanian	Macedonian	Albanian	Macedonian	Muslim Catholics	50 0
MKD Turks	Turks	Macedonian	Turkish	Macedonian	Muslim	82
MKD Serbs	Serbs	Macedonian	Serbian	Macedonian	Macedonian Orthodox	8
MKDVlachs[21]	Vlach	Macedonian	Vlachs	Macedonian	Catholics	9
MKD Mixed[22]	Mixed	Macedonian	Mixed	Macedonian	Mixes	2
Yug Gorans[23]	Goa	Serbian	Macedonian	Macedonian	Muslim	6
Total						80
						594

in the medical sector showed that above all the main problems the focus is directed at confidentiality, privacy, discrimination. Every breach of confidentiality could impact insurability, destroy family relationships and causes stigmatization or discrimination. This is recognized in some of the countries which were discussed above.

Therefore a general transparency of the activities of genetic data banking is necessary for the individual related with intensive information (existence, ownership, application and group of persons covered by the databank) maintaining at the same time a status of anonymity concerning the concrete data. It is obvious that the sampling, storage and examination of body material without knowledge and consent of the concerned person would be an offense against the right of informational self determination. Even the already formulated consent declarations are often not enough determined or temporarily limited. It has to be discussed, if it will be necessary to create documentation and reporting obligations disregarding the principle of appropriation. But it must be clear for the concerned person to know who, how far, for which purpose and where his genetic data will be used. The person must be able to revoke the consent.

The regulations of this law contain more or less general clauses and undetermined legal notions needing an interpretation for each single case. The Data Protection directive of the EU (Directive 95/46/EG from 24.10.1995, L 281, p. 21ff.) established new measures concerning the handling of patients' data in medicine and health care, forbidding the collection of data about health and sexual life, Art. 8 I. Those data are protected. Art. 8 III foresees several exceptions if the data are really necessary and an obligation to keep the identity of the concerned person secret. These exceptions have to be interpreted when transferring the directive into the national law. For example, in some new countries such as Estonia the EC directive was transferred immediately into national law.

In India, on the other hand, the formal legal situation is in progress, and the cause or the basis of the need for such formal and clear law is just beginning to be appreciated.

REFERENCES

Cambon-Thomson, (2003). *Populations and Genetics: Legal Social-Ethical Perspectives. The Netherlands* (Knoppers, B. M., Ed.). Leiden, Boston.

Crosbie, D. (2000). *Protection of Genetic Information: An International Comparison.* London: Report to the Human Genetics Commission.

Decoding Estonia. (2010). Retrieved from http://www.geenivaramu.ee/index.php?id=237. Accessed 7 March 2010.

Ethical Guidelines for Biomedical Research on Human Subjects. (2000). Delhi.

Failure to define law on privacy could cost society dear. (2001). *Times of India,* August 26, 2001.

Hirtzlin, I. (2003). An empirical survey on biobanking of human genetic material and data in six EU countries. *European Journal of Human Genetics, 11*, 475–488. doi:10.1038/sj.ejhg.5201007

HUGO Ethics Committee. (2002). *Statement on Benefit Sharing.* Retrieved from http://www.gene.ucl.ac.uk/hugo/benefit.html. Accessed 23 October 2002.

Indian Genome Variation Consortium, Brahmachari, Samir K. et al. (2005). The Indian Genome Variation database (IGVdb): a project overview. *Human Genetics, 118*, 1–11. doi:10.1007/s00439-005-0009-9

International Disease Classification (IDC). (2010). Retrieved from http://apps.who.int/classifications/apps/icd/icd10online/. Accessed 15 March 2010.

Kennedy, S. (2004). Women health biobank in India. *The Indian Journal of Medical Research, 120*, 131–132.

Kriari-Catranis, I. (2003). Genetic data and confidentiality: the Estonian experiment. *Law and the Human Genome Review, 19*, 147–157.

Kruuv, K., & Nomper, A. (2002). The Estonian Genome Project. Presented at the International Workshop "Law in Genetic Era" at the Central European University, Budapest, 7-9 June.

Mattick, J. S. (2003). The human genome and the future of medicine. *The Medical Journal of Australia, 179*, 212–216.

Metspalu, A. (2002). Biotechnology as an instrument of politics: the example of Estonia. In *Sinclair House Debates: Who Owns the Human Genome?* Bad Homburg v. d. Höhe: Herbert Quandt Foundation, The Foundation of Altana AG.

Metspalu, A. (2003). Workshop on "Biobanks for Health" in Oslo from January 28 - 31, 2003.

National Health and Medical Research Council (NHMRC). (1999). *Guidelines for Genetic Registers and Associated Genetic Material.* Issued by the National Health and Medical Research Council in accordance with the National Health and Medical Research Act, 1992 (Cth). Retrieved from http://www.nhmrc.gov.au/_files_nhmrc/file/publications/synopses/e14.pdf.

Norwegian Biobank Act. (2003). *Ministry of Health*, Law 2003-02-21, Oslo.

Redecker, N. v. (2003). The Estonian gene bank la: an example for Germany? In German-Estonian Legal Questions. Frankfurt: N. v. Redecker.

The Human Genetics Society of Australasia (HGSA). (1990). *Guidelines for Human DNA Banking*. July, 1990.

Waters, M. D., Selkirk, J. K., & Olden, K. (2003). The impact of new technologies on human population studies. *Mutation Research, 544*, 349–360. doi:10.1016/j.mrrev.2003.06.022

ENDNOTES

1 Using biobanks in an unforeseen way will not be discussed here (e.g., for forensic purposes if this is not the goal **of** the bank itself).

2 You can find this distinction in the new Norwegian Biobank Act. Ministry of Health,. The Biobank Act. (Biobankloven), Law 2003-02-21 no 12, Oslo 2003a.

3 New possibilities will come; methods will change (Waters, Selkirk and Olden, 2003).

4 Certain quality and security requirements must be followed, for example: the Australian Standard AS 4400-1995, Personal Privacy Protection in Health Care Information Systems.

5 The following Labs and institutions were involved: Institute of Genomics and Integrative Biology (IGIB), Delhi, Centre for Cellular and Molecular Biology (CCMB), Hyderabad, Indian Institute of Chemical Biology (IICB), Kolkata, Central Drug Research Institute (CDRI), Lucknow, Industrial Toxicological Research Centre (ITRC), Lucknow and Institute of Microbial technology (IMTECH), Chandigarh, Indian Statistical Institute (ISI), the Anthropological Survey of India[1], the Centre for Genomic Application (TCGA), Department of Science and Technology (DST), CSIR with the Chatterjee Group (TCG) SilicoGene Informatics Private Limited along with Lab-Vantage, India.

6 The biobank should also act as a national focus for training in the fields of molecular genetics, biostatistics and epidemiology.

7 If it will come into force, it might be very important for future research that the weight will be changed. The comparison of the actual sentences concerning the function of the council is as follows: "MCI may with the previous sanction of the central government make regulations to carry out the purposes of the Act" and the new sentences "Every regulation made shall be subject to Parliamentary oversight... central government may direct the MCI to make/ amend/ revoke regulations and if it fails to comply, the government may do so itself." This will change the spirit and empower the Council.

8 The law changes in some countries; for example not only the man can be registered as the owner of a house, but also his wife nowadays. You can see also a progress in the economic area. A woman now can open a bank account without the family`s or husband`s permission and on this basis she can get loans.

9 Viewing of pornography is a criminal act.

10 See Failure to define law on privacy could cost society dear. August 26, 2001. *Times of India.*

11 Esp. if the costumers view pornos in internet cafés or misused the internet for other criminal activities.

12 The National Task Force on IT and Software.

13 It show the Indian tradition of confidentiality.

14 The Human Gene Research Act of December 13, 2000 came into power as of August 1, 2001.

15 The Estonian name for the genome foundation is: Eesti Geenivaramu (EGV), http://www.geenivaramu.ee/index.php, as of 4/23/03.

16 The Ethics Committee of the Human Genome Organization recommends stating in the relevant contracts with drug companies a clause with a 1-3% profit sharing; see HUGO (2002).

17 Changed on 14.02.2007 (RT I 2007, 22, 111) comes enforce 1.04.2007.

18 RTI 2007, 22,111 – enforced since 01.04.2007.

19 Further indications as to the Estonian system of health security in Kruuv and Nomper, 2002, p. 7.

[20] Kriari-Catranis emphasized these rights in her article (2003), with a cross reference to the decision in favor of Finland by the European Court of Human Rights, http://dhcour.coe.fr/eng/z.html, Case Z. of Finland (9/1996/627/811), judgement of February 25, 1997, p. 18, where emphasis is placed upon the protection of medical data.

[21] This minority group is mostly unknown.

[22] As "Mixed" are Montenegriner, Jewish and others.

[23] The Gorans are living in the region Gora southwest from the Kosovo (Muslims).

[24] http://www.mlrc.org.mk/law/1004.htm

[25] http://www.mlrc.org.mk/list.htm

[26] http://www.hdnamkd.org.mk/donation.htm

Chapter 14
The Regulation of Genetic Testing and the Protection of Genetic and Medical Information in Singapore

Terry Kaan[1]
National University of Singapore, Singapore

ABSTRACT

In the decades since its independence in 1965, the transformation of Singapore's economy and its transition to a relatively developed economy has also in like manner transformed its health care system, and of the demands made of it. The emergence and availability of new medical technologies has put into sharp focus many novel legal, ethical as well as social issues. This chapter looks at how Singapore has attempted to respond to issues thrown up by genetic testing and screening technologies. A particular focus of this chapter will be the tension between privacy concerns, and the imperatives of access for biomedical research, given that biomedical research has been championed by the Singapore government as one of the future leading sectors of the economy of Singapore. This chapter also examines Singapore's approach to the question of "genetic exceptionalism:" Does genetic information possess special qualities or attributes that remove it from the realm of ordinary personal information, and which thereby demands special treatment and protection? In this context, the impact of the doctrine of genetic exceptionalism on industry (in this case the insurance industry) is examined.

THE BACKGROUND

On November 25, 2005, the Singapore Bioethics Advisory Committee released a Report entitled "Genetic Testing and Genetic Research" (The Singapore Bioethics Advisory Committee, 2005).[2] In the Report (the "Genetic Testing Report"), the BAC laid down ethical guidelines for an array of genetic testing procedures and for genetic research. A year and a half later, the BAC released a further Report entitled "Personal Information in Biomedical Research" (The Bioethics Advisory Committee, 2007) ("the Personal Information Report") which *inter alia* took up and expanded on some of the themes first examined in the Genetic Testing Report. Both the Genetic Testing

DOI: 10.4018/978-1-61692-883-4.ch014

Report and the Personal Information Report was subsequently accepted in full by the Singapore Government, and (through the mechanism which will be explained below) effectively assumed the status of binding professional guidance for the medical profession as regards genetic testing and screening procedures, and on matters on the use of genetic testing and genetic information used in research.

Some contextual background may be useful at this point for a proper appreciation of the Genetic Testing Report and the Personal Information Report, and on their impact and implications[3]. Towards the end of 2000, the Singapore Cabinet decided to establish a national-level agency to establish guidelines, and to make recommendations for supporting legislation and regulatory structures for the development of biomedical research in Singapore.

To this end, the Singapore Bioethics Advisory Committee was appointed by the Cabinet. Applying an approach commonly employed by the government in Singapore, its constitution was and continues to be that of a committee of experts[4] (the majority of which were drawn from the medical, legal and biomedical research sectors) without any separate official or legal identity of its own, reporting and making its recommendations to a committee of ministers charged with steering the Government's plans for the development of the biomedical industry and research sector in Singapore[5]. One of the impetus for the establishment of the BAC was no doubt the launch by the Singapore Government of the Singapore Biomedical Sciences Initiative (BMS Initiative) in June 2000; the vision was for Singapore to be the "Biopolis of Asia," with the BMS Initiative targeted at the development of the biomedical industry and research sector "as the fourth pillar of Singapore's industry cluster, alongside electronics, chemicals and engineering."[6]

That the BMS Initiative has had a profound impact on the reshaping of the present and future of the Singapore economy is not in dispute. For 2007, the biomedical sector's contribution to the manufacturing sector amounted to S$24 billion (US$17 billion as at January 2010 exchange rates), and accounted for 6% of Singapore's GDP (Agency for Science, Technology and Research, 2008). Currently, by far the great bulk of the contribution of the biomedical sector comes from pharmaceutical manufacturing, as opposed to biomedical research (including biomedical research services). The Singapore Government, however, recognizes that research is a fundamental support pillar of the biomedical sector, and to this end has poured considerable resources and funds into developing the biomedical research services sector. The most visible result of this push towards the development of biomedical research in Singapore has been the construction of the Biopolis complex at the new new one-north [sic] development situated in close proximity and designed as a complement to the two existing Science Parks, the Fusionopolis complex, the National University of Singapore and the National University Hospital.

From a regulatory and legal point of view, however, the breathtaking pace of these economic and physical infrastructural developments has not been matched by developments in the formal law and the regulatory structure. Perhaps surprisingly, Singapore did not until the establishment of the BAC in 2000 have a national-level body aimed specifically at the development and coordination of the ethical governance of biomedical research. Until the establishment of the BAC, that responsibility was in part addressed by the National Medical Ethics Committee (the NMEC).

The NMEC was established in 1994 by the Ministry of Health to identify and advise the Ministry on, among other things, "...the prevailing ethical issues relating to public health, medical practice and research in Singapore [and] ... on the potential ethical issues which may occur in Singapore based on the trends in other developed countries" (National Medical Ethics Committee, Singapore, 1998). Although most of the work of the NMEC dealt with matters of medical practice, it did have occasion to advocate law reform in areas that had considerable repercussions in the social and public spheres.[7]

The main focus of the NMEC, however, was on ethical issues arising in the context of medical practice in Singapore, and not issues concerning biomedical research. In the context of the practice biomedical research in Singapore, directions issued by the Ministry of Health on the advice of the NMEC to clinics, hospitals and medical practitioners made for the effective but indirect regulation of biomedical research in Singapore– but only because most biomedical research was carried out in hospitals (which are subject to the regulatory jurisdiction of the Ministry of Health),[8] or involve the recruitment of subjects or access to information held by such hospitals, or by medical practitioners (who are also subject to the regulatory jurisdiction of the Ministry of Health).[9] But this mechanism of indirect regulation by the Ministry of Health through its regulatory levers on the hospitals and the medical profession left open the question of how *future* biomedical research activities that did not involve hospitals or medical practitioners were to be regulated–n obvious development if Singapore were to develop itself as a biomedical research and development hub.

From a different perspective, there was also the difficulty that the interests of a body charged with the development of professional ethics within the context of medical practice did not necessarily coincide entirely with that of a body tasked with the development of guidelines for the ethical governance of research. Fundamentally, the relationship of physician and patient is not the same as that of the relationship of research and subject. Put in its bluntest form, there is a fundamental conflict involved in asking a patient to participate as a human subject in a research trial involving physical interference with the patient or with his treatment regime: except in very few cases, [10] there is arguably always *some* degree of risk of harm[11] involved in biomedical research trials involving human subjects.

The establishment of the BAC at the end of 2000 resulted in a neater separation of functions between the two bodies. The NMEC was freed to concentrate on matters falling within the sphere of the physician-patient relationship and professional practice; and the BAC on the development of a professional and regulatory framework for the ethical governance of biomedical research in Singapore.

Nonetheless, some of the basic framework for the ethical governance of biomedical research in Singapore had been put in place by the NMEC a few years before the establishment of the BAC. In September 1997, the NMEC had issued the first set of national guidelines for the ethical governance of biomedical research, entitled *Ethical Guidelines on Research Involving Human Subjects.*[12] As the NMEC explained, the objectives of the guidelines were specifically "...to ensure that the rights and the welfare of research subjects were protected. They were prepared to assist the ethics committees of the various hospitals in assessing the ethical acceptability of research proposals. ...The Guidelines were accepted by the Ministry and sent to all the hospital ethics committees and the National Medical Research Council for their reference" (NMEC, 1998, pages 8-9).

With the implementation of this Ministry of Health directive, the basic framework for the ethical governance of biomedical research was established. Each hospital now had an ethics committee or institutional review board (IRB) distinct from the Medical Board (which dealt with issues of professional and operational policy within the hospital) or the hospital ethics committee (which dealt with ethical issues arising from medical practice, such as whether new medical devices or forms of surgery or therapy should be permitted) which concentrated entirely on the ethical review of proposed research protocols[13].

Before any biomedical research of any kind could be carried in a hospital, or by researchers employed by or affiliated to the hospital, or carried on or involving patients of the hospital or their medical records or on the tissue collections held by the hospital, a research protocol proposal had to be drafted for review by the hospital research ethics committee or its IRB. The sole exception were clinical trials (pharmaceutical trials) involving the testing of drugs for their safety, efficacy, appropriate dosage and other effects: these were regulated on a separate track by the much older statutory regime laid down under the Medicines (Clinical Trials) Regulations 1978, [14] which applied a set of procedures designed to conform to international norms as represented by the ICH-GCP Guidelines.[15]

Several challenges of particular urgency confronted the BAC from the moment of its establishment. The first was that biomedical research other than clinical trials had far overtaken the traditional clinical trials in both volume and importance. As observed by the BAC, by 2002, "hospital ethics committees of the five main restructured hospitals reviewed nearly three times as many applications for such research as they did for pharmaceutical trials" (The Singapore Bioethics Advisory Committee, 2004).[16] The second challenge was that, unlike for clinical trials, there was no statutory scheme in place for the regulation of biomedical research other than clinical trials.

The third challenge was related to the second. Unlike for clinical trials (for which international consensus had been practically achieved in the form of the ICH-GCP documents), there were not any convenient international consensus on the body of rules to be applied to biomedical research apart from clinical trials. This situation was not surprising, given the bewildering range and diversity of research falling within the definition of biomedical research. Internationally, the situation as far as harmonization of national ethical codes was not much better: indeed, it should be remembered that it was not until 1993 did the CIOMS and the WHO bring to publication the first edition of the CIOMS International Ethical Guidelines for Biomedical Research Involving Human Subjects (CIOMS, 2002). Prior to the CIOMS document, there were essentially only the World Medical Association's Declaration Helsinki (World Medical Association, 2008) and the Nuremberg Code (1949) to go on.

THE BAC REPORTS

To date, the BAC has issued a total of six reports.[17] These examined and made recommendations for legislation and laid down practice guidelines for research involving human cloning and human embryonic stem cell research ("the Cloning and Stem Cell Report") (Bioethics Advisory Committee, Singapore, 2002a); on human tissue research ("the Human Tissue Report") (BAC, 2002b); on research involving human subjects, and on the kind of institutional infrastructure required for the ethical governance of research within hospitals and other research institutions ("the Human Research Report") (BAC, 2004); on genetic testing and research ("the Genetic Testing Report") (BAC, 2005); on the use of medical records and personal information in biomedical research ("the Personal Information Report") (BAC, 2007); and finally on the use of human eggs donated for biomedical research (BAC, 2008). Of these, the discussion in this chapter will concentrate on the guidelines laid down by the BAC in the Genetic Testing Report and the Personal Information Report.

One of the first tasks of the BAC was in effect to define its own terms of reference: it was charged with formulating and recommending policy to the Government in relation to biomedical research, but *what* was biomedical research? At the very least, it had to be clearly distinguished from clinical trials, which fell outside of the ambit of the BAC's terms of reference, if only because there was already in existence a clear statutory regime for the control of clinical trials. But unlike as with the ICH-GCP for clinical trials, there were no universally accepted international body of standards which could be readily and directly adopted. This new kind of biomedical research was of a much more amorphous and heterogeneous nature than clinical trials. Ultimately it came to be defined both in Singapore and elsewhere in negative terms: effectively, it was any biomedical research *not* covered by current rules governing clinical trials. So it was thus defined by Singapore's national Bioethics Advisory Committee as:

...human biomedical research that involves an interaction (whether direct or otherwise) with a human subject or human biological material, and therefore exclude any human biomedical research in relation to: (a) Genetically modified organisms, (b) Animals and their treatment; and (c) Economic, sociological and other studies in the disciplines of the humanities and social sciences ... (BAC, 2004, paragraph 3.2).

Inherent in this definition is the careful separation and teasing out of biomedical research as a distinct activity from the broader category medical therapy. The BAC recognized the importance of this distinction, and adopted an earlier guideline of the NMEC which distinguished between the two in the following terms:

Human research can be broadly defined as studies which generate data about human subjects which go beyond what is needed for the individual's well-being. The primary purpose of research activity is the generation of new information or the testing of a hypothesis. The fact that some benefit may result from the activity does not alter its status as "research."[18]

This careful distinction captures well some of the inherent tensions and potential conflicts that may occur when a biomedical researcher is a physician, and his subjects or potential subjects are the patients that he is treating as a physician. This tension is well demonstrated by the controversy over the changing texts of the World Medical Association's Declaration of Helsinki in its recent reiterations; there is increasing recognition that there are inherent conflicts of interests between that of a physician acting as a physician in relation to his or her patients (the physician-patient relationship), and that of the physician-researcher acting as a researcher-investigator in relation to his or her research subjects, who may be also the physician-researcher's patients (the researcher-subject relationship).[19]

This is the inherent and inescapable tension that occurs whenever a physician takes on two hats simultaneously, namely that of the physician to his or her patient, and that as a researcher with the same patient as his or her research subject. This duality of function of most biomedical researchers currently carrying out research in Singapore is a recurrent theme that emerges in all the BAC Reports, and is an important key to the understanding and interpretation of the Reports.

THE GENETIC TESTING REPORT AND THE PERSONAL INFORMATION REPORT

On November 25, 2005, the BAC released the fourth of its Reports, the Genetic Testing Report (Bioethics Advisory Committee, Singapore, 2005). Unlike any of the other BAC Reports, the Genetic Testing Report deals with ethical issues that straddled the boundaries between research and therapy. Indeed, when viewed as a whole, it might be argued that the guidelines and recommendations set out in the Genetic Testing Report deals overwhelming with therapy and medical practice than it does with biomedical research.[20]

The Report deals with ethical considerations arising from a diverse assemblage of procedures which had genetic testing as their common ground. In the Report, the BAC set out to consider the ethical implications and the social impact of genetic testing and screening technologies--defining the difference between a genetic test and otherwise ordinary medical diagnostic procedure that might yield genetic information directly or indirectly as well, laying out general principles in relation to genetic testing and screening for heritable conditions, considering the use of genetic testing technologies in assisted reproductive technologies, and deliberating whether genetic tests ought be allowed to be supplied directly to the public (for example, mail-order genetic tests) that cut out a medical intermediary.

Definitions and Genetic Exceptionalism

The unspoken link between all the diversity of themes in the Report was the question of genetic exceptionalism, i.e., the controversy over whether or not genetic information is so qualitatively different from ordinary medical information that it deserves or requires special protection and treatment (Green and Botkin, 2003). The answer to this question would determine whether or not the Genetic Testing Report itself was required. If genetic exceptionalism as a doctrine was rejected, then it would follow that genetic testing procedure and genetic information would fall to be bound by the same rules as any other kind of clinical procedure or medical information, and it would not be necessary to have a report dealing with special rules for genetic testing and genetic information.

To this first question, there was a corollary: If it were to be accepted that special protection and treatment ought to be accorded to genetic information, how was genetic information to be defined? The BAC sought to tackle this question first if only because it was necessary to define the very scope of the Report. If the question of genetic exceptionalism was to be tackled, then it was necessary first to define what might have to be given special status apart from ordinary medical information. But this second question is not as straightforward as it may appear at first glance. For example, should genetic information relate only to heritable mutations or germ-line mutations? If the answer is yes, then from this perspective, a test designed to confirm non-heritable somatic mutations giving rise to conditions such as Down Syndrome (trisomy 21) would not result in genetic information, even though the cause of the condition is manifestly genetic in the sense that it is an error of chromosomal replication.

On the other hand, if genetic information is to be defined in terms of information about heritable mutations of the human genome, the definition may become too vague or unwieldy because it would capture too much. For example, it is not necessary to carry out any kind of medical or clinical test (let alone a genetic test of any description) to elicit information about heritable conditions. For example, taking a family history is one of the most basic procedures known to physicians, and that family histories (if sufficiently complete and accurate) can provide powerful predictive information about the genetic inheritance of patients.

The difficult line between these two extremes was drawn by the BAC in the following terms. First, genetic testing:

Genetic testing is the analysis of human DNA, RNA, genes and/or chromosomes, or the analysis of human proteins or certain metabolites, with the primary purpose of detecting a heritable genotype, mutation, phenotype or karyotype (BAC, 2005, paragraph 2.1).

Then the BAC went on to restrict the definition and scope of genetic information by reference to its definition of genetic testing:

Genetic information broadly refers to any information about the genetic makeup of an individual. It can be derived from genetic testing as defined in paragraph 2.1 in either clinical or research settings or from any other sources, including details of an individual's family history of genetic diseases. This report is concerned with information about heritable conditions obtained by genetic tests whether it is for clinical purposes or for research (BAC, 2005, paragraph 3.1)." [emphasis added]

The difficulty is of course that many kinds of routine clinical investigations will throw up medical information which clearly has a genetic dimension. For example, a purely clinical diagnosis confirmed through pathological examination of tumor samples exhibiting the same kind of breast cancer in several closely-related women from the same family clearly has genetic implications, and would be properly regarded as qualifying as genetic information. But this kind of information would be excluded by the BAC definition. On the other hand, even simple clinical tests such as blood typing may throw up uncomfortable genetic probabilities (or improbabilities, as the case may be) such as the paternity of a child. In such cases, it is not clear such information which clearly pertain to the genetic inheritance (or non-inheritance or non-relatedness) of one individual to another would fall within the BAC definition.

The BAC then went on to consider the question of genetic exceptionalism. This doctrine is strongly opposed by some biomedical researchers (and the life insurance industry, as discussed below), who argue strenuously against it, both on point of principle and on grounds of pragmatism.

On the point of principle, the arguments against genetic exceptionalism are essentially that there is no logical or substantive basis to distinguish genetic information derived from genetic tests from any other kind of medical information pertaining to the individual; that many kinds of genetic information are patently obvious to any casual observer and cannot be hidden (e.g., eye or skin color, or any identifiably distinct ethnic or population-based characteristic) and therefore cannot and should not attract special privacy protection; that genetic information of predictive value has been collected for many years in the form of family histories without opposition for legitimate and fair purposes such as assessing the fair actuarial basis for the premium to be paid on an insurance policy; and finally that medical information (and other kinds of personal information) are already strongly protected by law in most developed countries.

On the grounds of pragmatism, there are fears that rules even more stringent that those applied to ordinary medical information will stifle participation by otherwise willing volunteers in biomedical research trials, and discourage biomedical researchers from engaging in much-needed research because of the high compliance administrative overheads imposed by laws based on genetic exceptionalism.

The BAC took the path of moderation, rejecting the strict legal exceptionalism institutionalized by legislation in some countries, notably the United States and Germany (The Genetic Information Non-Discrimination Act, 2008; HumanGenetic Examination Act of Germany, 2009). It preferred instead the more nuanced approach in the United Kingdom and in Australia, with their emphasis on flexible informal or indirect regulation rather than direct regulatory intervention in the form of substantive legislation (BAC, 2005, pages 20-22). The BAC agreed with the observations of the UK Human Genetics Commission in holding that there were undeniably some qualities which set genetic information apart from ordinary medical information.

In particular, the BAC noted that much of the sensitivity attaching to genetic information related to the *potential* or *perceived*[21] predictive power; that such predictive power is related not only to a given individual's future health, but also potentially to the individual's relations, children and other descendants; and that such information could be gathered from very small samples such as hair samples or cheek swabs. But the BAC concluded, in the very first guideline that it laid down in the Genetic Testing Report, that:

Genetic information derived from clinical genetic testing should be regarded as medical information and the usual standards in medical ethics apply in its derivation, management and use (BAC, 2005, page 22).

At casual glance, this would appear to be a very decisive nail in the coffin of genetic exceptionalism. But this first recommendation needs to be read in context. What the BAC was in effect saying was that, *at minimum, all* kinds of genetic information covered by the Report attracted the same kind of legal and ethical protection and obligations attaching to medical information. In other words, genetic information was to be protected as a species of medical information, if nothing else. It is in the recommendations following this seemingly surprising first statement that the intent of the BAC becomes clear: If genetic exceptionalism were to be completely rejected out of hand, then there would be little need for the recommendations which followed. In effect, the rest of the Genetic Testing Report deals with the special kind of consideration and protection which the BAC felt that genetic information merited over and above ordinary medical information.

One of the reasons advanced by the BAC for not adopting the strict formal exceptionalism as enshrined in the legal systems of some countries was its view that genetic information in itself did not give rise to any special quality that entitled it to special consideration or protection (again, over and above that already accorded to ordinary medical information) as a category. Instead, the BAC agreed with the Australian Law Reform Commission and the Australian National Health and Medical Research Council in noting that a "commensurate level of protection may be required where there is a likelihood of special threat to privacy or discrimination," a position echoed by the statements from the Council of Europe, and the Bioethics Committee of Japan (BAC, 2005, paragraph 3.5).

The idea, then, was that instead of applying a blanket regime with a protective threshold high enough to cover all sorts of genetic information, the level of special protection (*if* any was required) should be commensurate with the vulnerability of that information to abuse, and with the seriousness of the consequences of such abuse. In the BAC's words, the stringency of the special protection should be "proportional": this was a theme that it would eventually expand upon in the later Personal Information Report. We will take this up later on in this chapter, but first let us consider the application of the earlier issue of genetic exceptionalism in the context of an application in real life.

GENETIC EXCEPTIONALISM AND LIFE INSURANCE

It has long been accepted practice in Singapore, as elsewhere, for life insurers to require prospective customers to give their consent to the insurance companies contacting their medical providers or physicians directly for information about the prospective customers' state of health, and/or require their prospective customers to undergo a medical examination and medical tests carried out by physicians appointed by the insurance companies. In this way, insurers obtain medical information about their prospective customers. Should insurance companies by extension have the right to whatever genetic information (in the BAC sense of information derived from genetic tests, and not from family histories) that an individual may already possess, and should insurance companies have the right to require genetic testing of prospective customers?

One difficult point in this context is of course that from a national health policy perspective, genetic testing and screening is of benefit to both individuals and the state in at least some situations. For instance, screening for thalassemia in populations such as Singapore's with a relatively high incidence of such inherited conditions (so that informed decisions can be made as to whether or not to have a child), or for genes strongly linked to certain forms of breast cancer if there is a strong trend towards breast cancer in the women in a particular family (so that those confirmed to be carrying such genes can take the precaution of having themselves examined and tested more frequently). So there is a state interest in promoting such tests, both to prevent tragedy at the individual level, and the burden on public health care costs at the state level.

The problem is that the genetic information in the second situation would be the kind of information which life insurance companies in Singapore regard as vital information about the health of their prospective customers which their prospective customers ought to disclose. The life insurance companies, however, concede that they should not want to mandate genetic tests for any prospective customer as a condition of the acceptance of a proposal; the line is simply drawn between genetic information which the customer may already have, as opposed to genetic information which the customer does not at that point in time have, and the life insurance industry in Singapore urged the BAC not to bar access to genetic information already available to the prospective customer at the time of the proposal. [22]

The conflict put the government and individuals on one side (although, as will be explained, their interests do not necessarily coincide), and the insurance industry on the other. Understandably, the insurance industry would prefer to regard genetic information (to be fair to the industry, they have made clear that they are interested only in genetic information of proven, and not speculative, predictive value) as any other kind of medical information on which they may make an actuarial assessment of the risks involved in underwriting a policy. There is a strong argument in their favor of not allowing individuals to conceal relevant and reliable information about the nature of the risk to individuals–this may in at least some circumstances amount to fraud on the insurance companies involved.

There is also the argument that the insurance industry already require the disclosure (without any consumer opposition) information of a clearly genetic nature–for example, an individual's family history, particularly in relation to conditions such as cancers, heart disease, high blood pressure and diabetes. The argument here of course is that genetic information is already being asked for, and given, and that genetic information obtained from a genetic test is not substantially different from that derived from a family history because both kinds of genetic information are merely predictive of *possibilities* or *probabilities* in actuarial terms, and do not represent genetic *certainties*.[23]

From a public policy point of view, however, the more precise the information that insurance companies can extract from their customers, the better position insurance companies are in to cherry-pick their customers, with the effect of making insurance prohibitively expensive for those considered by the insurance industry to be at a particular risk because of their genes. Or, in the worst case, to be denied insurance altogether[24]. This runs against the principle of the pooling of risks, which governments and consumers arguably see to be main reason for having insurance in the first place.

From the consumers' perspective, there is also the added unfairness and danger inherent in the asymmetry of information; a given consumer will not be told of why his or her proposal has been rejected, or why the premiums asked for appear to be unduly loaded, or of the actuarial or statistical bases justifying such a loading. A consumer would have no means of knowing whether an insurance company may be assigning a speculative value (or a value unsupported by empirical evidence or actuarial experience) to a disclosed genetic trait.

From the public health care perspective too, giving the insurance industry free rein to demand genetic information derived from tests would have a severe chilling effect on voluntary preventive or prophylactic genetic testing by individuals at risk, on population screening programs, and on participation by volunteers in biomedical research trials. Why find out whether you may be at risk for X, Y or Z condition if such information might result in your being refused insurance, or charged a premium? Many would also point out that it would be counter-productive for the insurance industry to demand such information because it would simply result in those people who have private suspicions that they might be genetically at risk would simply defer testing themselves until *after* they had taken out the necessary insurance policy.

There is also the question of whether genetic test results form part of the disclosable family history of a potential customer: If X applies for an insurance policy, and has not undergone any kind of genetic testing, does X have to disclose that he or she is aware that siblings Y and Z have undergone genetic testing for a strongly familial and deadly kind of cancer, and that the results of those tests were positive?

Access by insurers to genetic information has been seen as an invasion of privacy by consumers in many developed jurisdiction. In the United States, for example, one response (if a somewhat belated and flawed one) was passage of the Genetic Information Nondiscrimination Act of 2008[25] by the US Congress, which prohibits insurers from requiring prospective clients to undergo genetic tests or from using genetic information to discriminate among policyholders.[26]

On the issue of whether any one should be compelled to undergo genetic testing as a condition of acceptance by an insurance company, the BAC came to the conclusion that "no one should be compelled to undergo genetic testing in order to obtain insurance coverage" (BAC, 2007, paragraph 6.11). This was perhaps the easier of the two decisions to make, since the life insurance industry in Singapore was not in any event asserting such a right. But what of the right that they *were* asserting, i.e., the right to genetic test results that an individual might already have? On this second point, the BAC chose[27] to defer the decision by adopting the position taken by the UK Government by recommending a moratorium on the use of genetic information by insurance companies until November 2011.

Under the terms of the Concordat and Moratorium agreed between the UK Government and the Association of British Insurers (HM Government and the Association of British Insurers, 2005), prospective clients may not be asked to submit to testing as a condition for acceptance of an insurance proposal, nor may insurance companies ask for disclosure of genetic test results unless the policy is for an amount in excess of £500,000 for life insurance, or £300,000 for critical illness insurance[28], *and* then only for predictive genetic tests for conditions approved by the Genetic and Insurance Committee (GAIC).[29, 30]

Critically, the Concordat and Moratorium makes clear that genetic information derived from participation in a research program (as opposed genetic information derived from a clinical genetic test) is not something that insurance companies may ask for under any circumstances; a prospective customer is not to "be asked to disclose any predictive or diagnostic genetic test results acquired as part of clinical research" (HM Government, 2005, Clause 16(iii)). This provision is important from a public health perspective, and perhaps also from the perspective of a country eager to promote and develop its biomedical research sector, if ordinary individuals are to be encouraged to participate in biomedical research programs.

The absence of such a provision might occasion a pause for thought for many people who may otherwise be willing to participate in biomedical research programs, because of the risk of the research results exposing a genetic condition which insurers may deem to be relevant to actuarial risks. Secondarily, there is also the concern that genetic testing carried out in the context of a research program may not be to standards or protocols appropriate to a clinical genetic test (which may, for example, require further or more sensitive confirmatory tests to eliminate false positives or negatives before a definitive diagnosis is offered).

The Concordat and Moratorium also makes clear that so-called genetic "relational information" (the best example being family history: information not about oneself, but about one's family members, and often information that is not obtained first hand but through other persons: e.g. information that the informant's grandfather X died of colon cancer at the age of 35 before the informant was even born) should not be asked for by insurance companies (HM Government, 2005, Clause 16(ii)).

The limitation by the GIAC of "relevant" genetic conditions to Huntington's disease also points to a different consideration. Apart from a small number of conditions such as Huntington's Disease which is a monogenic condition of full penetrance, the effects of the vast majority of known or suspected genetic risks are not well-known; we know that certain genes may predispose a person carrying that gene for a heritable disease, but in most cases, there is not the certainty as is tragically the case in Huntington's Disease (BAC, 2005, paragraph 3.6).

A good analogy might be that we know that it is well-established that heavy smoking and allowing oneself to get morbidly obese will increase one's chances of dying earlier of cancer or from heart disease, but there is no guarantee of an early demise through such factors even for the determinedly suicidal. At least in the case of heavy smoking and obesity, life insurance companies have assembled sufficiently large databases for actuarial tables, but there is much less information available for actuaries to work on when it comes to the effect of genes. There is a further level of uncertainty; even if a life insurance company has in place an actuarial estimate of the risks of some common genetic conditions, these actuarial assessments are not available to the public, nor are the basis of the actuarial calculations. These points further buttress the current argument that any actuarial weight given by life insurance companies to genetic conditions, except for monogenic conditions of full penetrance, fails the test of transparency, and is intrinsically unfair to the consumer.

Ultimately, however, it should be noted that in both Singapore and the United Kingdom, unlike the United States with its Genetic Information and Non-Discrimination Act, the recommendations in place merely postpone the substantive decision that has eventually to be made. The final recommendation by the BAC in this regard merely recommends a moratorium without specifying any particular time limit, to "allow time for both the insurance industry and the government to look into the substantive issues" (BAC, 2007, paragraph 6.14). For now, at least, given the representations of the life insurance industry to the BAC, it does not appear that the industry will be pressing the issue of access to genetic information. But with genetic tests (particularly those for "genetic markers") becoming more and more commonplace and forming part of ordinary clinical investigations, there may be pressure in the future for a clear and substantive resolution of this issue.

PROPORTIONALITY

In its Personal Information Report, the BAC enlarged on the concept of proportionality that it first hinted at in its Genetic Testing Report. It urged researchers and IRBs to "be mindful of possible public sensitivity towards certain kinds of research" in the taking of consent for participation in biomedical research trials, or to the use of personal (and thereby presumably also genetic) information for the same, noting that the UK Nuffield Council on Bioethics had identified "certain types of genetic research that may be of public concern, such as those relating to personality, behavioral characteristics, sexual orientation or intelligence" (BAC, 2007, paragraph 5.11). Here then was a recommendation that stricter measures (than for "ordinary" medical information) for the protection of privacy and for the taking of consent for the use of the information ought be applied depending on the *perceived* sensitivity of the information involved.

The BAC was careful to make a distinction between "actual and perceived risk of harm to the individual concerned (BAC, 2007, paragraph 5.13);" there was a recognition that some information might be regarded as being of a very sensitive nature for reasons other than purely scientific or medical ones.[31] It was not the scientific probability of something happening or a condition eventuating that determined whether or not information was sensitive. Instead, respect for the individual demanded that information be treated as sensitive if society generally, or a subset of it, should regard it as being so.

There is inevitably the question of how information which is regarded as being sensitive only by the person to whom the information pertains. In such a case, the general principle of proportionality when applied would appear to require the information to be accorded the level of sensitive sought for it by the person to whom the information pertains, although there are obviously practical and logical limits.[32]

GENETIC TESTING PROCEDURES

In this section we will consider the guidelines laid down by the BAC for specific genetic testing procedures. In these recommendations, we may perhaps observe more clearly the interplay of the doctrines of genetic exceptionalism and of proportionality. Essentially, the question being repeatedly asked in all of the cases boils down to a risk/benefit analysis. But unlike in biomedical research involving human subjects, or in clinical trials, where "risk" is taken in terms of the risk of physical to the subject, "risk" in the context of pure medical information or genetic information generally deals with emotional or social harm to the individual concerned resulting from a breach of his right to privacy and confidentiality of such information.

Most of the BAC's recommendations relating to genetic testing procedures are uncontroversial and are in agreement with those in most developed countries. So the BAC is at pains to insist that all genetic testing should be completely voluntary, and that it should be "conducted in a manner that is respectful of the welfare, safety, religious and cultural traditions and perspectives of individuals."[33] Not only did the BAC take the view that no testing should be done without the appropriate consent, it felt strongly enough to recommend that "the non-consensual or deceitful taking of human tissues for the purpose of genetic testing should be prohibited (BAC, 2005, page24);" and that the government should consider legislation similar to that in England criminalizing such conduct.[34]

Genetic Testing: Children and Minors

But as in many areas of medical law, the main difficulties lie not in the main body but at the interstices and cracks of the law. It is easy enough to say that full and informed consent must be obtained before a genetic test is to be carried out, but what if the person to be tested is an incompetent adult (for example, unconscious, or insane), or a child or a minor? In none of these cases would the person concerned have the necessary legal capacity to give full and informed consent. A particularly difficult situation is the situation where parents seek to have their minor children tested for genetic conditions which the parents may know or fear that they themselves have, and which they fear may have been inherited by their children.

It would certainly be useful in *some* situations, and directly beneficial to the children concerned, for the parents to have clear information about whether or not a child has a genetic inheritance that will give rise to a high probability or certainty that the child will develop a certain hereditary genetic condition, especially if there are available medical interventions and treatment which can be started while the child is still a child or minor to the improvement of the child's health, quality of life or life expectancy. But there is also the situation where nothing can be done for the child even if an inherited disease is diagnosed upon genetic testing. Or it may be an adult-onset genetic condition which will affect the child or minor only later in adult life.

The BAC's response was that it did not support "the broad use of genetic testing on children and adolescents" as a general rule, and that genetic testing "should generally be deferred until the child is mature or when required to make reproductive decisions." This was on the basis of respect for the right of a person *not* to know about his or her genetic makeup: Once told, the person loses that right to remain in deliberate blissful ignorance of any dreadful genetic inheritance. For many, the prospect of the likelihood of an early death through an incurable genetic condition is information that they would rather not have.

The exception which the BAC allowed was the situation where genetic testing "for genetic conditions where preventive intervention or treatment is available and beneficial in childhood." In this situation, where genetic testing would be directly and immediately of benefit to the child or minor, the BAC held that genetic testing *should* be permitted, and indeed encouraged.

To this general principle that genetic testing ought not be carried out in children or minors where no effective intervention was available, the BAC however conceded a loophole: Genetic testing in such situations where testing might not be normally recommended on the principles outlined above might, under certain circumstances, be permitted if "compelling interests of other family members or public health interests exists, and the physician should be able to decide, together with the parents, whether or not determine the carrier status of the child" (BAC, 2005,paragraphs 4.8 – 4.14). Whether this get-out clause will eventually prove to be the exception swallowing the rule remains to be seen, but the need for such a provision may be more understandable when seen in the context of very successful populational genetic screening programs for hereditary conditions common in the local population such as that for thalassemia which have greatly reduce the incidence of the inheritance of such diseases.[35]

One possible situation in which this loophole may be used might be by parents seeking to test their children for what effectively amounts to their own interests, rather than the children. For example, if both parents are carriers of X recessive gene, there is a one-in-four chance that a child born to them will have full-blown X genetic double recessive condition. Assume that such a child born with such a condition is doomed to an early death in childhood, and there is no treatment that can help the child. Would the parents be entitled to have their child tested, with a view to helping them plan for another child (which they would otherwise not have had) to replace the doomed one?

On a different but related point, the BAC is clear that no one should be given genetic about himself or herself against his or her will: There is a "right not to know" (BAC, 2005, paragraphs 4.24 - 4.26). But what the BAC does not make clear is whether a person has the right not to know about genetic results that *may* have a bearing on his or her own genetic inheritance. For example, if siblings W, X and Y have had themselves tested (with positive results) for a gene strongly linked to the development of a specific kind of cancer in later life, do they even have the right to tell their sibling Z of their results, let alone urge Z to have himself tested so that he can take prophylactic measures?

Genetic Selection: Testing the Unborn

The discussion so far has dealt with genetic testing of children and adults. But what of the unborn? One of the issues which the BAC had to grapple with in its Genetic Testing Report was whether a number of new developments in assisted reproductive technologies (ART) ought to be allowed. All of them involved some form or other of genetic testing, not just of the fetus, but of the unimplanted embryo. The new ART technologies, collectively labeled "Preimplantation Genetic Testing" (PGT) by the BAC, encompassed specific procedures such as Preimplantation Genetic Diagnosis (PGD), Preimplantation Genetic Screening (PGS), and Preimplantation Tissue Typing (PTT).

All these technologies involve the same central procedure. Early embryos are created *in vitro* outside the womb through standard IVF procedures using sperm and eggs outside the womb. But unlike in standard IVF procedures, genetic tests are carried out on the early embryos *before* implantation, with a view to choosing which embryos should be implanted. In PGD, the objective is to select an embryo that is free of a deadly genetic flaw that of which one or both of the contributing parents are known carriers. In PGS, the objective is similar, but the screening is carried out for parents who have no known specific genetic condition but who have had difficulties having a child through natural means. In PTT (the "savior sibling" genetic selection procedure), the objective is to select an embryo that is immunogenetically compatible with that of an existing sick sibling, with a view to bringing about a child who can be a source of lifesaving cord blood or bone marrow to the sick sibling.[36]

Fertility is a matter of consuming public interest and national concern in Singapore. One of the unwelcome consequences of economic prosperity in recent decades has been a precipitous decline in the total fertility rate, which fell from 4.66 per female at independence in 1965 to just 1.28 in 2008–far below replacement rate for the population (Singapore Department of Statistics, 2009). The law on abortion in Singapore certainly has had an impact. The Termination of Pregnancy Act (Chapter 324, enacted December 27, 1974), enacted at a time in Singapore's history when it was struggling to control a burgeoning population, permits abortion on demand up till 24 weeks of pregnancy. It has been reported that in 2006, the number of abortions exceeded 12,000 (Perry, 2008) in a year for which only 38,317 live births was reported (Singapore Department of Statistics, 2009, page. 29). Against this background, any help for parents desperately seeking to have a child was to be encouraged. But PGT went beyond the bounds of ordinary ART procedures; the process of selection for implantation through genetic testing raised uncomfortable issues of whether it was ethically acceptable for parents to select for desired characteristics in this way.

Following a review of the position in other countries, the BAC came to the cautious conclusion that PGD and PGS ought to be allowed, but should be strictly "limited to preventing to preventing serious genetic conditions," for example, if there was a high risk of a child being born with a serious or eventually lethal genetic condition because of the parents' carrier status (BAC, 2005, paragraphs 4.27-4.43). It further held that the "use of preimplantation genetic testing for the selection of desired traits or gender for non-medical reasons should not be allowed" (BAC, 2005, paragraphs 4.44-4.46) and selection for "social" considerations such as gender was out of the question. [37] A much tougher ethical nut to crack was the question of PTT: Should genetic testing for the selection of a "savior sibling" be permitted? The BAC gave a cautious yes, but on the condition of strict licensing on a case-by-case basis. (BAC, 2005, paragraphs 4.47-4.50).

One question which might be asked, following on the principles explored in the discussion on the genetic testing of children and minors, is whether the fact of such selection by genetic testing, together with the relevant genetic information about their parents' genetic heritage as well as their own, ought be disclosed to children born of such PGT procedures. In the case of children born as a result of PGD or PGS procedures, there is arguably no objection to children being informed of what they *do not* have, as opposed to what they might have. For example, if both parents are carriers for thalassemia, and have a child selected through PGD to be free of any gene for thalassemia, there seems to be no ethical difficulty in letting the child know, because the disclosure is ultimately about a gene of concern which the parents have, and which the child does not.

The situation is a bit more clouded in the situation where the parents settled for a child (also selected through PGD) who retains one "bad" gene from one of the parents (and hence carrier status, but who in consequence will not suffer full-blown thalassemia major, and will therefore be of normal health). If the child is told in this case, he or she will be effectively given information about his or her own genetic makeup, which the BAC has ruled that the child has a right *not* to know.

Here lies the difficulty in the PTT and the child or minor genetic screening cases: If a deleterious gene is uncovered during the course of the screening, at what stage of the child's life should he or she be told? If he or she has a right *not* to know, then how is the subject to be broached at all? Merely asking someone whether he wishes to have some information about his genetic inheritance (particularly if that offer comes from a parent) alerts that person to the fact that potentially important information is available which could affect his or her health, or his or her progeny.

If the genetic information relates to an adult-onset disease which only manifests itself in later life, then it seems clear according to the BAC guidelines that the child or minor be informed upon the age of majority because such information would be "required to avert serious harm" (BAC, 2005, Recommendation 7). Again, it is the silent carrier status situation that presents the difficulty. Information about the silent carrier status of the child will have no impact on the future health of the child himself or herself. But when the child grows to maturity and has children of his or her own ("the grandchildren"); his or her carrier status may have a direct (and devastating) impact on the health and survival of the grandchildren. Here again, it seems that the avoidance of "serious harm" recommendation would probably apply, but again on the proviso that the information is only offered when the child attains the age of majority.

THE SINGAPORE REGULATORY LANDSCAPE

One obvious question which might be asked is: What kind of regulatory or legal force does the pronouncements of the BAC have in Singapore? The BAC is not a government department, and has no legislative constitution or even a distinct legal identity of its own, beyond that of an advisory committee to the Singapore government. Technically, it reports to and advises only the Singapore government that appointed it. Nor has, apart from a few notable exceptions, [38] the main body of the BAC recommendations and guidelines been enacted into law despite the acceptance of the Reports by the Singapore government.

Despite this, however, the BAC guidelines and recommendations have from the practical and professional perspective the force of mandatory practice rules backed by powerful sanctions through a remarkable indirect regulatory mechanism. The Ministry of Health (or through its delegate, the Director of Medical Services) has regulatory jurisdiction and direction over all hospitals and clinics [39] and all registered physicians [40] in Singapore. By a directive issued on January 18, 2006, by the Director of Medical Services addressed to "all registered medical practitioners," it was announced that the "Singapore Medical Council would, in evaluating appropriateness of a doctor's actions in research involving human subjects, rely on the BAC's recommendations as the standard for ethical conduct." [41] Under the Medical Registration Act, the Singapore Medical Council is the statutory body charged with professional discipline of registered physicians in Singapore.[42]

Under this indirect regulatory scheme, a registered physician who breaches any of the guidelines laid down by the BAC would be liable to disciplinary action by the Singapore Medical Council. The catch, of course, is that the Singapore Medical Council's jurisdiction extends only to registered physicians. Although most if not all biomedical research in Singapore is currently carried out by registered physicians, or in hospitals subject to the regulatory oversight of the Ministry of Health under the Private Hospitals and Medical Clinics Act, or involve the recruitment of patients from such hospitals or information from such hospitals or their patients, there will eventually come a time when biomedical research without such links will emerge in Singapore.

Without direct legislation governing biomedical research (as currently exists for clinical trials), it would be possible (theoretically, at least) for a biomedical researcher who is not a registered physician to carry out biomedical research outside of the regulatory jurisdiction of the Ministry of Health, provided that the research did not involve any collaboration with a registered physician or with any hospital or clinic subject to the jurisdiction of the Ministry of Health.

In practice, such deliberate attempts to evade the jurisdiction of the Ministry of Health in order to carry out research which might not be approved under the BAC guidelines is highly unlikely, if only for the fact that such researchers might face considerable difficulties in obtaining the all-important peer and industry acceptance of the results of their research. Indeed, the concern of researchers in precisely such a situation is likely to be the very opposite, that they should not be left in a legal vacuum, and have the comfort of clear law backing their efforts.

In 2003, the Ministry of Health did float a draft Regulation of Biomedical Research Act 2003, [43] but this was subsequently shelved. No new draft has yet been announced.[44] Coming as it did after only the second of the BAC's report (and particularly before the BAC's Human Subject Research Report released November 23, 2004), the draft Act was perhaps premature. In its report following public feedback on the draft Act, the Ministry of Health stated that it had "decided to adopt a step-by-step approach to regulating biomedical research activities" (Ministry of Health, Singapore, 2010).

But with a considerable body of recommendations and practice guidelines now in place as a result of a total of six Reports, and with the maturity of biomedical research in Singapore, it may be time again for the Singapore government to look into the matter once more and consider afresh a more direct approach to the regulation of biomedical research in Singapore. Just as was done for to facilitate the development of clinical trials so many decades ago, it is now perhaps time to do the same for biomedical research through the introduction of direct and universal regulation, complete with rules on legislative basis.

REFERENCES

Association of British Insurers. (2005). *Insurers and government reach agreement on genetic testing and insurance (*March 14, 2005). Retrieved from http://www.abi.org.uk/Media/Releases/2005/03/Insurers_and_Government_reach_agreement_on_genetic_testing_and_insurance.aspx. Accessed January 9, 2010.

Bioethics Advisory Committee. Singapore (BAC). (2002a). *Ethical, Legal and Social Issues in Human Stem Cell Research, Reproductive and Therapeutic Cloning: A Report from the Bioethics Advisory Committee, Singapore* (June 21, 2002). Retrieved from http://www.bioethics-singapore.org/.

Bioethics Advisory Committee. Singapore (BAC). (2002b). *Human Tissue Research: A Report by the Bioethics Advisory Committee, Singapore* (November 12, 2002). Retrieved from http://www.bioethics-singapore.org/.

Bioethics Advisory Committee. Singapore (BAC). (2004). *Research Involving Human Subjects – Guidelines for IRBs: A Report by the Bioethics Advisory Committee, Singapore* (November 23, 2004). Retrieved from http://www.bioethics-singapore.org/.

Bioethics Advisory Committee. Singapore (BAC). (2005). *Genetic Testing and Genetic Research: A Report by the Bioethics Advisory Committee, Singapore* (November 25, 2005). Retrieved from http://www.bioethics-singapore.org/.

Bioethics Advisory Committee. Singapore (BAC). (2007). *Personal Information in Biomedical Research: A Report by the Bioethics Advisory Committee, Singapore* (May 7, 2007). Retrieved from http://www.bioethics-singapore.org/.

Bioethics Advisory Committee. Singapore (BAC). (2008). *Donation of Human Eggs for Research: A Report by the Bioethics Advisory Committee, Singapore* (November 3, 2008). Retrieved from http://www.bioethics-singapore.org/.

Epps. (2006). Singapore's multi-billion gamble. *The Journal of Experimental Medicine*, 203, 1139-1142.

Genetic Information Non-Discrimination Act of 2008, Public Law 110-233, 12 Stat. 881, May 21, 2008.

Green, M., & Botkin, J. (2003). "Genetic exceptionalism" in medicine: clarifying the differences between genetic and nongenetic tests. *Annals of Internal Medicine*, *138*, 571–575.

HM Government and the Association of British Insurers. (2005). *Concordat and Moratorium on Genetics and Insurance* (March 2005). Retrieved from http://www.abi.org.uk/Information/Codes_and_Guidance_Notes/528.pdf.

Human and Fluss. (2001). *The World Medical Association's Declaration of Helsinki: historical and contemporary perspectives* (World Medical Association: 24 July 2001). Retrieved from http://www.wma.net/en/20activities/10ethics/10helsinki/draft_historical_contemporary_perspectives.pdf. Accessed 20 November 2009.

Human and Fluss. (2003). Dismantling the Helsinki Declaration. *Editorial Canadian Medical Association Journal,* Nov. 11, 169.10.

Human Genetic Examination Act of Germany. (Gesetz über genetische Untersuchungen bei Menschen (Gendiagnostikgesetz - GenDG). Enactment No. 374/09 of the German Federal Parliament (Bundestag), April 24, 2009, original German text and a parallel English translation available from EuroGentest at http://www.eurogentest.org/uploads/1247230263295/GenDG_German_English.pdf, and accessed January 10, 2010.

Janson-Smith, D. (2002). *The genetics and insurance committee*, July 19, 2002. Retrieved from http://genome.wellcome.ac.uk/doc_WTD021010.html. Accessed January 10, 2010.

Kaan, T. (2009). *Walking the tightrope: the regulation of biomedical research in Singapore.* Paper presented at the 2009 ILST Conference on Innovation, Competition and Regulation, Taiwan on December 3-4, 2009, organized by the Institute of Law for Science and Technology, National Tsing Hua University, Hsinchu, Taiwan.

Ministry of Health Singapore. (2003) Public consultation on the draft regulation of Biomedical Research Bill - summary of feedback received 10 November 2003 - 30 November 2003. Retrieved from http://www.moh.gov.sg/mohcorp/econsultationpast.aspx?ecid=81. Accessed January 10, 2010.

Normile. (2007). An Asian tiger's bold experiment. *Science,* 316(5821), 38-41.

Perry, M. (2008). Channel NewsAsia report, "Sexually active women advised to use the Pill to reduce abortion rate", posted 21 May 2008 1950 hrs, from http://www.channelnewsasia.com/stories/singaporelocalnews/view/349189/1/.html, re-accessed 10 January 2010.

(1801-1803). Reynolds. (2000). Declaration of Helsinki revisited. *Journal of the National Cancer Institute*, *92*(22).

Singapore Department of Statistics. (2009). *Yearbook of Statistics Singapore 2009*. Retrieved from http://www.singstat.gov.sg/pubn/reference.html. Accessed 10 January 2010.

The Agency for Science, Technology and Research, with the Ministry of Health. (2008). Joint Media Release: "13th BMS IAC meeting announces key achievements in Translational & Clinical Research efforts in Singapore," 17 October 2008. Retrieved from https://www.nmrc.gov.sg/corp/uploadedFiles/NMRC/News,_Views_And_Events/News/IAC%20Media%20Release.pdf. Accessed 20 November 2009.

The Council for International Organizations of Medical Sciences (CIOMS) in collaboration with the World Health Organization. (WHO). (2002). *International Ethical Guidelines for Biomedical Research Involving Human Subjects*. Current edition, Geneva. Retrieved from http://www.cioms.ch.

The International Conference on Harmonization of Technical Requirements for Registration of Pharmaceuticals for Human Use. (1996). *Guideline for Good Clinical Practice E6 (R1), 10 June 1996.*

The National Medical Ethics Committee (NMEC). Singapore. (1998). *National Medical Ethics Committee: A Review of Activities, 1994-1997.* Ministry of Health Singapore, 18 March 1998. Retrieved from http://www.moh.gov.sg/mohcorp/publicationsreports.aspx?id=2406.

The Nuremberg Code: Directives for Human Experimentation. (1949). Trials of War Criminals before the Nuremberg Military Tribunals under Control Council Law No. 10, Vol. 2. Washington, D.C.: U.S. Government Printing Office. Retrieved from http://ohsr.od.nih.gov/guidelines/nuremberg.html.

The World Medical Association. (2008). *The Declaration of Helsinki (Ethical Principles for Medical Research Involving Human Subjects).* Current edition by the 59th General Assembly of the WMA, Seoul, 22 October 2008. Retrieved from http://www.wma.net.

Williams. 2008. The Declaration of Helsinki and public health. *Bulletin of the World Health Organization*, 86(8).

Yeoh, K. C. (2008). Singapore's biomedical sciences landscape. *Journal of Commercial Biotechnology*, *14*(2), 141–148. doi:10.1057/palgrave.jcb.3050083

ENDNOTES

1. Terry Sheung-Hung Kaan is Associate Professor at the Faculty of Law, National University of Singapore, where he teaches Tort and Biomedical Law & Ethics. Correspondence address: Faculty of Law, NUS, Eu Tong Sen Building, 469G Bukit Timah Road, Singapore 259776. Email: lawterry@nus.edu.sg. The views expressed in this paper are the author's own personal opinions, and do not reflect the views of the NUS or that of any committees or bodies of which he may be or have been a member, or with which he is or may have been affiliated.
2. . The Singapore Bioethics Advisory Committee, *Genetic Testing and Genetic Research* (November 25, 2005) (BAC, 2005). The full text of the Report together with all its appendices is available from the website of the BAC at http://www.bioethics-singapore.org/.
3. Some parts of this paper, and in particular the summary of the economic background, the history of establishment of the BAC and its relationship with the National Medical Ethics Committee, and the general overview of the

regulatory framework for ethical governance are adapted from and first presented in an oral presentation based on an unpublished draft (Kaan, 2009).

4. The author was a member of the BAC from January 2001 until December 2006. During most of this time, he was also the Chair of the BAC's Human Genetics Sub-Committee (HGS), which spearheaded initial work on 3 of the first 4 Reports eventually issued by the BAC: the BAC reports on *Human Tissue Research* (November 12, 2002), *Research Involving Human Subjects: Guidelines for IRBs* (November 23, 2004), the Genetic Testing Report, and *Personal Information in Biomedical Research* (May 7, 2007). The author wishes to make clear that all the views presented in this paper are entirely the author's own personal opinion, and do not reflect that of the BAC or any other committee or body with which he is currently or may have a member of or have been affiliated to or associated, including the BAC and the National Medical Ethics Committee (NMEC).

5. Initially, the BAC reported to the Government's Life Sciences Ministerial Committee. Currently, it reports to the Steering Committee on Life Sciences, led by Dr Tony Tan, a former Deputy Prime Minister, and currently Chairman of the National Research Foundation.

6. See Yeoh, 2008, p. 142. (Mr. Yeoh is the Executive Director of the Economic Development Board's Biomedical Sciences Cluster). For other perspectives of Singapore's ambitions in the biomedical industry and research sector, and the BMS Initiative, see also Epps, 2006 and Normile, 2007.

7. As for example, its reports on living wills and on the principles governing human organ transplantation. See NMEC, 1998.

8. Under the Private Hospitals and Medical Clinics Act (Chapter 248).

9. Under the Medical Registration Act (Chapter 174).

10. For example, Phase 1 clinical trials involving experimental cancer drugs which double as a experimental therapy of last resort for cancer patients who have exhausted every other proven therapeutic option, or biomedical research involving the use of wholly and irreversible anonymized and aggregated information.

11. Which may be physical harm in direct effects, or the loss of opportunity for a better therapeutic option; or social or emotional harm in the case of biomedical medical research involving personal information if that information is made public or made available to an unauthorized party.

12. The full text of these guidelines are set out in Annex IV/D (NMEC, 1998).

13. In recent years, further kinds of ethics committees have been added to the mix–hospitals which carry out human organ transplants are now required to have transplant ethics committees under the Human Organ Transplant Act (Chapter 131A).

14. S54/1978, promulgated as subsidiary legislation by the Minister for Health on 27 March 1978 in exercise of his powers under the parent statute, the Medicines Act (Chapter 176).

15. The International Conference on Harmonization of Technical Requirements for Registration of Pharmaceuticals for Human Use (1996). This document forms the basis of the Singapore Guideline for Good Clinical Practice referred to in the Medicines (Clinical Trials) Regulations.

16. BAC, 2004. Full text and appendices accessible at http://www.bioethics-singapore.org/.

17. As at the end of 2009.

18. NMEC, 1998, paragraph 2.2.1, adopted by the BAC at paragraph 3.21 of the BAC's *Human Research Report* (2004).

19. See, for example, Reynolds, 2000; Williams, 2008; Human and Fluss, 2001; and Human and Fluss, 2003.

20. A chapter on genetic testing in the context of biomedical research proper is relegated to a brief 3½ pages (pages 50-53) right at the end of the Report. Obviously, most of the principles articulated for genetic testing and screening in the context of the physician-patient relationship, especially those in relation to consent and privacy, can be applied directly to genetic testing in the research context.

21. Not the same thing, as uninformed lay people may have completely unfounded assumptions about the actual predictive power of genetic tests and their results.

22. See the written responses received by the BAC from the Life Insurance Association, Aviva Ltd and the Chief Actuary, Great Eastern Life Assurance Co Ltd, Annex F, The Personal Information Report, at F-68, F-157 and F-158, as well as the Position Paper entitled "Genetics and Life Insurance: A Position Paper by the Life Insurance Association, Singapore" at page A-4-1 of the Personal Information Report. Although representations had already been received on this point from the life insurance industry following the call for feedback and representation on the Consultation Paper issued by the BAC on April 5, 2005 which eventually formed the basis of the Genetic Testing Report on November 25, 2005, the BAC deferred their recommendations on this thorny issue to their subsequent Personal Information Report issued on May 7, 2007.

23. Except in the uncommon situation of a full-penetrance monogenic genetic condition such as Huntington's Disease, as discussed below.

24. It should be borne in mind that a rejection of an underwriting proposal by any one insurance company has a significant impact on the insurability of a given individual, as insurance companies commonly require potential customers to disclose rejections by other insurance companies.

25. Public Law 110-233, 12 Stat. 881, May 21, 2008.

26. Sec. 101 (b).

27. The BAC made its recommendations in 2007 before the passage of the Genetic Information Non-Discrimination Act in May 2008.

28. The Concordat and Moratorium, Clauses 16 and 20 (HM Government, 2005). The Concordat states that these policy value thresholds would exclude 97% of all policies underwritten in the UK in 2004.

29. Currently, only Huntington's Disease has been approved (Association of British Insurers, 2005).

30. "The GAIC was set up by the [UK] government in 1999 in response to growing public concern about how the industry wanted to use genetic test results, and the potential for discrimination through genetic screening. ...The concern is whether we know enough about the relationship between genes and health to say whether insurers are justified in using the results of such tests. ...[and the GAIC] is an independent review body set up to evaluate the use of genetic tests by the insurance industry" (Janson-Smith, 2002).

31. An example might be the case of an individual, without any scientific basis whatsoever, that being of X blood type may genetically doom him to social incompatibility with persons of Y or Z blood type, thus diminishing in his eyes (and others with like beliefs) his chances for matrimony, if information about his blood type were to be leaked to the public.

32. For example, if a person should irrationally insist on "keeping secret" patently and publicly observable characteristics such as the color of his eyes, or his apparent ethnicity.

33. For this part generally, see BAC, 2005, pp. 23-41.

34. Section 45 of the UK Human Tissue Act 2004. No proposal for such a provision has to date been put before the Singapore Parliament.

35. For accounts of these programs in Singapore, see the Commissioned Position Papers in Annex C of BAC, 2005.

36. The descriptions of these procedures are adapted from the review of these procedures in BAC, 2005, pp. 30 – 37.

37. Gender imbalance is not an issue in Singapore: as at June 2008, there were approximately 1,839,700 female to 1,803,000 male Singapore Residents (Singapore Department of Statistics, 2009, p. 26).

38. Notably the Human Cloning and Other Prohibited Practices Act (Chapter 131B), which implemented some of the recommendations in the BAC's Human Cloning and Stem Cell Report; and the National Registry of Diseases Act 2007 (Chapter 201B) in relation to the BAC's Personal Information Report.

39. Through the Private Hospitals and Medical Clinics Act (Chapter 248).

40. Through the Medical Registration Act (Chapter 174).

41. Ministry of Health Directive 1A/2006.

42 . Parts II and VII, Medical Registration Act.

43. Available from http://app.reach.gov.sg/Data/adm05/c6/p984/draft_BMR_bill.pdf, last accessed January 10, 2010.

44. As at January 10, 2010.

Chapter 15
Legal Aspects of Bioethics in Tajikistan

Firuza Nasyrova
Tajik Academy of Sciences, Tajikistan

ABSTRACT

The chapter presents a detailed account of the available legal mechanisms on biotechnological issues in her country. Among the issues discussed are biosafety regulations, especially ones concerning the Cartagena Protocol as well as the governance structure in Tajikistan on these issues.

LOCATION

Tajikistan is a mountainous landlocked country in Central Asia, bordering with Afghanistan, Uzbekistan, Kyrgyzstan, and China. The total area of Tajikistan Republic is 143.1 thousands of square km of which 93% is Himalayan mountainous terrain, and population more than 7 million people. Tajikistan consists of 4 administrative divisions: 2 provinces (Sughd in the north and Khatlon in the south), 1 autonomous province (Gorno-Badakhshan - Pamir), and disrtricts of republican subordination around Dushanbe.

Tajikistan is an agrarian-industrial country, exporting cotton, vegetables and fruit.

DOI: 10.4018/978-1-61692-883-4.ch015

HISTORICAL INSIGHT

The Tajik people have a rich and ancient culture. The history of Tajik culture, philosophy, mentalities, traditions, behavior, and especially ethic aspects roots back to interpretations in pre-Islamic tradition and in the Koran. Religious and philosophical ideas of Tajik ancestor's in pre-Islamic time have been given in the most ancient book in "Avesto": the holy book-encyclopedia of Zoroastrians containing not only religious dogmas but also ideas on cosmogony, philosophy, morals and law. Together with information on history, religion, language, philosophy, geography, economics, literature, etc. there are most valuable information on medicine in "Avesto." At this source in telling about diagnostics of diseases, the ways to treat them, treatment of sick by means of word (mantra), curative plants and grasses, and surgeon knife as well. Issues of private hygiene, tasteful and healthy meal, cleanness of drinking water and other beverages, cleanness of environment, i.e., ecology, personality of physician, different methods of treatment, of healing of sick etc. are included in the separate chapters of one of the part of "Avesto" "Vandidat." According to "Vandidat" the supreme God of Zoroastrian religion, Akhura Mazda, created a large number of curative plants, grasses and donated them to the people for them taking treatment and were healthy (Soatov, 2003, pp. 210-216).

The characteristic feature of Zoroastrians is dualism, which means that there are two fundamental concepts exist, such as light (good) and darkness (evil), and the struggle between these is the pivot and content of the universe. For Zoroastrians the main fundamental philosophy of their life have been formulated in ternary ideal of the righteous man ("good thoughts," "good words" and "good deeds") opposed to the ternary ideal of the infidels ("wicked thoughts," "wicked words" and "wicked deeds"). The Zoroastrian idea about infinite time as an initial substance gave rise to zervanism–a teaching in the framework of which a materialistic trend developed that denied the creation, and God as the creator of the universe, and affirmed the belief in the eternity of the world. During feudalism period Mazdakism (Mazdak, the end of the V–the beginning of the VI century) became very influential. Those philosophical trends assimilated the Zoroastrian ideal about the struggle between light (good) and darkness (evil). He believed that the light was absolutely free in its actions while ignorance limited darkness. On social level, synthesizing economics, philosophy, and theology, Mazdak preached a communistic doctrine espousing communal ownership a property and of women, his social doctrine proclaimed ideas of justice and equality. In that period many written monuments of religious, political and scientific thought were created, many of which were destroyed during the invasion of the army of the Arab Caliphate (VII-VIII centuries) (Bashiri. 2003, pp. 1-98; Achrorova. 2007, pp. 283-301).

During the VIII-XV centuries philosophical, social and political ideas were developing in the atmosphere of Arabian conquests and a forcible spreading of Islam. Spiritual cultures of Iranian and Arabic peoples became most closely linked.

During the Samanids period (IX-XI centuries), the Tajiks developed as a nation with its State system and literary language–Dari (Tajik) language. This period promoted the creation of Central Asian Islamic culture by the completing of the process of Islamization in Central Asia that began immediately after Arabic conquest. Possibility to return to Zoroastrian religion was still preserved. But the Samanids, having become the rulers of Central Asia, didn't follow that reactionary way. They started the cultural mechanisms of Islamization, and successfully promoted Islam to be the ideology of their state, as well as the social consciousness of the masses. Thus appeared the real possibility to forming Central Asian Islamic culture. At this period Central Asia become one of the most important centers of oriental science. That was the time when astronomical observatories, "Houses of Wisdom" and libraries were built. Scientists from Central Asia translated and commented the scientific heritage of ancient Greece and India, and wrote original works on mathematics, astronomy, mineralogy, applied mechanics, physics, chemistry and medicine (Dinorshoev, 1999, p. 130-133).

Ethics as a scientific trend had a significant place in works of prominent oriental thinkers, and Ali Ibn Sina (Abu Ali al-Husain ibn Sina-e Balkhi (Latinized Avicenna)) was one of them; a philosopher, musician, poet, physician, the follower of Aristotle, he was a scientist of encyclopedic knowledge and one of the greatest minds of the middle Ages. Roger Bacon was certainly right when he wrote that Ibn Sina was "the greatest teacher of philosophy after Aristotle" (UNESCO, 2006, p. 1-18). Tajik by birth, he was born in the suburb of Bukhara (Afshona) in 980, lived in different cities of the Central Asia and Iran, served various rulers as a court physician and vizier, and died in 1037 near Hamadan. Abu Ali Ibn Sina wrote in the Arabic and Tajik languages. He left a rich scientific heritage–about 300 works, and among those "Book of Healing," "Book of Knowledge," "Book of Directions and Remarks," "Book of Salvation" and "Canon of Medical Science." The latter had been considered as one of the fundamental medical guidelines during five centuries.

With regard to ethical principles, "Canon of Medical Science" is a literary source containing elements of bioethics. It differs from its antiquity analogues in the attempt to characterize medicine not only as a system of medical and biological knowledge, but also as a canon of spiritual and ethical basis of the entire medical science. There were many reasons why "Canon" excited much interest in different historical periods including the European Renaissance. It has a powerful spiritual and moral potential for the development of medical science the subject of which is not just a human being but also his/her lifestyle with all the variety of cultural, legal, religious and other features. According to Ibn Sina, it is essential to humanize the process of bringing up healthy people and to find the ways for maintaining and developing spiritual harmony throughout a person's life.

Ethical views of Ibn Sina are still topical nowadays. He saw a physician as a patient's friend, tutor and helper. To fulfill this noble mission, the physician, apart from professional knowledge and experience, should have many positive qualities–mercy, respect for human dignity, readiness to self-sacrifice in a patient's interests, etc.

In 2003, the Islamic Republic of Iran and UNESCO instituted the Avicenna prize for the development of modern bioethics in science. The role of Avicenna in the development of current bioethical views runs all through the book *Avicenna and the Ethics of Science Technology Today* published by UNESCO in 2006. Avicenna's life and works invite us to think about the ethics of science (Abdullkhodjaeva, 2007, pp. 323-349).

Starting from the second half of the XIX century, the Russian, and western cultures as well have been asserted influence on the philosophical and sociopolitical development of the Tajiks and other nations in the former Soviet Union. There were also new studies in the field of sociology, ethics and aesthetics.

The late eighties and early nineties were marked with complicated and contradictory events, the most significant of which was a disappearance of the USSR from the map of the world. Instead, about 15 new independent states appeared, of which one was Tajikistan. Thus, within a short historical interval Tajikistan moved from one historical period to another; the state and power structure changed, as well as its attributes. The former political system was destroyed, which resulted in the change of ownership patterns and social relations.

CURRENT SITUATION

The current scientific and technological evolution in the field of biomedicine is characterized by the growing importance of the role of universal ethical values relating to the protection of human rights and dignity. This tendency becomes particularly essential in the course of deep social and cultural changes that have been happening during the last decades in the post-Soviet space. The development and implementation of the new countries' national policies in the field of ethics and bioethics by bringing up initiatives and consolidating activities in lawmaking, education, creation of the system for the ethical review and international cooperation is presently one of the topical issues.

Currently, in Tajikistan, like all over the world, the need for biomedical research is growing steadily, as well as the concern for such human values as a person's health, rights and dignity. Therefore, people of different professions, religions and nationalities have come together to form medical ethics committees that would perform ethical review of each research involving human subjects.

The development of Tajikistan as an independent republic in the post-Soviet space was accompanied with economic, political and social instability. A significant support from the international community was directed to the needs associated with the civil war (1992-1995) and its consequences. Emergency measures of providing medicines, vaccines, bandaging material taken by the international community made an essential contribution into the solution of health care problems. At the same time, there was a need in more thorough structural changes that would help to overcome existing problems and facilitate the development of safe mechanisms for providing health care and safeguarding patients' rights.

The involvement of Republic of Tajikistan, as an independent state, in international research projects requires ethical and bioethical review to ensure the compliance with international regulations (Achrorova, 2007, pp. 283-301).

Currently there are two Bioethics Committees in Tajikistan:

1. Bioethics Council, established in 2007, consists of leading scientists, medical doctors, teachers, lawyers, etc. under the Academy of Sciences of the Republic of Tajikistan (RT). The Council deals with education, biotechnology, and other related topics, and publishes manuals on ethical review in medical research. The national legislative committees are also involved in the development of laws relating to biosafety and GMOs. They are still developing a charter of the committee and adopting different documents on bioethics, biotechnology, and others. The Head of the Bioethics Council is Prof. Karimov Kh., Vice President of the Tajik Academy of Sciences. In cooperation with the Ministry of Agriculture, the Council reviews the ethical content of PhD theses in veterinary science and biology. Under financial support of UNESCO Bioethics Council have been provided different regional workshops, seminars on bioethics, biotechnology and GMO impacts to the human health and environment.
2. Medical Ethics Committee (MEC) under Ministry of Health of Tajikistan was established in 2004. Professor Ahrorova Zuhra has been the first Head of MEC. The Committee is made up of leading specialists in ophthalmology, dermatology, gynecology, psychology, path physiology and two lawyers. MEC was created, ratified and adapted such important documents as the Statement of MEC, Clinical Practice and Laboratory Practice rules. MEC has prepared a draft of the law on protection of human rights and dignity, which now is under consideration of the Government of Tajikistan. A study of medical doctors conducted by MEC showed limited knowledge of patient and doctors rights.

LEGAL REGULATIONS

Health is one of few human values without which many others comforts and values lose their significance. Today every democratic country acknowledges the importance of human rights protection in all spheres of social life. The ethical review of biomedical research involving human subjects is the important component in rights, interests and dignity of human subjects and in the same time is the necessary condition for further development of medical science and technology. Many legal norms prohibit actions that are hazardous to the society and human health.

In Tajikistan biomedical research and protection of human rights in this sphere is regulated by a number of national legal acts, and first of all by Constitution. According the principal Law, Tajik residents have the right to health protection, free health care provision by a network of state health care institutions. The moral and ethical components in biomedical research should correspond to legislative principles referring to the protection of the patient's rights and dignity. One of the rights guaranteed by the Constitution is the right for health protection implying the following:

- Medical care and social protection;
- Safe environment, food products and drinking water;
- Qualified medical and sanitary care including a free choice of a physician and healthcare institution;
- Safe and healthy living and working conditions, as well as safe and healthy conditions for rest, education and upbringing;
- Sanitary and epidemiological well-being in the territory where a person lives;
- Truthful and timely information about an individual's health including existing and potential risks and the degree of risk;
- Participation in measures on health protection and public expertise with regard to these issues;
- Possibility to create public organizations for the purpose of facilitating health protection and protection of human rights;
- Legal assistance in any case of discrimination referring to health condition;
- Compensation for a caused harm;
- Making an appeal in the event of wrongful decisions or actions of healthcare professionals;
- Independent medical expertise in case of discordance with the decision of the State expertise.

Currently, the Committee on Medical Ethics at the Health Ministry of Republic of Tajikistan (HM RT) has worked out a number of important documents adapted to the needs of Tajikistan:

- Statement "On the Committee on Medical Ethics" (HM RT Order of No 118, 10 of March 2005);
- "Ethical Aspects of Regulations for Clinical Practice in Republic of Tajikistan" (Supplement to the HM RT Order No 118, 10 of March 2005);
- "Guidelines for Laboratory Practice in Republic Tajikistan" (Supplement to the HM RT Order No 118, 10 of March 2005) (Achrorova, 2007, pp. 283-301).

To protect both the individual and the community, these legislative acts state that ethical and scientific review should precede a biomedical research including the evaluation of inform consent to participate in the research and special procedures aiming to protect interests of individuals who cannot consent. In developing national regulative documents, we follow international guidelines and documents of the Forum for Ethics Committees in the Commonwealth of Independent States such as International Ethical Guidelines for Biomedical Research Involving Human Subjects and Guidance on Good Clinical Practice developed by the World Health Organization and International Conference on Harmonization of Technical Requirements got the Registration of Pharmaceuticals for Human Use, materials of FECCIS international conferences (2003-2005). Compliance with these documents ensures the protection of rights, honor, dignity, safety and well-being of research participants, as well as reliability of research findings. We were also guided by the Constitution of Republic of Tajikistan and by Tajikistan Laws "On the Protection of the Population Health," "On Pharmaceutical Products and Pharmaceutical Practice," "On Narcotic Drugs, Psychoactive Drugs and Precursors," "The Concept of Health Care Reformation in Republic of Tajikistan" (2002). These documents are focused on the international practice; implementation of new, more efficient organization approaches, improving the quality and availability of medical and sanitary care and a further development of international cooperation (Kubar and Henk ten Have, 2007, pp. 7-8; Mikirtichian et al., 2007, pp. 223-282).

The health care legislation in Tajikistan includes Constitution norms, Tajikistan Law "On the Protection of Population Health" and other national legislation acts, international regulations adopted by Tajikistan, international treaties and normative documents issued by state structures. All these documents state that the society and government are responsible to contemporaries and future generations for the health level of the Tajikistan population. The current legislation is regulating social relations in the sphere of health care covering a wide range of issues from a harmonious physical and spiritual development and the improvement of heredity to social and legal protection of the patient. The laws aim at improving conditions of work, life and rest, resolving ecological problems, developing the quality of medical care and promoting a healthy lifestyle.

The first steps of the independent Tajikistan were accompanied with economic crisis in all fields including health care. Economic, political and social recession affected indices of the population health. Most urgent were problems concerning food, water supply, sanitary situation, the growth of infectious diseases (typhoid, malaria, tuberculosis, diphtheria, acute viral hepatitis) acute respiratory diseases, anemia, diseases caused by iodine deficit and reproductive health. Despite instability in political and socioeconomic life during the first years of independence, Tajikistan Government took comprehensive measures on the protection of the population health. During that period, Tajikistan ratified important international documents referring to the protection of human rights such as the Convention on the Elimination of All Forms of Discrimination against Women (ratified in 1993), Convention on Children's Rights (ratified in 1993). Presently, the following documents form the legislative basis in health care:

- Constitution of Republic of Tajikistan (RT) (1994);
- RT Law "On AIDS Prevention" (No 904; 23 of December 1993);
- RT Law "On Donorship of Blood and its Components" (No 901; 23 December 1993);
- RT Law "On the State Sanitary Inspection" (No 987; 20 of July 1994);
- RT Law "On the Protection of Population Health" (No 419; 15 of May 1997);
- RT Labor Code (1997);
- RT Family Code (1997);
- RT Law "On Narcotic Drugs, Psychotropic and Precursors" (No 874, 10 of December 1999);
- RT Law "On Pharmaceutical Products and Pharmaceutical Activity" (No 39; 6 of August 2001);
- RT Law "On Psychiatric Care" (No 90, 2 of December 2000);
- RT Law "On Reproductive Health and Reproductive Rights" (No 72; 2 of December 2002);
- RT Law "On Private Medical Practice" (No 60; 3 of December 2000);
- RT Law "On Salt Iodination" (No 85; 2 of December 2002);
- Decree of the RT Government "On the Approval of the National Program for the Control of Tropical Diseases (Malaria) in Republic of Tajikistan in 1997-2005" (No 342, 4 of December 1997);
- Decree of the RT Government "On the Approval of the National Program for the Control of Viral Hepatitis "B" in Tajikistan in 2000- 2007" (No 100; 11 of March 2000);
- Decree of the RT Government "On the Approval of the Concept of Healthcare

Reformation in Republic of Tajikistan" (No 94; 4 of March 2002);
- Decree of the RT Government "On the Approval of the National Program for the Prevention and Control of HIV/AIDS/STD in Republic of Tajikistan in the Period to 2007" (No 475; 27 of November 2002);
- Decree of the RT Government "On the RT Strategy for Preventing the Menace of AIDS Spreading for the Period of 2002-2005" (No 389; 1 of October 2002);
- Decree of the RT Government "On the Approval of the RT Strategy for the Protection of Population Health in the Period to 2010" (No 436; 5 of November 2002)
- Decree of the RT Government "On the Approval of the National Program for Tuberculosis Control for the Period 2003-2010" (N° 524; 31 of December 2002);
- Decree of the RT Government "On Family Medicine" (2000);
- Decree of the RT Government "On the Approval of the National Program for the Development of a Healthy Life-Style in the Period to 2010 (No84; 3 of March 2003).

The problem of protection of research subjects' health and rights is reflected in strategic documents "The RT Strategy for the Protection of the Population Health in the Period to 2010", "The RT Strategy of Reducing Poverty (2002)" and "The RT Strategy of Achieving Goals of the Development of the Millennium". The documents aim at reducing mortality of children under 5 years of age by 2/3 and mortality of mothers by ¾; at improving access to the service of reproductive health, at preventing HIV/AIDS spreading and at eliminating sex discrimination in the field of primary and secondary education. To ensure the national demographic policy, the Government approved and adopted the program "The Concept of RT State Demographic Policy for the Period 2003-2015" (2002) (Achrorova, 2007, pp. 283-301).

Legislation initiatives in the sphere of protection of patient "rights," activity directed to adaptation of international guidelines, as well as transformation of the positions of international declarations and conventions in the sphere of biomedicine into domestic legal acts, give us the reason to expect the strengthening and development of the institute of ethical review in Tajikistan.

BIOSAFETY REGULATION IN TAJIKISTAN

Cartagena Protocol on Biosafety

New technologies often have great potential and promise much, but also needs to be assessed adequately in order to ensure that they are safe, as well as environmentally and socially sustainable. Genetic engineering (GE) and genetically modified organisms (GMOs) are example where, despite promises and expectations of benefits, concerns remain over their potential risks to human health and the environment.

The Cartagena Protocol on Biosafety is the first international law to specifically regulate genetic engineering, and this largely reflects the global climate of concern about safety, health and environmental risks of GMOs, along with the wider political and socio-economic implications of this technology. The Protocol recognizes that GMO may have biodiversity, human health and socio-economic impacts, and that these impacts should be risk assessed or taken into account when making decision on GMOs. Precaution is the basis for the Protocol itself, and is operationalized in decision-making and risk assessment.

The Cartagena Protocol on Biosafety entered into force on 11 September 2003. It is legally binding international agreement under the United Nations Convention on Biological Diversity (CBD).

The entry into force of the Protocol was an important defining moment in global biosafety regulation. At this moment more than 130 countries have been adopted the Protocol. The Protocol's entry into force means that it is legally binding in the international legal system and in the legal system of countries that have ratified, approved, accepted, or acceded to it.

Since most countries do not have adequate laws or regulations on biosafety, and lack the capacity, technological and financial resources to regulate genetic engineering, the Protocol is necessary for developing countries.

The central objective of the protocol is to regulate the international (i.e. transboundary) movement of GMOs in order to "derive maximum benefits from biotechnology while at the same time protecting biodiversity and human health from potential risks posed by GMOs." Its guiding principle is the *precautionary approach* and its central directive is that the import of GMOs into signatory country for release requires the *advance informed agreement* of the country authority (Khor, 2007, pp. 13-14).

Ethical Aspects of GMOs

Different applications of GE and GMO usage represent various types of risk. For instance, with GE medical applications such as GM vaccine, a GM drug or somatic cell therapy, the beneficiary coincidently carriers the potential risks. For germ cell line therapy, however, diseases may be cured by genetic "surgery," and the "improved" genome will be passed on as new genotype in the next generations. Accordingly, the risk of harm may be transferred to future generations. Issues that present putative risks across generation gap, raise questions concerning of human needs today against the opportunities for future generations to fulfill their needs. The situation becomes even more complex when society and the environment may experience the risk. For instance, we do not know with certainty if GM crops will promote general welfare by providing more nutritious food or help to ensure on non-target organisms or threaten biodiversity.

Ethically responsible decision making must be based on the best available knowledge, but also on conception of missing knowledge. This requires awareness of the relevant scientific uncertainties and knowledge gaps involved. Research on such topics has made significant progress during the last decade, valuable and useful instruments to represent ethical principles need to be established. Ethical aspects relate directly to the scientific description of the risk assessment and management of GE, taking into account the adverse effects and unexpected effects that need to be avoided, as well as the benefits we need to achieve. This may initiate creative thinking about design of risk-associated research. Truly creative thinking must include proper monitoring of the promised benefits and potential health and environmental risks as well as social, ethical and cultural issues that the communities find important to protect (Myhr and Traavik 2007, pp. 123-135).

International Commitments of the Republic of Tajikistan in the Field of Biosafety

The policy of the Republic of Tajikistan in the field of biosafety is part of its policy in the field of health care and environmental protection from the standpoint of the sustainable development concept. The state policy in these fields is based upon the fundamental principles, stated in a number of international agreements, which Tajikistan is a Party to, as well as upon the relevant national legislative acts. The main objective of the Republic of Tajikistan in the field of biosafety is, on the one hand, to create an enabling environment for deriving maximum benefit from the achievements of modern biotechnology, to foster development of genetic engineering as one of the priority research areas of focus and, on the other hand, to ensure human and environmental health when carrying out genetic engineering activities, implementing new biotechnologies and consuming their products.

The Republic of Tajikistan is a Party to a number of international conventions, related to safe use of achievements of modern biotechnology. The most important of them are the following:

1. Convention on Biological Diversity.
2. Cartagena Protocol on Biosafety to the Convention on Biological Diversity.
3. Convention on Access to Information, Public Participation in Decision-Making and Access to Justice in Environmental Matters (the Aarhus Convention).

National Biosafety Framework *(2004, p. 68)*

National Biosafety Framework for Tajikistan (NBFT) is an outcome of UNEP-GEF Project implementation related to establishment of National Biosafety Framework. Project is based on technical and legal potential of Tajikistan related to monitoring and regulation of safety of the environment, human health, biosafety, and on the results of consultations and workshops organized in the project frames. Active work on establishment of NBF, involvement of politicians, Parliament members and local community to the issues of biosafety management within the project, has promoted to ratification of Cartagena Protocol on biosafety by Tajikistan in October 2003. Cartagena Protocol is a supplementary agreement to the Convention on Biodiversity, Tajikistan being the Party of, and it identifies biosafety as one of the main issues in biodiversity conservation. Protocol mainly regulates transboundary transfer of GMOs and provides international procedures ensured by the Parties. Currently, besides the developed NBF, there are still no any legislative, regulative or instructive documents on biosafety within the country. While developing NBF the experience of a number of countries has been studied related to biosafety regulation and best practice has been applied for its development, considering the country's peculiarities. This will also ensure the effectiveness of NBF application as a transient mechanism for realization of Cartagena Protocol. The project is the basis for the development of a stable framework and in the process of its implementation changes and improvements of a number of components are proposed for more effective activity. At this stage the National Biosafety Framework describes the following components:

- Biosafety policy;
- Regulatory framework;
- Administrative system;
- Monitoring and Enforcement;
- Public awareness and participation in decision-making.

NBF comprises mechanisms of nonconflict cooperation of authorized governmental institutions. NBF is proposed to ensure clarity, transparency and predictability of decisions on biosafety in Tajikistan. Main principle of relation to the use of genetically modified organisms–a *precautionary approach* has been outlined as a result of discussion of the present biosafety framework. Development of national legislation has been launched in the process of NBF preparation. The Law on biosafety is submitted to the Parliament for approval. The prepared law prescribes regulations of GMOs treating, provides potential risks management and stipulates the need of maximum benefit from biotechnology products. The country's legislation will comprise regulations on GMO decision making, identify control state authorities on biosafety and their cooperation. General principles and minimum supervision status for relevant stakeholders are already included into National Biosafety Framework.

The *National Biodiversity and Biosafety Center (NBBC)*, established according to the Government resolution No. 392 on January 9, 2003, is an authority responsible for biosafety and is developing a system of biosafety regulation in cooperation with stakeholders. The Center also provides commitments of Republic of Tajikistan under the requirements of the Convention on Biodiversity and Cartagena Protocol on Biosafety. Managing of works and coordinating of activities on the implementation of the Cartagena Protocol on Biosafety and the present NBF in the country is authorized on *CBD and CPB National Focal Point* of Republic of Tajikistan, Chairman of the National Biodiversity and Biosafety Center of Republic of Tajikistan, Dr. Neimatullo Safarov.

Current State

The Constitution of the Republic of Tajikistan provides rational use of nature resources and protection of human rights for safe and favorable environment, with biological safety being an integral part of. The Republic of Tajikistan bases on the need of wide and effective international cooperation to conserve nature resources and providing biosafety on its territory and at a global scale as well, therefore takes active part in all international initiatives. Government of the Republic of Tajikistan shares a world concern in the need to provide biosafety. Thus, January 2002, the letter (No. 19/1-4) was signed expressing the willingness of the republic to join the Cartagena Protocol, and October 22, 2003 Tajikistan ratified the Cartagena Protocol on Biosafety (Government Statement No. 932). Tajikistan admits the need in its legislation improvement and development of new laws and regulation acts to provide biosafety and meeting the international requirements, and has started to develop the Law on Biosafety. Government of the Republic of Tajikistan puts efforts to provide the conservation of genetic resources in the country. Biodiversity issues are reflected in several environmental policies of the Republic. The National Strategy and Action Plan on conservation and sustainable use of biodiversity (NBSAP) that was approved by Government Decree N°. 392 from September 1, 2003 places high emphasis on biosafety issues. As priority actions the Action Plan includes development of legislation on genetically modified organisms, establishment of Center on gene pool and ratification of Cartagena Protocol.

National Report on sustainable development (approved by Government July 13, 2002, No. 297) includes chapter on ecologically safe use of biotechnologies (chapter 16). The policy is focused on increase of productivity of agricultural crops, increase of food stuff production and development of pharmaceutical industry on the base of ecologically pure biotechnologies. However under biotechnologies one does not mean the use of genetic engineering methods and its monitoring. A number of organizations are involved to some extent in the process of ensuring the country biosafety, as State Committee on Environment Protection and Forestry; Ministry for Health; Ministry of Agriculture; Customs Committee of Ministry on State Income and Taxes; Academy of Science and Institutes of Higher Education, etc.

Thus, the most important objective of biosafety policy is the formation of coordinated activities of all relevant competent authorities and most stakeholders. The main priority is capacity building, personnel training, and development of adequate mechanisms of information exchange to realize the terms of national biosafety mechanisms. Tajikistan acknowledges the importance of National Biosafety Framework development to create and support the effective national and international biosafety system. The Cartagena Protocol serves as an international platform for providing biosafety.

Prospects and Needs

Development of the National Biosafety Framework (NBF) for Tajikistan is extremely valuable and helps expedite the establishment of biosafety policy principles in the country. This is considered to be perspective for further development and improvement of policy, mainly with the support of UNEP-GEF that will surely expedite the realization of biosafety framework in Tajikistan. In the present National Biosafety Framework the key principles of biosafety policy are based on prevention and reduction of possible adverse impacts on the environment, mainly related to biodiversity conservation and human health and to guarantees of safe use and application of modern biotechnology. Implementation of the National Biosafety Framework has to provide development of national institutions, establish scientific basis for risk assessment, improve control and management systems, and provide training on administrative and legal backgrounds on biosafety for all stakeholders. This may be provided only through long-term processes of capacity building. Biosafety policy in the country is under development. However, considering natural climatic and social-economic conditions of Tajikistan, in terms of deficit of high-quality seed material of agricultural crops and a plenty of food products, a policy on principles of the environment protection has been formulated during the last decades, as well as regulation policy of introduction new varieties and phytosanitary risk and other biosafety principles. The development of modern biotechnology is making an ever-increasing influence on agriculture, public health and industries, and the economic strategy as a whole, all over the world. Using biotechnology ideas and methods is urgent in Tajikistan as well, particularly in providing an increase of agricultural production. In the recent decade, almost all plant-growing branches of the country show a decrease of crop productivity. Cotton-growing, potato-growing and cereal growing need new technologies to be developed and introduced. However, considering that Tajikistan is a country of origin of many biodiversity species and possesses rich genetic resources, the traditional methods of agriculture will be more preferable and not the methods of modern biotechnology. Government intends to take steps on development of applying the environmentally safe alternative actions to increase yields, prevent land degradation, etc. The important issue is also the conservation of indigenous animal species and agricultural plant species. Tajikistan, having ratified the Cartagena Protocol on Biosafety, considers it as an international platform for biosafety and will apply all mechanisms stipulated by Protocol to create a biosafety framework. Thus the formation of new views on modern biotechnology will be based on current country's positions stipulated in the Constitution and acting law of the Republic of Tajikistan. All the decisions made will be discussed with stakeholders and organizations having access to information and chance to take part in decision making on biosafety.

Regulatory Framework

Until the present time in Tajikistan the mechanism of decision making and biosafety regulation has not been developed in view of lack of permission on importing GMO to the country from other producing countries. NBF regulatory regime for the country will be based on international legislation on biosafety, regulated by Cartagena Protocol on Biosafety to the UN Convention on Biodiversity, as Tajikistan is a Party to the Convention on Biodiversity and Cartagena Protocol. These international principles have become the basis for adopting national legislation on biosafety regulation. Development of legislative base is not easy and is a long-term process. In the process of development, agreement and adoption of legislative base, Tajikistan will be guided by principles of the present National Biosafety Framework and Protocol as national legislation envisages the application of international legislation while lacking of national one. The Cartagena Protocol, acting legislation and standards, will provide the basis for decision making at the first phase.

National Legislation

National legislation regulating environmental policy, food safety and quality, is presented by:

- The Constitution of RT (06.11.94);

- The RT Law of Nature Protection (1.02.1996 No. 223) determines the RT policy in the area of environmental protection;
- The RT Law on Nature Protected Areas (13.12.1996, No. 329) regulates a conservation of unique nature components, environment, and genetic resources;
- The RT Law on Plant Quarantine (12.05.2001, No. 25) determines the principles of legal regulation in providing of plant quarantine in Tajikistan;
- The RT Law on Human Health Protection (15.05.1997, No. 420) determines and regulates the relations in the area of human health protection;
- The RT Law on Production and Service Certification (13.12.1996) regulates the need of observing the current standards of foodstuffs;
- The RT Law on Consumers' Rights Protection (15.05.1997) provides a realization of people's rights for safe and healthy foodstuffs;
- The RT Law on Food Products Quality and Safety (10.05.2002) regulating product quality and safety through registration, licensing, and certifying food products;
- The RT Law on achievements of agricultural crops selection (04.11.1995, No. 119) regulates the state register of seeds permitted for application in Tajikistan.

And other legislation documents provide sanctions for violation of safety rules for the environment and human health, for violation of veterinary regulations, for violation of standards on food products, etc.

However, the current standards cannot identify the food products safety to the full, as well as food raw materials and agricultural fodder in view of lack of qualitative and quantitative requirements on biosafety. However, the acting national legislation does not reflect the legal aspects on regulation of genetic engineering in the republic. Legislative documents are not a regulation basis on GMO use, release to the market, import and export of GMOs. There is a lack of GMO register, public access to information on GMOs is limited, there are no set up methods of GMO safety assessment, as well as lack of state control on biosafety, including the borders of the Republic.

Prospects and Needs

Considering the current situation within the country and lack of any regulatory regime on biosafety, the most important objectives at the first phase of National Biosafety Framework are:

- Adopting the RT Law on Biosafety.
- Development and introducing amendments into the acting legislation.
- Development and adopting of relevant legislative documents on realization of Law on Biosafety to ensure implementation of the legislation developed.
- Preparation of guidelines for the national competent institution and authorized agencies.
- Development of inter-institutional guidelines on cooperation in the process of decision making.
- Development of instructive documents on inter-institutional procedures of biosafety regulation.
- Development of marking system for GMO products.
- Attraction of investments.

Development of national legislation has been launched in the process of NBF preparation. The main goal of the Law is the creation of a legislative base for regulation of the activity attracting genetically modified organisms, and protection of human health and the environment. The Law prescribes the activity related to the use of GMO in contained use, their release into the environment, at the market and their import/export. For further activity is required to develop relevant guidelines, technical norms and standards among which:

- Inter-institutional guidelines on cooperation mechanisms in decision making,
- Regulation documents on inter-institutional procedures on biosafety.
- Mechanism of protection the confidential information in the work of inter-institutional committees on biosafety.
- Instructions on the control and preparation of the documents for export/import of goods containing GMO.
- Instructions on mechanisms of the process of deliberate release of GMO into the environment and at the market. Thus, detailed instructions and methodologies are required to carry the above procedures, monitoring and enforcement,
- Instructions on transportation and packaging of GM-products in case of transit through the country's area.
- Instructions on standardization and certification of GM-products.
- Identification of the classification criteria of the used GMO on risk classes.
- Development of various safety measures, management regulations and other requirements per each risk class.
- Development of marking rules and commercialization of food products containing GMO.
- Establishment of GMO register, permissible for using in the country.
- Development of mechanism for information on quality and safety of products.

In 2004 Tajikistan Parliament adopted the Law "On Biosafety." The Law created the National Committee on Biosafety and requirements on GMO labeling. The competent authority of decision making is the National Biodiversity and Biosafety Center, acting since January 2004 and is now establishing relevant commissions, expert board and other sub-divisions. National Biosafety Commission (decision making) and expert board (risk assessment) acting as inter-institutional 'authority in implementing the National Biosafety Framework.

STRUCTURE OF THE ADMINISTRATIVE BIOSAFETY SYSTEM OF THE REPUBLIC OF TAJIKISTAN

Public Awareness and Participation

Among the effective components of the National Biosafety Framework project activities is public awareness and participation in decision making on biosafety issues in Tajikistan. Public awareness and participation in decision making on biosafety is required for establishing contacts and possibility to use overall efforts and potential of the stakeholders. Participation promotes to the involvement of all stakeholders in decision making, ensures transparency and reporting and provides equal process and outcomes for all stakeholders. The basis for public participation in decision making, access to information on the environmental issues is a number of international conventions and agreements.

The Cartagena Protocol being the key document regulating and providing biosafety also commits its Parties to assist and promote to public awareness and education, and its participation in biosafety ensuring. Public participation at all stages in decision-making involving genetic engineering and GMOs is critical for the effectiveness of regulatory framework and to inform policy and decisions. The Cartagena Protocol on Biosafety places clear obligations in Article 23 on Parties to promote and facilitate public awareness, education and participation, including access to information, and also requires mandatory public consultation and disclosure of results of decisions to the public in the decision-making process. Different countries will have different local environment and agro-ecosystems, in which there may be no previous fields release and hence experience with the GMO in question. Participation by local people with knowledge of local conditions thus becomes important. Public participation can help provide information in diverse local conditions, practices and cultures that are relevant for the assessment to risk and impacts of GMOs. The authorities have to know which ecological and social systems the technology is going to interact with, and have to engage with the people who live in those conditions, as they will have a key role in evaluating the risks and impacts of GMOs. Assessing risks requires knowledge of what people value, why they value what they do, and who decides upon the value. It is thus not possible to evaluate the impacts of new technologies without reference to social concerns (Ching, 2007, pp. 555-567).

Tajikistan has ratified a number of Conventions and expressed its willingness to join the Convention on access to information, public participation in decision making and access to justice on the issues related to the environment (Aarhus Convention) which envisages the conditions for public access to information on the environment state and its participation in decision making. There is no special legislation on the public awareness mechanisms and access to information. Only few legislation acts envisage public awareness mechanisms on various sectors. The Law RT "On nature protection" (1993) establishes the base for access to information on the environment state and public participation. However, at the state level, the Tajikistan Government provides the public involvement in national strategy and action plan development, including the most active NGO representatives in the governmental working groups. Public involvement in policy development is becoming the practice. The National Plans of Actions on Climate Change, Combating Desertification, and Biodiversity Conservation, developed by Tajikistan, elaborated the issues of raising the public awareness, envisaged mechanisms of raising the educational level of the population on these issues, partially developed mechanisms of raising the public awareness and promoting the public involvement. However, the above program does not touch biosafety issues.

At present time on behalf of some NGOs there is a high public concern in relation to GMO used in production of food products and forage that is reflected in periodicals at workshops with public involvement. In particular, a number of NGO meetings have been organized concerning GMO and biosafety issues, with the main objective of wide-scale involvement of public organizations in the process of discussions and decision making on genetically modified organisms.

CONCLUSION

The National Biosafety Framework has to create an enabling environment for deriving maximum benefit from the achievements of modern biotechnology, to foster development of genetic engineering as one of the priority research areas of focus and, on the other hand, to ensure human and environmental health when carrying out genetic engineering activities, implementing new biotechnologies and consuming their products. Its implementation raises different types of issues which should be addressed at the appropriate time and level. Strengthening and/or developing adequate regulatory mechanisms and authorities should base on sound scientific data and clear correct facts. The most important output of the project is that now we have National Biosafety Framework for the Republic of Tajikistan, which can serve as a basic guide to the implementation of the biosafety system in our country. The involvement of different ministries and several stakeholders in the preparation of this document ensures that different views were taken into account, that stakeholders in NBF now recognize their roles much better. It is very important for future capacity building when the Law of the Republic of Tajikistan "On Biosafety" will come into force.

REFERENCES

Abdullkhodjaeva, M. (2007). *Ethical Review of Biomedical research in CIS Countries (Social and Cultural Aspects)*. Saint-Petersburg.

Achrorova, Z. (2007). *Ethical Review of Biomedical Research in CIS Countries (Social and Cultural Aspects)*. Saint-Petersburg.

Bashiri, I. (2003). *From the Hymns of Zarathustra to the Songs of Borbad.* Dushanbe.

Ching, L Li. (2007). *Biosafety First.* Tapir Academic Press.

Dinorshoev, M. (1999). *The Contribution of the Samanids Epoch to the Cultural Heritage of the Central Asia: The Proceedings of the International Colloquium.* Dushanbe: UNESCO.

Khor, M. (2007). *Biosafety First*. Tapir Academic Press.

Kubar, O., & ten Have, H. (2007). *Ethical Review of Biomedical research in CIS Countries (Social and Cultural Aspects)*. Saint-Petersburg.

Mikirtichian, G., Nikitina, A., Sozinov, A., Guryleva, M., & Malaysheva, E. (2007). *Ethical Review of Biomedical Research in CIS Countries (Social and Cultural Aspects)*. Saint-Petersburg.

Myhr, A. I., & Traavik, T. (2007). *Biosafety First*. Tapir Academic Press.

National Biosafety Framework. 2004. Dushanbe. Available at http://www.biodiv.tojikiston.com.

Soatov, I. (2003). *From the Hymns of Zarathustra to the Songs of Borbad.* Dushanbe.

UNESCO. (2006). *Avicenna and the Ethics of Science and Technology Today*. Paris: UNESCO.

Chapter 16
Genetic Testing and Protection of Genetic Privacy:
A Comparative Legal Analysis in Europe and Australia

Sergio Romeo-Malanda
University of Las Palmas de Gran Canaria, Spain

Dianne Nicol
University of Tasmania, Australia

Margaret Otlowski
University of Tasmania, Australia

ABSTRACT

Progress in the field of biomedical science has made it possible to obtain greater knowledge of the human genome and the nature of genetic disorders. Thanks to these advances, doctors now have the tools to diagnose certain disorders, and to carry out genetic tests to determine increased risks of developing other illnesses and of passing them on to future generations. In addition to the classic single gene disorders (like hemophilia and sickle cell anaemia), susceptibility genes are also being identified for genetically complex diseases, including many types of cancer, Alzheimer's disease, diabetes and other illnesses (House of Lords, 2009, p. 8). We can look toward a future where genetic test results are an important part of every healthy person's medical file.

INTRODUCTION

Currently, genetic testing has developed to the extent that doctors can pinpoint missing or defective genes. However, in practically all cases, treatments for those diseases are still far off.

DOI: 10.4018/978-1-61692-883-4.ch016

Despite the paucity of available treatments for genetic conditions at the current time, genetic testing still has an important social value in informing people about genetic risk factors for them and their offspring (both present and future). For some, such as those with an increased risk of cancer (e.g. colon or breast cancer), regular surveillance can be undertaken which can be beneficial in ensuring early diagnosis to maximize the effectiveness of available treatments. Genetic testing of individuals can bring to light important personal and family information about current and future health, including mental health, even though this may be limited to giving advance warning of a propensity or predisposition to certain disorders, or information on reproductive capacity and the future health of offspring. Knowledge of such matters gives individuals the capacity to plan for the future and to avoid lifestyle choices that may exacerbate the risk of adverse health outcomes. And, of course, those who receive a negative test result will have their concerns about developing a genetic condition in the future or passing on genetic risk factors to their offspring alleviated.

But genetic testing could also have significant detrimental social impacts. Increasing knowledge of the human genome and expanding opportunities for genetic testing raise new challenges, even though diagnosis of genetic risk factors is not of itself a new field (having long been available by means of mapping family histories for genetic diseases). What has changed is the *means* by which genetic information is available and also the *extent and quality* of information which can now potentially be obtained as a result of the advancements in relation to genetic testing (Otlowski, 2002b). While family history remains an important consideration in assessing the risk of developing a genetic disease, or a disease which has genetic as well as environmental triggers, genetic testing provides higher quality information in the sense that it can actually identify the genetic mutations that provide those genetic triggers. Hence, while genetic testing does not provide certainty about the onset of disease–neither its inevitability nor its severity–it does enhance the capacity to measure risk. Such information provides knowledge of highly personal aspects of the tested individual and directly affects his or her innermost sphere. The personal nature of this information means that particular care needs to be taken in establishing appropriate regulatory frameworks for dealing with it.

Unless appropriate procedures are put in place to protect this information, there is a risk that individuals may simply choose to refuse testing, even though it might otherwise provide significant benefits to them and their families (Keogh, 2009). The role of the law in this area should be to provide an appropriate balance between protecting the interests of individuals who undergo testing and taking into account other legitimate interests.

The aim of this article is to point out the main dilemmas in the field, with specific focus on the issues associated with genetic privacy, centered around three key questions:

1. How is the protection of genetic privacy recognized by law?
2. Does a person who undergoes genetic testing have a right to know test results as well as a right not to know?
3. To what extent can genetic test information be used outside of the confidential doctor-patient relationship by third parties including:
   - Other healthcare providers;
   - Other family members; and
   - Insurers and employers.
4. What new issues does direct-to-consumer testing raise?

The article starts by considering the available evidence on the extent to which these dilemmas are affecting members of the public in their decision-making about genetic testing and use of their genetic information. The article then examines the solutions that have been offered to these dilemmas within the international, European and Australian regulatory frameworks.

The rationale for examining these issues in the European context is that the European framework for personal data protection was established much earlier than in other jurisdictions.[1] and has undergone much greater scrutiny at the policy level. As such, the European approach provides one possible model that might be adopted much more broadly, especially if we take into consideration that the key legislative development in the field of genetic data is the recognition of the human right to privacy (Romeo-Malanda & Nicol, 2007).

Inclusion of the Australian perspective to this issue is also of particular interest because of the considerable attention given to the protection of genetic information in this country in recent years. Whilst the *Privacy Act* 1988 (Cth) has been in force for over 20 years, it is only in more recent times that the procedures for dealing with genetic information have come under scrutiny and are being reformed. In particular, there has been a major national inquiry into the protection of human genetic information jointly conducted by the Australian Law Reform Commission (ALRC) and the Australian Health Ethics Committee (AHEC) of the National Health and Medical Research Council (NHMRC). The ensuing report, *Essentially Yours* 2003 (Australian Law Reform Commission and Australian Health Ethics Committee (ALRC/AHEC), 2003), made numerous recommendations for reform, some of which have already been implemented with other changes in progress. These recent developments thus represent a useful counterpoint for analysis for the issues with regard to protecting genetic data.

PUBLIC VIEWS ON GENETIC TESTING AND USE OF GENETIC INFORMATION IN EUROPE AND AUSTRALIA

The extensive debate at the policy level about protection of genetic and other personal data has not been matched by a detailed analysis of public concerns and needs with regard to the protection of their data. The absence of such knowledge, and the lack of public debate has forced regulatory and advisory bodies to make assumptions about what the public might find acceptable (Academy of Medical Sciences, 2006).

In Europe, "Eurobarometer" surveys of public opinion on issues in biotechnology have been undertaken for a number of years. The 2005 survey included, for the first time, specific questions relating to genetic testing and use of genetic information. In response to the question "Would you be willing to take a genetic test to detect any serious disease that you might get?," 32% of participants said that they definitely would and 32% said they probably would, whereas only 16% said they definitely would not and 15% said they probably would not (leaving 5% undecided) (Gaskell et al., 2006). The authors of the study conclude that while the European public is not overwhelmingly supportive of the use of genetic data for personal genetic diagnosis, nevertheless, there is majority support for such use. Conversely, and perhaps not surprisingly, use by government agencies and for commercial insurance purposes does not have the same level of broad support (Gaskell et al., p. 54).

There is little available data on the reasons why members of the public might be less than willing to undergo genetic testing. Privacy is generally thought to be of paramount concern by legislators, policy makers and academic commentators. In Australia, a study on the impact of privacy legislation on various stakeholders was undertaken on behalf of the NHMRC in 2004 (National Health and Medical Research Council (NHMRC), 2004; Whiteman, Clutton and Hill, 2006). While the study did not focus specifically on genetic data, but considered health information more generally, it provides useful information on public attitudes towards privacy in the health care context. The study illustrates that there is a low level of knowledge about privacy legislation in Australia and some confusion about the distinction between privacy and professional confidentiality. However, the results also suggest that general public trust in health care professionals is "sufficient warranty that their privacy rights would be protected" (NHMRC, 2004, p. 4). Privacy laws also reassure members of the general public that their health information will be treated confidentially, but health consumers tend to be much more cautious in this regard, and a substantial minority of participants in the study expressed concerned about misuse of their health information (NHMRC, 2004, p. 6).

In conclusion, what data there are on public trust in relation to genetic testing and use of genetic and other health information suggest that individuals do have some confidence that their privacy and confidentiality will be respected, and that they are willing to undergo testing and allow their health information to be used on this basis. In moving forward, it is important that this level of trust is not compromised. On this basis, adequate protection of privacy and confidentiality are likely to be pivotal issues.

REGULATORY FRAMEWORKS FOR THE PROTECTION OF GENETIC PRIVACY

The regulatory framework that is relevant to the present discussion comes within the human rights domain. In the specific field of biomedicine, human rights discourse emphasizes the link between new technological advances and the rights of the individual, with particular focus on privacy. The concept of privacy is complex and has many meanings. Allen, for example, identifies four dimensions (Allen, 1997, pp. 81ff; Otlowski, 2002b, p. 88; Setoyama, 2005, pp. 81ff): (a) informational privacy; (b) physical privacy; (c) decisional privacy; and (d) proprietary privacy. This work focuses on informational privacy interests in controlling access to and use of personal information (Rothstein, 1997, p. 453; Otlowski, 2002b, p. 88).

International Regulatory Framework for the Protection of Genetic Privacy

The right to privacy is recognized at the international level in various binding and non-binding legal documents. Article 12 of the Universal Declaration of Human Rights of 10 December 1948 states that "no one shall be subjected to arbitrary interference with his privacy... Everyone has the right to the protection of the law against such interference or attacks." The binding force of this right to privacy is recognized in three ways. First, the Declaration was adopted in the United Nations Charter for the purpose of defining the meaning of the words "fundamental freedoms" and "human rights." Secondly, the Declaration is widely regarded as forming part of customary international law (Cancado-Trindade, 1948). Thirdly, this same right to privacy is recognized in other binding United Nations human rights documents, for example Article 17 of the International Covenant on Civil and Political Rights of 16 December 1966. This provision has binding force in the countries of Europe and in Australia, on the basis that they have all ratified the Covenant.

Other legal documents further develop the right to privacy in international law in respect of medicine and genetics. In particular, the United Nations Educational, Scientific and Cultural Organization (UNESCO) adopted the Universal Declaration on the Human Genome and Human Rights (UDHGHR) in 1997. The UDHGHR was adopted by the General Assembly of the United Nations on 9 December 1998, emphasizing the importance of this document, particularly because this is a matter without precedent (Gros-Espiell, 1997, pp. 131ff). The UDHGHR uses the language of confidentiality rather than privacy in defining the rights of individuals with regard to their genetic data. Article 7 states that "genetic data associated with an identifiable person and stored or processed for the purposes of research or any other purpose must be held confidential in the conditions set by law".

Subsequent to the adoption of the UDHGHR by UNESCO in 1997, the International Declaration on Human Genetic Data (IDHGD) was adopted on 16 October 2003. The IDHGD provides further clarification on the status of genetic data. The key principles of genetic privacy, non-disclosure to third parties and anonymity can all be drawn from Article 14 of the IDHGD, which provides that:

(a) States should endeavor to protect the privacy of individuals and the confidentiality of human genetic data linked to an identifiable person, family or, where appropriate, group, in accordance with domestic law consistent with the international law of human rights.
(b) Human genetic data, human proteomic data and biological samples linked to an identifiable person should not be disclosed or made accessible to third parties, in particular, employers, insurance companies, educational institutions and the family, except for an important public interest reason in cases restrictively provided for by domestic law consistent with the international law of human rights or where the prior, free, informed and express consent of the person concerned has been obtained provided that such consent is in accordance with domestic law and the international law of human rights. The privacy of an individual participating in a study using human genetic data, human proteomic data or biological samples should be protected and the data should be treated as confidential.
(c) Human genetic data, human proteomic data and biological samples collected for the purposes of scientific research should not normally be linked to an identifiable person. Even when such data or biological samples are unlinked to an identifiable person, the necessary precautions should be taken to ensure the security of the data or biological samples.
(d) Human genetic data, human proteomic data and biological samples collected for medical and scientific research purposes can remain linked to an identifiable person, only if necessary to carry out the research and provided that the privacy of the individual and the confidentiality of the data or biological samples concerned are protected in accordance with domestic law.
(e) Human genetic data and human proteomic data should not be kept in a form which allows the data subject to be identified for any longer than is necessary for achieving the purposes for which they were collected or subsequently processed.

While neither the UDHGHR nor the IDHGD create binding legal obligations, their force lies in providing moral guidance to states. According to some academics, these international Declarations, though not coercive legal rules in a strict sense, should be integrated into general principles of domestic law, as has been seen with the Universal Declaration on Human Rights (Romeo-Casabona, 2002, p. 42). Many of these principles have been shaped by other international laws, and they are of paramount importance in achieving adequate state regulation in relation to the biomedical sciences.

European Regulatory Framework for the Protection of Genetic Privacy

Europe has a complex regulatory framework because it is made up of a number of sovereign states. In this paper, attention is focused on the collective European legal system rather than the legislation of each individual European country. Reference is made to the two key arms of the European legal system: the Council of Europe and the European Union.

The Council of Europe is an international organization of 47 member states in the European region, among whose purposes is the protection of human rights.[2] This organization has produced several documents with regard to the right to privacy and data protection, with special reference to privacy in the fields of medicine and genetics. The Convention for the Protection of Human Rights and Dignity of the Human Being with regard to the Application of Biology and Medicine: Convention on Human Rights and Biomedicine (CHRB), of 4

April 1997 is particularly important because, as a Treaty, it has binding legal force.[3] Two Additional Protocols to the CHRB concerning Biomedical Research (APBR), of 25 January 2005, and Genetic Testing for Health Purposes (APGT), of 27 November 2008[4] are also of particular relevance to the present study.

The European Union has 27 member states.[5] It was established by the Treaty of Maastricht in 1993. Legislation of the European Union that is relevant to this study includes the Charter of Fundamental Rights of the European Union, of 7 December 2000, and Directive 95/46/EC of the European Parliament and of the Council, on the Protection of Individuals with Regard to the Processing of Personal Data and on the Free Movement of Such Data, of 24 October 1995. The latter is especially relevant because it generates in all the EU Member States an effective obligation to act according to the terms stipulated in it. Directives are legislative acts of the European Union which require member states to achieve a particular result without dictating the means of achieving that result. They can be distinguished from European Union regulations which are self-executing and do not require any implementing measures. Directives normally leave member states with a certain amount of leeway as to the exact rules to be adopted and can be adopted by means of a variety of legislative procedures depending on their subject matter.

The right to privacy in relation to health is recognized by the Council of Europe in Article 10 CHRB, which states that "Everyone has the right to respect for private life in relation to information about his or her health". Article 25 APBR provides further that "Any information of a personal nature collected during biomedical research shall be considered as confidential and treated according to the rules relating to the protection of private life" and Article 16 APGT provides that "Everyone has the right to respect for his or her private life, in particular to protection of his or her personal data derived from a genetic test".

The Charter of Fundamental Rights of the European Union recognizes the "protection of personal data" as a right in itself (Article 8). Directive 95/46/EC further develops this right. Its object, as stated in Article 1.1, is to "protect the fundamental rights and freedoms of natural persons, and in particular their right to privacy with respect to the processing of personal data". This Directive states in Article 2(a) that *personal data* "shall mean any information relating to an identified or identifiable natural person ('data subject')". Article 29 of Directive 95/46/EC established a Data Protection Working Party, which is an independent advisory body to the European Union on data protection and privacy. Its tasks are laid down in Article 30 of Directive 95/46/EC and in Article 14 of Directive 97/66/EC, of 15 December 1997, concerning the processing of personal data and the protection of privacy in the telecommunications sector.

The Working Party's Opinion 4/2007 on the Concept of Personal Data is of particular interest. According to the Working Party's interpretation of the definition of personal data in Article 2(a) of Directive 95/46/EC, there are three distinct categories of data depending on the likelihood of identification of the person from whom they are obtained (Data Protection Working Party, 2007, pp. 12ff): (a) data relating to an identified person; (b) data relating to an identifiable person; and (c) anonymous data. Data relating to an identified person can be defined as data that appear clearly and directly linked with the person from whom they were obtained. Data relating to an identifiable person (known as "dissociated data") are not directly attributable to that person, but may be linked to them by diverse procedures, which can be easily carried out.[6] Finally, anonymous data can be considered as data where the identity of the data subject is not known, and identification is not possible because the data were anonymous when collected, or were later anonymized (Romeo-Casabona, 2004, p. 38). To be classified as anonymous, the process of anonymization should be irreversible, that is to say, that the data cannot be returned to the form taken previously.

Article 2(a) of Directive 95/46/EC does not mention the category of anonymized data explicitly. However, Recital 26 limits the reach of Article 2 (a) in stating that "to determine whether a person is identifiable, account should be taken of all the means likely *reasonably* to be used either by controller or by any other person to identify the said person"[7] (emphasis added). This provision indicates that when unreasonable means are required to identify a person, the data will move into the category of anonymous data. However, what will amount to "reasonable means" is not easy to say.[8]

The principles of data protection do not apply to data rendered anonymous in such a way that the data subject is no longer identifiable. This implies that protection will not be given to *any* data that have been subjected to an anonymization process (Romeo-Casabona, 2004, p. 34). However, if these anonymous data were processed and it became possible to identify the data subject again, they would regain the status of personal data and the principles of data protection would again be applicable to them (Romeo-Casabona, 2004, p. 42).

These definitions are particularly important in the research context, because they place limitations on the ways in which data can be obtained and uses to which it can be put. The principle of autonomy requires that each person is entitled to decide to whom, when, and to what extent personal information relating to him or her can be processed. On this basis, identified data can only be used for research purposes if consent has been obtained, but truly anonyimzed data can be used without consent. According to Article 8.2.a) of Directive 95/46/EC, a data subject must give his or her explicit consent to the processing of personal data concerning health.[9]Genetic data undoubtedly fall within the scope of this provision (Data Protection Working Party, 2004, pp. 5 and 7). In addition, Article 9 APGT states that "a genetic test may only be carried out after the person concerned has given free and informed consent to it." The person concerned may freely withdraw consent at any time.

Australian Regulatory Framework for the Protection of Genetic Privacy

In Australia, the regulatory framework is also complicated by virtue of Australia's status as a federation, with some legislative powers vested in the Commonwealth, and others devolved to the States and Territories. Privacy is an area of overlapping jurisdiction and this has represented a significant challenge to efforts to create uniform privacy obligations across the multiple Australian jurisdictions. Initially, the *Privacy Act* 1988 (Cth) only applied to Commonwealth government agencies, but the legislation was significantly expanded in 2001 to include the private sector with the introduction of the National Privacy Principles (NPPs) to augment the Information Privacy Principles (IPPs) applying to the public sector. The amending Act (Privacy Amendment (Private Sector) Act 2000 (Cth)) drew on OECD (Organization for Economic Cooperation and Development) and European regulatory instruments in respect of privacy and brought about more comprehensive privacy regulation for Australia albeit still fragmented with separate coverage for public and private sectors and divisions between Commonwealth and States and Territories (Guidelines for the Protection of Privacy, 1980; Regulation (EC) 45, 2001).

Recent reform recommendations of the ALRC which have since been largely endorsed by the Commonwealth Government (Australian Government First Stage Response, 2009), have called for consolidation of the separate privacy principles currently applying in the public and private sectors (IPPs and NPPs) into Unified Privacy Principles regulating the collection, storage and use and transfer of information which would also form the basis of the proposed co-operative scheme involving the states (ALRC, 2008). For the purposes of this paper, attention will focus on the federal privacy scheme in Australia as this is the dominant regime. As in the European context, duties of confidentiality which arise in particular relationships such as doctor/patient also play an important role in protecting the interests of patients in having their personal information kept confidential. Breach of this duty may result in damages in tort, contract or for equitable breach of confidence, and may also be the basis for disciplinary proceedings. The question of whether or not a common law tort of privacy exists in Australia has still not been fully canvassed by the courts (ALRC, 2008, Chapter 74).

The focus of protection under the Commonwealth *Privacy Act* 1988 is on "personal information" defined in the Act as: "information or an opinion (including information or an opinion forming part of a database), whether true or not, and whether recorded in a material form or not, about an individual whose identity is apparent, or can reasonably be ascertained, from the information or opinion" (Section 4). The Act provides for a category of "sensitive information" to which special protections apply including the need for consent of the individual before such information can be collected subject only to the statutory exceptions provided.[10] "Sensitive information" is defined as including "health information" about an individual.[11] Genetic information had been understood to fall within this category, however, further to the recommendations of the ALRC/AHEC *Essentially Yours* Report, genetic information is now expressly included in the Act under the definitions for health information and sensitive information.[12] The purpose of this amendment has been to clarify and raise awareness as to the applicability of the *Privacy Act* to this form of personal information and also to ensure that forms of genetic information which are not also health information (e.g. parentage or kinship information) are covered by the Act.

Notably, however, the protection of the legislation only extends to genetic information, not the tissue samples themselves from which such information can be derived. Whilst the ALRC/AHEC *Essentially Yours* Report did recommend the expansion of federal privacy principles by redefining personal information as including genetic samples, this recommendation was not supported by the (then) Commonwealth government and has not been implemented (Government Response to Recommendations, 2009). There had been some precedent for reform of this kind: one of the Australian states, New South Wales, has adopted this approach in its privacy legislation which explicitly applies information privacy principles to genetic samples (Privacy and Personal Information Protection Act 1998 (NSW) and Health Records and Information Privacy Act 2002 (NSW)). The failure to implement such a provision in the Commonwealth privacy legislation arguably leaves genetic samples without adequate protection. There is some prospect of mitigation of this situation, however, as other ALRC/AHEC recommendations for the introduction of an offence for non-consensual genetic testing along similar lines to the provisions in section 45 of the UK *Human Tissue Act* 2004 were supported and a Discussion Paper has been developed and circulated for comment (Model Criminal Law Officers' Committee, 2008).

Also to be noted are the ethical guidelines which apply in the research setting. In 2007 the NHMRC issued a revised *National Statement on Ethical Conduct in Human Research* (National Health and Medical Research Council, 2007). This includes chapters on databanks, human tissue samples and human genetics. In defining "data" the *National Statement* distinguishes between "individually identifiable data," "re-identifiable data" and "non-identifiable data" (National Health and Medical Research Council, 2007, p. 29). According to the *National Statement*, "individually identifiable data" is where the identity of a specific individual can reasonably be ascertained. "Re-identifiable data" is data from which identifiers have been removed and replaced by a code, but it remains possible to re-identify specific individuals by, for example using the code or linking different data sets. "Non-identifiable data" is defined as data which have never been labeled with individual identifiers or from which identifiers have been permanently removed and by means of which no specific individual can be identified (National Health and Medical Research Council, 2007, p. 29).

Essentially these are the same categories of data that are identified in European Directive 95/46/EC. The main difference is that in Europe this Directive is a legal document but in Australia the equivalent provisions are found in ethical guidelines. There is ongoing debate in Australia as to the legal status of the *National Statement* (Nicol, Otlowski and Chalmers, 2001).

THE RIGHT TO KNOW AND NOT TO KNOW

As a matter of principle, everyone who undergoes genetic testing has a right to know their test results. There is increasing recognition of a right *not* to know, including the right not to undergo genetic testing. This right is important because most hereditary diseases cannot be treated or even prevented at present; in many cases, the only certain prediction is that the disease will develop and that nothing can be done to prevent or delay its onset.[13]

International Law on the Right to Know and Not to Know

According to Article 10 IDHGD, "the person concerned has the right to decide whether or not to be informed of the results." This provision provides not only a clear reference to the right to know, but also to the right "not to know." This right is also protected in other international instruments including the UDHGHR (Article 5).

European Law on the Right to Know and Not to Know

European legislation provides that everyone is entitled to know any information collected about their health (Articles 10.2 CHRB, 16.2 APGT and 26 APBR). This is, however, subject to the so-called "therapeutic exception,"[14] which may restrict a person's right to know in the interests of their health. In cases where this exception applies, it might be necessary to communicate the withheld information to a next of kin, in order to make the right decision.[15]

In the research setting, there are at times debates as to the extent of researchers' obligations in connection with large-scale projects to feed back clinically relevant information to participants; in the context of emerging biobanks it has been argued that there may be not only an *ethical* duty but also a *legal* duty to disclose, arising under Article 2 of the *European Convention on Human Rights* (Johnstone and Kaye, 2004).

In addition to the right to know, European legislation also recognizes the right not to know. Articles 10.2 CHRB and 16.3 APGT state that the wishes of individuals not to be informed shall be observed. Individuals may have their own reasons for not wishing to know about certain aspects of their health and these wishes must be observed.

Australian Law on the Right to Know and Not to Know

There is no specific legislation in Australia which guarantees the right to know or not to know one's genetic information. However, as with any medical procedure, the law protects the autonomy of competent individuals to decide whether to undertake genetic testing and to accept medical treatment or advice about lifestyle changes arising from such testing.[16]

In the Australian research setting, the *National Statement* is not prescriptive as to whether clinically relevant information must be provided to participants but it does require researchers to prepare and follow an ethically defensible plan to disclose or withhold that information (National Statement 3.5.1.).

At the same time, the *National Statement* recognises (paragraph 3.5.2(a)(i)) care must be taken to ensure respect for the participant's right not to know. However, the *National Statement* expressly states that even if participants or relatives indicate that they prefer *not* to receive genetic information that is important for their health, they must nevertheless be advised that they will be approached to confirm this decision when the results of the research are available (see paragraph 3.5.2(c)).

PRIVACY, CONFIDENTIALITY AND DISCLOSURE TO THIRD PARTIES

As a starting point in this section, it is important to distinguish the terms "privacy" and "confidentiality." Although they are connected in some sense, they are not synonyms (Cavoukian, 1995). Privacy includes the right to control one's own personal information (this perspective is also known as "informational self-determination").[17] Confidentiality, on the other hand, is but one means of protecting private information, by creating a legal obligation to keep the information confidential. Essentially, it imposes an obligation on the recipient of personal information about another person not to use that information for any purpose other than that for which the information was given (Otlowski, 2001, pp. 66 and 72). Obligations of confidentiality can arise in a whole range of circumstances, but they most clearly exist in certain recognized relationships, particularly the doctor-patient relationship. As such, confidentiality is an important but not absolute safeguard in protecting the privacy of health information (Elizalde, 1995, p. 307). While the right to privacy (and with it the duty of confidentiality) has been accepted as a fundamental human right,[18] it is also recognized that there is a tension between privacy and other aspects of human life, including personal health and welfare, the rights of family members to be informed of genetic risk factors and the rights of insurers and employers to be informed of matters that have actuarial or occupational health and safety implications.

In some limited cases, it may be necessary for a medical practitioner to share information about a patient with other professionals involved (directly or indirectly) in the patient's care. In this respect, medical data should only be communicated to those who are subject to the rules of confidentiality incumbent on health-care professionals, or comparable rules of confidentiality. We are referring here to the so-called notion of "shared confidentiality" (Romeo-Casabona, 2001, p. 780).

In the research context, consent requirements are pivotal. In some circumstances consent may not be required, when the information has been anonymised in such a way that the data subject is no longer identifiable (Romeo-Casabona, 2004, p. 34). However, it is also important to be aware that, given modern advances in genetic technology, the feasibility of fully anonymizing genetic data is open to question (National Statement 3.5.8.).

In the family context, because genetic features are transmissible from one generation to another, if a person carries a gene responsible for a certain disease it is quite likely that the gene will also be found in other members of that person's close family. If an individual refuses to inform relatives of relevant genetic data, questions inevitably arise as to whether their medical practitioner: (a) is at liberty to disclose relevant information to genetic relatives in the absence of consent from the patient; and (b) has a duty to make such disclosure. These are frequent dilemmas in medicine and there are no universal answers to them. Two different approaches have emerged (Skene, 1998): the *legal model,* which takes into account the need to pass on information in some circumstances and the *medical model,* which focuses on the duty to inform. At the core of the legal model is the individual's right to privacy, stemming from the idea that each person has the right to control their own body and genetic information, while the medical model rejects the language of individual rights and instead stresses the need to warn patients who have risks of a genetic nature.

In countries where the "legal model" has been adopted, doctors may only disclose relevant information with the express consent of the patient, even if a family member is at risk. Nevertheless, even where this legal model applies, there may still be the possibility of disclosing this information in very closely defined circumstances. This is not because a duty to forewarn exists, but because breach of confidentiality is justified in all the circumstances (Moniz, 2004; European Standards on Confidentiality and Privacy in Healthcare, 2006, paragraph 3.4.2). Pursuant to the medical model, which has regard to the familial and shared nature of genetic information, relevant information must be shared between family members and under no circumstances can this be prevented by the patient. Effectively, doctors are under a duty of disclosure.

International Law on Privacy, Confidentiality and Disclosure to Third Parties

According to Article 14. (b) IDHGD, human genetic data should not be disclosed or made accessible to third parties, particularly employers, insurance companies and educational institutions, except for an important public interest reason in cases restrictively provided for by domestic law or where the prior, free, informed and express consent of the person concerned has been obtained.

The right of family members to be informed of genetic risks is recognized in Article 4.a) IDHGD. This provision states that human genetic data "may have a significant impact on the family, including offspring, extending over generations." Whenever a gene is discovered in a given individual, therefore, the question arises as to whether this information should be communicated to the individual's relatives for the purposes of diagnosis, prevention or therapy, etc.

European Law on Privacy, Confidentiality and Disclosure to Third Parties

In Europe there exists no express recognition of the duty of non-disclosure to third parties in a binding legal text. However, The European Court of Human Rights has held that it is implicit in the right of privacy, guaranteed by Article 8 of the Convention:

the protection of personal data, not least medical data, is of fundamental importance to a person's enjoyment of his or her right to respect for private and family life as guaranteed by Article 8 of the Convention (art. 8). Respecting the confidentiality of health data is a vital principle in the legal systems of all the Contracting Parties to the Convention. It is crucial not only to respect the sense of privacy of a patient but also to preserve his or her confidence in the medical profession and in the health services in general. Without such protection, those in need of medical assistance may be deterred from revealing such information of a personal and intimate nature as may be necessary in order to receive appropriate treatment and, even, from seeking such assistance, thereby endangering their own health and, in the case of transmissible diseases, that of the community.[19]

In the family context, the legal model has prevailed to date in Europe, allowing medical practitioners to disclose relevant information to genetic relatives in the absence of consent from the patient. However, the CHRB offers no precise guidance to medical practitioners in deciding the circumstances in which genetic information can or cannot be disclosed to family members. It provides only general guidance in Article 10.1, which gives everyone the right to privacy in relation to their health, in conjunction with Article 26, which allows certain very limited restrictions to be placed on the exercise of the rights and protective provisions in the Convention, including the protection of the rights and freedoms of others.

The Data Protection Working Party states that "given the complexity of the issues described above, the Working Party takes the view at this stage that consideration should be given to a case by case approach in deciding how to address possible conflicts between the interests of the data subjects and those of their biological family" (Data Protection Working Party, 2004, p. 9).

Romeo-Casabona and others argue that the protection of essential rights of third parties can be achieved on grounds of the *conflict (or collision) of duties*[20] or, as the case may be, the *state of necessity,*[21] both of which concepts exist in European legislation (Romeo Casabona, 2001, p. 784; Nys, Romeo-Casabona and Desmet, 2002). However, Opinion 4/2007 puts this assertion in question. In this Opinion, the Working Party provides guidance on the way in which the concept of personal data in Directive 95/46/EC and related community legislation should be understood and how it should be applied in different situations. The Working Party's analysis is based on the four main "building blocks" that can be distinguished in the definition of "personal data" in Article 2(a) of Directive 95/46/EC: i.e., "any information," "relating to," "an identified or identifiable," "natural person."

One of the key findings of the Working Party is that the second building block ("relating to") has been too often overlooked, and that it should play a crucial role in determining the substantive scope of the concept as a whole, especially in relation to objects and new technologies (Data Protection Working Party, 2007, p. 9ff). The Opinion suggests that there will be sufficient connection between data and a particular person simply if it can be established that the person may be treated differently from others as a result of the processing of such data ((Data Protection Working Party, 2007, p. 11). On this basis, genetic data could be considered to be personal data, not only of the source of the sample, but also of members of their biological family. The difficulty with this proposition is that it could mean that all the members of the biological family of the tested subject (parent, children, siblings, etc.) must be considered as "data subjects." The logical consequence of this proposition is that family members not only have access to all data, but also must consent every act of processing. This is generally not practicable.

Another area of possible conflict is that related with the use of genetic information for insurance and employment purposes. There is no binding legal text at the European level dealing specifically with this issue. According to the Explanatory Report of CHRB (Paragraphs 84-86), because there is an apparent risk that use is made of genetic testing possibilities outside health care (for instance in the case of medical examination prior to an employment or insurance contract), it is of importance to clearly distinguish between health care purposes for the benefit of the individual on the one hand and third parties' interests, which may be commercial, on the other hand.

Article 12 CHRB prohibits the carrying out of predictive tests for reasons other than health or health-related research, even with the assent of the person concerned. Therefore, it is forbidden to perform predictive genetic testing as part of pre-employment medical examinations, whenever it does not serve a health purpose of the individual. This means that in particular circumstances, when the working environment could have prejudicial consequences on the health of an individual because of a genetic predisposition, predictive genetic testing may be offered, but only for the purpose of protecting the individual's health (Opinion of the European Group, 2003).[22] In addition, this Explanatory Report also points out that insofar as predictive genetic testing in the employment or private insurance contexts does not have a health purpose, it entails a disproportionate interference with the individual's right to privacy. An insurance company will not be entitled to require a person to undertake a genetic test for the purpose of entering into or modifying an insurance policy. Nor will it be able to refuse a policy on the ground that the applicant has not submitted to a test, as entry into the policy cannot reasonably be made conditional on the performance of an illegal act.

Australian Law on Privacy, Confidentiality and Disclosure to Third Parties

On the question of disclosure to other family members, this is an area where Australian law has undergone significant change in recent times. Previously, doctors' confidentiality obligations together with statutory privacy principles precluded them from disclosing a person's genetic information to genetic relatives without the person's consent, irrespective of how important that information may be to them. In response to calls for sharing of relevant genetic information amongst at-risk relatives, the ALRC/AHEC Inquiry recommended an amendment to the *Privacy Act* 1988 (Cth) to allow such disclosure without the consent of the person in circumstances where disclosure is necessary to lessen or prevent a serious threat to an individual's life, health or safety, even where the threat is not imminent. This recommendation has since been implemented in the *Privacy Act* 1988 (Cth).[23] The Act specifies that such disclosure must be conducted in accordance with guidelines relating to such disclosure which are to be issued by the NHMRC and approved by the Privacy Commissioner.[24] Notably, the amendments do not make it *obligatory* for health practitioners to make such disclosure, but rather, legally permits them to do so if they see fit. Further, the changes only apply to a health practitioner in the private sector The enabling guidelines have now been finalized and took effect on 15 December 2009: [25] it remains to be seen, however, the extent to which the provision is used in practice given that it does not compel disclosure.

There is considerable disquiet about the use of genetic test information in non-medical contexts, such as by insurers or employers, particularly with increasing computerization and linkage of health records. As noted, it is beyond the scope of this paper to consider *misuses* of genetic information for non-medical purposes, so the focus is on circumstances in which Australian law permits third parties such as insurers or employers to access and use a person's genetic test information.

These issues have been most prominent in Australia with regard to life insurance, for which there is underwriting based on individual risk assessment (unlike Australian health insurance which is community rated). When applying for life insurance, individuals are required to disclose family history and the results of any genetic tests and insurers are entitled to take this information into account for the purposes of underwriting for life insurance and related products. Insurers are exempt from disability discrimination legislation[26] but must be able to justify the way in which they use the genetic information with regard to actuarial, statistical or other data.

The ALRC/AHEC Inquiry into the protection of genetic information spent considerable time canvassing this issue. Whilst the ensuing Report recommended retention of the disclosure obligation of individuals to ensure symmetry of information between applicants and life insurers and to protect insurers from adverse selection, they did recommend introduction of a process of vetting of genetic tests to determine which are suitable for use in life insurance underwriting, having regard to their scientific reliability, actuarial relevance and reasonableness.[27] This recommendation had received government endorsement but is yet to be implemented. In relation to privacy issues, the Report acknowledged that the practice of insurers of collecting genetic information from applicants about their genetic relatives for use in underwriting insurance polices in relation to the applicants potentially involved breach of the privacy of genetic relatives but was accepted to be a necessary practice and accordingly a recommendation was made for a Public Interest Determination to be sought under the *Privacy Act* 1988 (Cth).[28] The peak life insurance body, IFSA (Investment and Financial Services Association Limited), has responded to concerns about maintaining genetic privacy of applicants. The policy on genetic testing which has since become an industry standard, seeks to make clear that applicants will only be assessed on the basis of the information that they have provided, and information provided about genetic relatives will not be used to assess applications from those people.[29]

There is little evidence of genetic testing presently being used in the employment setting. Up until recently, there were no laws to prevent an employer from seeking this information, or even requesting genetic testing as part of their recruitment, although under disability discrimination legislation, there are some constraints on the use that could be made of this information (Otlowski, 2002a). The real challenge is to ensure that legitimate uses of genetic test information are permitted, such as screening for susceptibility to workplace hazards that cannot otherwise be avoided, but to protect employees and job seekers from unfair discrimination motivated by employer expediency and profit (Otlowski and Williamson, 2003, p. 5). The proposals advanced by the ALRC/AHEC seek to strike a proper balance to allow uses of genetic testing which are consistent with occupational health and safety interests, but prohibit other uses.[30] There has been some progress towards implementation.

There has been an amendment to the *Disability Discrimination Act* 1992 (Cth) to strengthen existing protection against genetic discrimination in employment by amending the definition of 'disability' to include a genetic predisposition to a disability, as well as prohibiting employers from requesting or requiring genetic information from a job applicant or employee, except where reasonably required for purposes not involving genetic discrimination (*Disability Discrimination and Other Human Rights Amendment Act* 2009 (Cth)). The effect of this provision is to put the onus on employers to provide evidence that none of the purposes of the request was that of unlawfully discriminating.

GENETIC PRIVACY AND DIRECT-TO-CONSUMER TESTING

As a general rule, genetic testing is carried out in the medical sector upon referral by a doctor. However, in recent years genetic testing has increasingly been offered directly to consumers (Pearson, 2008). Although the access to these tests can be organized by over-the-counter sales in pharmacies or drugstores, the main channel is the Internet. The main concern regarding direct-to-consumer genetic testing is that the services offered (whether via internet or over the counter in pharmacies) cannot match the high professional standards of medical and genetic consultation required for normal genetic testing (Hogarth, Javitt and Melzer. 2008). There are also challenges for protecting privacy given the potential for specimens to be sent for genetic testing against the will of a person or without their knowledge.

Because choices in relation to obtaining information about one's genetic status are very personal decisions which can have significant implications, appropriate genetic counseling is seen as playing an integral role in ensuring informed decision-making. The objective of genetic counseling is to educate people at risk of a genetic disorder about their genetic status and about the possible implications and complications of their genetic condition. Genetic counseling helps people make decisions about their future lives with respect to diagnostic, therapeutic, ethical and practical factors.

One form of supplying direct-to-consumer genetic testing that is under discussion as being particularly problematic is the supply of complete self-testing kits that allow the customer to directly read the positive or negative result of the test from the kit at home, comparable to a common pregnancy testing kit. In this scenario, it may be that there is no provision for counseling at all except for the written advice on the web page. In the alternative, counseling may be offered as an additional special service at extra cost and at the customer's request, or it may be that the customer is invited to contact a doctor or health practitioner from the company by phone for counseling. In other cases, the customer may be recommended to consult their own doctor on receipt of test results.

International Framework on Direct-to-Consumer Testing

Internationally, there is no specific framework on direct-to-consumer testing. As such, general principles relating to autonomy and privacy apply. Article 2 of the IDHGD defines genetic counseling as "a procedure to explain the possible implications of the findings of genetic testing or screening, its advantages and risks and where applicable to assist the individual in the long-term handling of the consequences; It takes place before and after genetic testing and screening."

European Framework on Direct-to-Consumer Testing

In Europe, there are no legal or other binding regulations that explicitly deal with direct-to-consumer genetic testing. However, some legal texts from both the European Union and the Council of Europe could be of significance.

As to concerns related to clinical validity and usefulness of direct-to-consumer tests, Directive 98/79/EC of the European Parliament and of the Council of 27 October 1998 on in vitro diagnostic medical devices is of interest. This Directive applies to in vitro diagnostic medical devices and their accessories in order to guarantee their security and quality. In the context of direct-to-consumer genetic testing, it is not clear that the Directive applies to laboratory-developed tests performed by laboratories outside of Europe. There is also a lack of clarity as to the extent to which the Directive takes into account the clinical validity or usefulness of a medical device. Finally, even if direct-to-consumer testing falls under the scope of Directive, it is not clear which types of tests would be covered, because the Directive does not require pre-market review for "low risk" items (Hogarth and Melzer, 2007; Hogarth, Javitt and Melzer. 2008, pp. 173ff). In order to overcome these concerns, Directive 98/79/EC is currently undergoing review by the European Commission, and public consultation has commenced.

The CHRB and APGT may address certain matters relating to the lack of proper genetic counseling in the context of direct-to-consumer genetic testing (Kaye, 2008), although it is not definitive. According to Article 12 CHRB, "tests which are predictive of genetic diseases or which serve either to identify the subject as a carrier of a gene responsible for a disease or to detect a genetic predisposition or susceptibility to a disease may be performed only for health purposes or for scientific research linked to health purposes, *and subject to appropriate genetic counseling*" (emphasis added). In the same sense, Article 7.1 APGT states that "a genetic test for health purposes may only be performed under individualized medical supervision." Use of the term "individualized" stresses the need for personal genetic counseling in order to enable an informed decision to be made. This clearly excludes the practice of indirect or remote counseling conducted by many direct-to-consumer genetic testing companies.

However, the practical effect of the APGT and the CHRB remains unclear (Borry, 2008, p. 182), in part because they have not yet been signed by a number of European countries. Moreover, both allow member states to make an exception to general rules for counseling mentioned above (Articles 26 and 7.2 respectively). Finally, because they apply to tests which are carried out for health purposes it may be possible to argue that direct-to-consumer tests are not done for health purposes.

Concerns relating more closely with the right to privacy (which might arise, for example, where the person asking for the test is not actually the source of the sample), remain unanswered in the European context. Specific regulation of direct-to-consumer genetic testing at European level may be needed to deal with these issues. The UK Human Genetic Commission has recently released a draft common framework of principles for direct-to-consumer testing (Human Genetic Commission, 2009), which is likely to provide important guidance in formulating a more general European response.

Australian Framework on Direct-to-Consumer Testing

Regulating access to genetic testing was another of the issues addressed by the ALRC/AHEC national inquiry with a number of recommendations made (ALRC/AHEC, Chapter 11). There has been considerable progress with regard to establishing a sound framework for the manufacture, sale and promotion of direct-to-consumer genetic tests within Australia. This is governed by a combination of consumer protection laws, (in particular, the *Trade Practices Act* 1974 (Cth)), privacy legislation, regulation of laboratories through accreditation requirements, and regulation of therapeutic goods further to which a new regulatory framework has recently been developed for the comprehensive regulation of in-vitro diagnostic services in Australia.[31] The real challenge lies in the regulation of companies operating outside of Australia which offer direct-to-consumer genetic testing services through the internet with testing conducted by overseas laboratories. A number of concerns have been raised including the questionable validity and/or utility of some of the direct-to-consumer tests that are available, and concerns about the impact of obtaining this information on individuals, particularly in the absence of genetic counseling. One issue on which at least preliminary action has been taken relates to non-consensual genetic testing: as noted, a Discussion Paper has been circulated for the introduction of an offence for non-consensual genetic testing which potentially will apply to persons who obtain or use bodily material for a genetic test without the consent of the person (Model Criminal Law Officers' Committee, 2008). Whilst there would inevitably be difficulties with enforcement of this legislation beyond Australia, it would at least ensure that any such activities undertaken within the jurisdiction would be punishable.

CONCLUSION

This chapter has shown that the right to privacy in the context of genetic testing and access to genetic data is treated very seriously in the international legal framework, and in Europe and Australia. In Europe, both the Council of Europe and the European Union have enacted general privacy legislation and specific legislation dealing with privacy in the context of genetic data. Moreover, the Data Protection Working Party has provided valuable guidance on how to interpret these instruments.

Similarly, in Australia, the *Privacy Act 1988* (Cth) provides general protection of the privacy of personal information. The ALRC/AHEC Report provides an essential guide and recommendations with regard to the specific issues associated with genetic testing and access to genetic data. Responses to ALRC/AHEC recommendations, in the form of amendments to the *Privacy Act 1988* (Cth) and changes to the *National Statement* provide an additional specific layer of protection for genetic privacy.

For the most part, however, the approach taken in Australia is quite different from that in Europe. The approach in Europe is clearly based on recognition of human rights, particularly the human right to privacy. Although Australia is signatory to a number of international human rights instruments, there is little in the way of concrete, legally binding, human rights obligations in this jurisdiction.

Rather, recognition of human rights is implicit in Australian legislation and case law and in ethical guidelines in the research context. This is not to say that genetic privacy is inadequately protected in Australia, but that it is one of various regulatory options for achieving such protection, which all rely on the ongoing international human rights discourse for their legitimacy.

Both the European and the Australian regulatory frameworks are likely to be challenged with increasing availability of and demand for direct-to-consumer testing. It is important that a proactive approach is taken to these new challenges in both regions, against the backdrop of further policy debates and regulatory developments internationally.

REFERENCES

Academy of Medical Sciences. (2006). *Personal Data for Public Good: Using Health Information in Medical Research.*

Allen, A. (1997). Genetic privacy: emerging concepts and values. In Rothstein, M. A. (Ed.), *Genetic Secrets: Protecting privacy and Confidentiality in the Genetic Era*. New Haven, CT: Yale University Press.

Australian Government First Stage Response to the Australian Law Reform Commission Report 108. (2009). *For Your Information: Australian Privacy Law and Practice*. Retrieved from http://www.dpmc.gov.au/privacy/alrc.cfm. Accessed 25 October 2009.

Australian Law Reform Commission (ALRC). 2008. *For Your Information: Privacy Law and Practice* (Report 108).

Australian Law Reform Commission and Australian Health Ethics Committee (ALRC/AHEC). (2003). *Essentially Yours, the Protection of Human Genetic Information in Australia, Report 96.*

Beyleveld, D., & Townend, D. (2004). When is personal data rendered anonymous? Interpreting Recital 26 of Directive 95/46/EC. *Medical Law International*, *6*, 73.

Borry, P. (2008). Europe to ban direct-to-consumer genetic tests? *Nature Biotechnology*, *25*(7), 736. doi:10.1038/nbt0708-736

Cancado-Trindade, A. A. (1948). Universal Declaration of Human Rights, Paris, 10 December 1948. *Audiovisual Library of International Law.* Retrieved from http://untreaty.un.org/cod/avl/ha/udhr/udhr.html. Accessed 19 September 2010.

Cavoukian, A. (1995). Confidentiality issues in genetics: the need for privacy and the right 'not to know.'. *Law and the Human Genome Review*, *2*, 53.

Data Protection Working Party. (2004). *Working Document on Genetic Data, 5 and 7.*

Data Protection Working Party. (2007).

Elizalde, J. (1995). Confidentiality, privacy and genetic data. In *The Human Genome Project: Legal Aspects, Vol. I.* Bilbao: Fundación BBV.

European Standards on Confidentiality and Privacy in Healthcare. (2006). Paragraph 3.4.2.

Evans v. The United Kingdom, 7 March 2006.

Z. v.Finland, 25 February 1997.

Gaskell, G. (2006). Europeans and biotechnology in 2005: patterns and trends. *Final Report on Eurobarometer*, *64*(3), 51.

Glass v. The United Kingdom, 9 March 2004.

Government Response to Recommendations. (2009). Retrieved from http://www.alrc.gov.au/inquiries/title/alrc96/response.htm. Accessed 24 September 2009.

Gros-Espiell, H. (1997). El proyecto de Declaración Universal sobre el Genoma Humano y los Derechos de la Persona Humana de la UNESCO. *Revista de Derecho y Genoma Humano. Law and the Human Genome Review*, 7.

Guidelines for the Protection of Privacy and Transborder Flows of Personal Data. (1980).

Hogarth, S., Javitt, G. and Melzer, D. (2008). The current landscape for direct-to-consumer genetic testing: legal, ethical, and policy issues. *Annual Review of Genomics and Human Genetics,* 9, 165 ff.

Hogarth, S., & Melzer, D. (2007). The IVD Directive and genetic testing problems and proposals. Retrieved from http://www.eshg.org/documents/BriefingIVDDirectiveFINAL_july_20071.pdf.

House of Lords Science and Technology Committee. (2009). *Genomic Medicine. 2nd Report of Session 2008-09.*

Human Genetic Commission. (2009). Common framework of principles for direct-to-consumer geentic testing services. Principles and consultation questions. Retrieved from http://www.tga.gov.au/ivd/forthcoming.htm. Accessed 29 September 2009.

International Covenant on Civil and Political Rights. *(*1966*).*

Johnstone, C., & Kaye, J. (2004). Does the UK biobank have a legal obligation to feedback individual findings to participants? *Medical Law Review, 12*, 239. doi:10.1093/medlaw/12.3.239

Kaye, J. (2008). The regulation of direct-to-consumer genetic tests. *Human Molecular Genetics, 17*, 182. doi:10.1093/hmg/ddn253

Keogh, L. (2009). Is uptake of genetic testing for colorectal cancer influenced by knowledge of insurance implications? *The Medical Journal of Australia, 191*, 255.

Model Criminal Law Officers' Committee of the Standing Committee of Attorneys-General. (2008). Non-Consensual Genetic Testing. Discussion Paper.

Moniz, H. (2004). Privacy and intra-family communication of genetic information. *Revista de Derecho y Genoma Humano. Law and the Human Genome Review, 21*, 111.

National Health and Medical Research Council. (2007). *National Statement on Ethical Conduct in Human Research.* Canberra: NHMRC.

National Health and Medical Research Council (NHMRC). (2004). *The Impact of Privacy Legislation on NHMRC Stakeholders: Comparative Stakeholder Analysis.*

NHMRC. (2009). *Use and Disclosure of Genetic Information to a Patient's Genetic Relatives under section 95AA of the Privacy Act 1988 (Cth).* Guidelines for Health Practitioners in the Private Sector.

Nicol, D., Otlowski, M., & Chalmers, D. (2001). Consent, commercialisation and benefit-sharing. *Journal of Law and Medicine, 9*, 80.

Nys, H., Romeo-Casabona, C. M., & Desmet, C. (2002). Legal aspects of prenatal testing for late-onset neurological diseases. In G. Evers-Kiebooms, M. W. Zoetewelj and P. S. Harper (Eds.), *Prenatal Testing for Late-onset Neurological Diseases* (84-109). Oxford: BIOS Scientific Publishers.

Opinion of the European Group on Ethics in Science and New Technologies to the European Commission. (2003). Ethical aspects of genetic testing in the workplace.

Otlowski, M. (2001). Protecting genetic privacy: an overview. In *Regulating the New Frontiers: Legal issues in Biotechnology*. Hobart, Melbourne: Centre for Law and Genetics.

Otlowski, M. (2002a). Employers' use of genetic test information: is there a need for regulation? *Australian Journal of Labour Law, 15*, 1.

Otlowski, M. (2002b). Protecting genetic privacy in the research context: where to from here? *Macquarie Law Journal, 2*, 91.

Otlowski, M. (2007). Disclosure of genetic information to at risk relatives: recent amendments to the *Privacy Act* 1988 (Cth). *The Medical Journal of Australia, 187*, 398.

Otlowski, M. (2007). The use of legal remedies in Australia for pursuing allegations of genetic discrimination: findings from an empirical study. *International Journal of Discrimination and the Law*, *9*, 3.

Otlowski, M., & Williamson, R. (2003). Ethical and Legal Issues and the "New Genetics.". *The Medical Journal of Australia*, *178*, 4.

Paulsen, J. S., Ferneyhough Hoth, K., Nehl, C., & Stierman, L. (2005). Critical periods of suicide risk in Huntington's Disease. *The American Journal of Psychiatry*, *162*, 725–731. Retrieved from http://www.huntington-assoc.com/Critical%20ab05.pdf. doi:10.1176/appi.ajp.162.4.725

Pearson, P. (2008). Genetic testing for everyone. *Nature*, *453*, 570–571. doi:10.1038/453570a

Regulation (EC) 45/2001 of the European Parliament and of the Council of 18 December 2000 European Union Directive on the Protection of Individual with Regard to the Processesing of Personal Data and on the Free Movement of such Data. (2001).

Romeo-Casabona, C. M. (2002). *Los genes y sus leyes. El derecho ante el genoma humano*. Bilbao, Granada: Comares.

Romeo-Casabona, C. M. (2004). Anonymization and pseudonymization: the legal framework at a European level. In Beyleveld, D., Townend, D., Rouillé-Mirza, S., & Wright, J. (Eds.), *The Data Protection Directive and Medical Research Across Europe*. England: Ashgate.

Romeo-Malanda, S., & Nicol, D. (2007). Protection of genetic data in medical genetics: a legal analysis in the European context. *Revista de Derecho y Genoma Humano. Law and the Human Genome Review*, *27*, 97–134.

Roscam Abbing, H. D. C. (1995). Genetic Information and third party interests. How to find the right balance? *Law and the Human Genome Review*, *2*, 39.

Rothstein, M. A. (1997). Genetic secret: a policy framework. In Rothstein, M. A. (Ed.), *Genetic Secrets: Protecting Privacy and Confidentiality in the Genetic Era.*

Setoyama, K. (2005). Privacy of genetic information. *Osaka University Law Review* 52,81 ff.

Skene, L. (1998). Patients' rights or family responsibilities? two approaches to genetic testing. *Medical Law Review*, *6*, 1. doi:10.1093/medlaw/6.1.1

M.S. v.Sweden, 27 August 1997.

Thiele, F. (2003). Genetic tests in the insurance system: criteria for a moral evaluation, *Poiesis Prax*, pp. 193 f.

Whiteman, D. C., Clutton, C., & Hill, D. (2006). Australian public's views on privacy and health research. *British Medical Journal*, *332*, 1274. doi:10.1136/bmj.332.7552.1274-a

ENDNOTES

1 The Convention for the Protection of Individuals with regard to Automatic Processing of Personal Data, of 28 January 1981, was the first international legally binding text on data confidentiality.

2 The main success of the Council of Europe was the European Convention on Human Rights in 1950, which serves as the basis for the European Court of Human Rights. The Council of Europe is not to be mistaken with the Council of the European Union or the European Council, as it is a separate organization and not part of the European Union.

3 However, unlike the European Convention on Human Rights which applies to all EU Member States, the Convention on Human Rights and Biomedicine has not yet been signed or ratified by many States, including most of the larger States. In spite of it not

applying directly to many EU States, it is nevertheless significant in that it has been drawn upon by the European Court of Human Rights in making judgments involving States who are not parties to this Convention. See in this respect, *Glass v. The United Kingdom*, 9 March 2004 (paragraph 58); *Evans v. The United Kingdom*, 7 March 2006 (paragraph 40).

4 Although this legal document also has the force of a treaty, and is therefore binding for the States that ratify it, it is still in an early stage of ratification and has not yet entered into force. According to Article 25 APGT, "This Protocol shall enter into force on the first day of the month following the expiration of a period of three months after the date on which five States, including at least four member States of the Council of Europe, have expressed their consent to be bound by the Protocol."

5 These are: Austria, Belgium, Bulgaria, Cyprus, Czech Republic, Denmark, Estonia, Finland, France, Germany, Greece, Hungary, Ireland, Italy, Latvia, Lithuania, Luxembourg, Malta, The Netherlands, Poland, Portugal, Romania, Slovakia, Slovenia, Spain, Sweden, and United Kingdom.

6 Directive 95/46/EC gives a definition of identifiable data in the following terms: "An identifiable person is one who can be identified, directly or indirectly, in particular by reference to an identification number or to one or more factors specific to his or her physical, physiological, mental, economic, cultural or social identity" [Article 2(a)].

7 It should be noted that the recitals in the Directives do not have a statutory power, but they are very useful to interpret the rules contained in the legal text.

8 Recommendation No. R (97) 5 of the Council of Europe on the Protection of Medical Data, of 13 February 1997 is a little more concrete in this respect, as it states that "an individual shall not be regarded as 'identifiable' if identification requires an unreasonable amount of time and manpower" (principle 1). See Data Protection Working Party (2007), pp. 15 ff. See also Beyleveld and Townsend, 2004, pp. 73ff.

9 However, domestic legislation of the Member States can foresee some exceptions to this principle. For instance, when the processing is necessary to protect the vital interests of the data subject or of another person where the data subject is physically or legally incapable of giving his consent. See Directive 95/46/EC, Article 8.2 and 3.

10 See Schedule 3 of the *Privacy Act* 1988 (Cth) Principle 10 dealing with sensitive information.

11 Schedule 3 of the *Privacy Act* 1988 (Cth) Principle 10 dealing with sensitive information.

12 Schedule 3 of the *Privacy Act* 1988 (Cth) Principle 10 dealing with sensitive information; "health information" now includes reference to "genetic information about an individual in a form that is or could be predictive of the health of the individual or a genetic relative of the individual. Sensitive information now expressly includes "genetic information about an individual that is not otherwise health information."

13 A study developed in the US concludes that suicide rate for Huntington's disease sufferers is much greater than the national average (Paulsen et al., 2005).

14 In general, such a therapeutic exception is only justifiable in very exceptional circumstances (Roscam Abbing, 1995).

15 This is the case, for example, in the Spanish legislation (Article 5.4 of the Law 41/2002 on the autonomy of the patient and the rights and obligations with regard to clinical information and documentation).

16 Although in theory, individuals should be free to decide whether to undergo genetic

testing, or whether to access results of genetic testing already undertaken, a number of cases have come to light in the course of empirical research in Australia into genetic discrimination that indicate inappropriate pressure is sometimes put on individuals (Otlowski and Williamson, 2003; Otlowski et al., 2007).

17 "The right to informational self-determination (...) should enable an individual to make decisions concerning personal data without being exposed to the coercion of a third party" (Thiele, 2003, p. 193).

18 See, for example, Article 17 of the International Covenant on Civil and Political Rights 1966.

19 *Z. v. Finland*, 25 February 1997 (paragraph 95). Cf. also, *M.S. v. Sweden*, 27 August 1997 (paragraph 41).

20 In order to use the conflict of duties solution, it is necessary to first determine the obligation of information towards others on the part of the person who performed or proposed the tests (e.g. a doctor in relation to his or her patient). That is to say, if the doctor is simultaneously doctor to other family members of the patient in question, then he or she may have the duty to communicate the fact to them.

21 When the clinician is doctor only to the patient having the test. In these cases, the doctor does not have to violate confidentiality, but when he or she does so, taken into account certain clearly defined circumstances, his or her conduct is not consider illicit.

22 See also, Opinion of the European Group on Ethics in Science and New Technologies to the European Commission. 2003. Ethical aspects of genetic testing in the workplace.

23 See National Privacy Principle 2.1, as amended by the *Privacy Legislation Amendment Act* 2006 (Cth). For discussion (Otlowski, 2007).

24 Section 95AA. At the time of writing, these guidelines had been developed by the NHMRC and are now required to be approved by the Federal Privacy Commission before the amendment can be used: *Use and Disclosure of Genetic Information to a Patient's Genetic Relatives under section 95AA of the Privacy Act 1988 (Cth): Guidelines for Health Practitioners in the Private Sector*. It is anticipated that they will be released later in 2009.

25 NHMRC and Office of the Privacy Commissioner, Use and Disclosure of Genetic Information to a Patient's Genetic Relatives Under s95AA of the Privacy Act 1988 (Cth): Guidelines for Health Practitionera in the Private Sector.

26 *Discrimination Act* 1992 (Cth) s 46.

27 ALRC/AHEC, Recommendation 27-1.

28 ALRC/AHEC, Recommendation 28-3.

29 IFSA Standard No. 11, Genetic Testing Policy 2005, 10.7.

30 ALRC/AHEC, Recommendations 30-1 – 32-6.

31 See Therapeutic Goods Administration, *Regulatory Framework for IVDs*. Retrieved from http://www.tga.gov.au/ivd/framework.htm. Accessed 19 September 2010.

Chapter 17

Property, Personality Rights and Data Protection with Regard to Biobanks: A Layered System with Germany as an Example

Jürgen Simon
Leuphana University Lüneburg, Germany

Jürgen Robienski
Lawyer in Hannover and Müden/Aller, Germany

ABSTRACT

This chapter is a discussion and analysis of the concept of property in the context of genetic research and biobanking. The authors are proposing a novel way of conceptualizing the concept of property, one which could unravel many difficulties that beset bioethical and legal debates on this issue. Once a certain part of the body is taken out, to what extent could the person from whom the tissue has been separated claim property right over the tissue? If the tissue is completely anonymized, then the person loses her right to claim the tissue to be her property. The linking could then be done by way of pseudonymization. Furthermore, the trustee model where the tissue is safeguarded and pseudonymized by a trustee could act as the middle path between complete retention of personal and property rights of the alienated tissue on the one hand, and complete disregard of the property or ownership rights to the tissue on the other

INTRODUCTION

Human body materials have been gathered and evaluated either for diagnostic or scientific-medical research purposes; due to this fact, until today there are a multitude of human biological material and data collections, some of them very large (Freier, 2005; Stellungnahme der Zentralen Ethikkommission, 2003, A-1632). Collecting human biological materials has been mostly done on the basis of an existing diagnosis and treatment contract or post mortem due to a proper donor declaration of the deceased, or today, more and more, from free donors. However, independent of the manner in which physicians or researchers

DOI: 10.4018/978-1-61692-883-4.ch017

receive the materials, whether it is with or without consent, any further use was to the benefit of others, which means for medical research purposes; otherwise the material was destroyed.

As options of biotechnological research continue to increase, in particular as the biotechnological industry is showing an increasing interest in scientific and commercial research as well as in any commercial utilization which may follow, a broad discussion is taking place at all levels on the questions of access to or collecting the needed body materials.

In this respect, the legal position of man is being discussed as a person who is, as either patient or donor, "providing" his human biological material (Laurie, 2005).

In this discussion process, there have been increasing criticisms to the effect that the institutional parties involved in this process, i.e., science, the market (biotechnological industry) and the law, keep reducing human biological resources to mere biological information. This allows them to come up with the opinion that anonymization on the one hand and "informed consent" on the other are sufficient to properly honor the rights and interests of individual donors or patients, in particular the right to informed self-determination and the dignity of the individual (Laurie, 2005). In this case the existence of further reaching rights, derived from the position of owner, in particular participatory rights, is usually negated (Laurie, 2005).

This is best shown in the "Moore case," known worldwide, in which the Californian Supreme Court held that any claims made by the plaintiff Moore were not to be considered, because individual property rights regarding body materials would hinder medical research by means of limiting access to necessary raw material and because a decision in a different direction would destroy the economical drive to do important medical research.[1]

The question whether the individual donor has property rights to his body materials and, thus, consequential rights, is being answered and justified actually not on the basis of legal-dogmatic ideas, but rather exclusively on the basis of interest-governed, economical thoughts.

The significance of this discussion on society carries varies greatly among countries. While countries in the southern hemisphere and those which are strongly religious have more public discussion on the matter, in Germany this discussion process seems to take place mostly at the academic level.

The general population does not, or hardly, take part in this discussion. Here is a new example. In fall of 2003, the project "Chromosoma" was initiated under the direction of the Federal Center for Political Education. In the center of the northern German city of Bremen, among the best shops in the area, a store opened with the same name. It offered potential customers a multitude of gene technological services such as "genetic horoscope," wish-embryo or performance increase through gene maximization. The interest was great. Particularly the welcome action was appreciated. Passersby were given a little package with some chewable mass in connection with the promise of a gift if the used chewing gum was returned. Upon its return one received an apple and an egg, a German picture or synonym for giving something away particularly cheaply. Many people followed the request and voluntarily provided their personal data like the name, address, birth date etc., although an information leaflet informed that "Cromosoma" wanted the saliva and, thus, the genetic material.

Person-identifiable genetic material was given for an apple and an egg. On the first day, only one passer by complained after she had thought about the matter and the meaning of her action at home (Zinkant, 2003, p. 65).

Particularly because the general public is not sufficiently sensitized with regard to the problem of use and continued use of genetic materials and its economical consequences, further explanations and enlightenment are necessary and the discussion process needs to be continued.

In the following we shall try to approach the question about property rights regarding body materials in a legal and dogmatic way. At the same time we want to see which legal positions the donor/patient or supplier of human biological biomaterial may adopt.

PROPERTY IN HUMAN BIOLOGICAL MATERIAL

According to the opinion which is ruling not only in Germany, it can be stated without doubt that the body of a living person is considered a *res extra commercium* (a non marketable good), and the body along with organs and firmly attached body or supporting parts (such as pacemakers or gold fillings) are not considered things in the sense of the law. Therefore, there can be no property in this regard (Spranger, 2005; NJW, 2005; Halàsz, 2004; Palandt/Heinrichs; Taupitz). However, human beings have a "property-like right to determination" instead as a result of their general personality rights (Halàsz, 2004, p. 40).

It is a different matter with regard to separated body components. It is an internationally very controversial issue which legal quality these have. These are considered – just like the human corpse – as a *res extra commercium,* or as a *res* nullum (Laurie, 2005; Tallachini). The German Civil Code Book does not provide for an explicit regulation about the legal position of separated body parts.

According to almost all opinions represented in the German legislation and in literature, individual body components become mobile objects after their separation from the body (BGH/NJW, 1994; Spranger, 2005; Halàsz, 2004, p. 20). The characteristic of being an object or thing comes into effect with the separation from the body analogously as per § 953 Civil Code, and this constitutes in itself the property of the separated body material belonging to the person from whose body the separated material comes (BGH/NJW, 1994, p. 128; v. Freier, 2005, p. 322; Freund and Weiss, 2004, p. 346; Lippert, 2004, p. 158; Taupitz AcP 1291, 208; Schünemann, 1985, pp. 86ff).

Only few opinions in Germany hold against this, that an analogous application of §953 Civil Code is excluded, since this very paragraph is applicable only for objects and since the separated body part was exactly not a thing prior to its separation. Although this opinion recognizes that separated body parts become objects after their separation, these parts first are without an owner and "simply" were subject to a privileged right to take ownership by the previous carrier according to § 958,2 Civil Code (Halàsz, 2004, p. 23; Schröder and Taupitz, 1991, pp. 37ff).

This opinion is constructed not only in a very complicated manner, it is also controversial as regards a unified natural point of view and the legal definition of an ownerless object. According to § 959 Civil Code a mobile object is only ownerless if "the owner gives up ownership with the intention to withdraw from his property."

Precondition for the applicability of this paragraph is first of all that the thing has been owned and that this ownership is to be given up by a conscious wilful decision of the owner (Freund and Weiss, 2004, p. 316). Only certain objects such as the wild animals running around in freedom as mentioned in § 960 Civil Code are ownerless by nature as they have not been attached to any legal subject.

The human body as such and therefore also all of his components, however, have come under a property-like attachment by way of the general personal right. Property and personal or personality rights both have the function to protect absolute rights against infringement by third parties. Therefore, the former carrier of separated body parts must receive the same authority as per § 953 Civil Code as does the owner of an object from which components are separated. The legal relationship of man to his body is, after all, much more intensive than the property right of an owner regarding his object. Now, if already separated object parts are to be the property of the object owner, this must certainly apply to separated body parts. So the personal right decreases in the direction of the property law (BGH/NJW, 1994, p. 128; v. Freier, 2005, p. 322; Halàsz, 2004, p. 40).

However, this does not mean that the personal right disappears completely in every case. Rather, the property law continues to be overlaid by the general personal right. The intensity of this layering depends on how explicit one may draw conclusion about the person of the former carrier of the body materials. If no conclusion may be drawn as to the person of the carrier, the general personal right is almost meaningless.

Even the fact that human biological body materials which become superfluous during diagnostic or treatment measures (e.g., during an operation) are qualified as waste according to the trash regulation book (number 180102 of the appendix EAKV) does not change the assessment of the law. On the contrary, German legislators define here explicitly the object characteristic of the separated body material, because precondition for confirming the term "waste" is that we are dealing with a movable object (V. Lersner). Applying the term "waste" simply gives the superfluous body material which is not intended for any special predefined use a special object characteristic, that of waste or trash. The qualification as waste does not have any object-related effects so that the body material qualified as waste becomes ownerless or a public matter. Waste continues to be the property of the previous owner, the qualification as waste only has meaning as per the KrW-/AbfG, i.e., Waste Disposal Regulation (Fluck).

Thus, the human being remains the owner of the body material after its separation (Freund and Weiss, 2004, p. 316; Breyer, 2004, p. 661; Lippert, 2001, p. 407). There is no automatic transfer of ownership for the collected sample to a third party (Lippert, 2001, p. 408).

Even if the separation is not final but done with the purpose of later "re-incorporation," the Federal Supreme Court does not confirm the object characteristic of the separated body material. A re-incorporation could be done with a donation of one's own blood or donation of one's own sperm for a later insemination or with a donation of tissue in order for transplantation after severe burns, for example (BGH/NJW, 1994, p. 128).

We can follow from this that any biotechnological research using human biological body materials is legally permissible only if the researching institution/biobank acquires the property or ownership of the body materials.

GENERAL PERSONAL RIGHT

As shown already, the property law is overlaid by the general personal or personality law. With regard to body materials which are not intended for re-incorporation, this law does not cover the material as such, but the information provided by the material, for example the genetic information which could be derived from it. The intensity with which the personal right overlays the property right depends on whether any conclusion may be drawn as to the person of the former carrier.

In the case of a complete anonymization–and we are not going to deal here with the question whether a complete anonymization is at all possible–this means that separated body materials are not overlaid by the general personality right since a relationship to a person does not exist any longer.

Only the property right of the donor would continue to exist. The property right requires, though, that an object may be assigned to a person. With objects which do not contain person-related information like a car or something like that, the assignment is without any relevance for the general personal right and the right to informed self-determination resulting from it. Body materials, however, always contain information which allows one to draw conclusions based on sensitive person-related data. Only a complete anonymization could prevent this. Once a complete anonymization is accomplished, the option of assigning the body materials to a person who is the owner is lost. Therefore, in case of a complete anonymization, the property right is actually lost, and the body materials actually become ownerless.

Separated human body materials are characterized in that the property right regarding them is inseparable from the general personal or personality right. Therefore, we can hardly speak of a weakening of the general personal right. Either the general personal right is affected or it is not affected. The Federal Supreme Court also does not explain how to imagine weakening of the general personal right. One could consider a weakening only if one understands a reduction in range of the personal reference by means of pseudonymizing.

This means for the former carrier of the body materials that he can keep his property rights only if he permits that person-related data are stored along with his body materials. In order to warrant also a high protection of the general personal rights, it is necessary to execute pseudonymization and maintain a high data protection as well.

This means further that the former carrier of body materials may only preserve his isolated property rights while preserving at the same time his general personal rights by way of a trustee.

Table 1 could be applied on the basis of the German law:

Perhaps one can picture this table like an onion:

The nucleus is made up by the human body materials. The first layer above it is the general personal right. The second layer is the property right. Now, the second layer may be removed without touching the first layer. But removal of the first layer is only possible once the second layer is gone. Protection for the first layer may be accomplished by providing a new layer on top of it.

CLAIMS FOR PROPERTY OF HUMAN BIOLOGICAL MATERIALS

Does it make sense for a donor to want to maintain property rights of his human biological materials?

If the property rights remain with the former carrier of body materials, he continues to have the

Table 1.

Type of Material	Property Right, Art. 14 GG	Personality Right, Art. 1 and 2 GG	Data Protection
Human Body	*Res extra commercium,* no significance	very high significance	very high significance
Human biological material for the purpose of re-incorporation	Object in the property of man, but small significance	great significance	very high significance
Anonymized human biological material	Object theoretically property of man (waste), but due to anonymization ownerless!	no significance	no significance
Pseudonymized human biological material, which is anonymized by trustees and transmitted	Object property of man	high significance	very high significance, between man and trustee
Pseudonymized human biological material	Object property of man	high significance	very high significance
Person related human biological material	Object property of man	very high significance	very high significance

sovereign or total rights of an owner, though he could grant the rights to the researching institution. He could, however, withdraw the granted user rights and apply his right to demand handing over and destroying the body materials or to prohibit any further research.

Only if he continues to hold property rights of body materials could one consider the case that he is successful in enforcing individual sharing rights, provided his body materials were the basis for substances which may be used commercially and are a success.

Then again, people in Germany are granted far reaching property and personal rights to their body materials. The infringement of these rights has so far not led to a ruling in favor of a commercial benefit sharing or even just a payment of damages. Already in 1974, the Federal Supreme Court rejected damages claimed for improper use of body materials. It held that an infringement of the law had occurred; however, damages were not to be paid since no commercialization was to happen (BGH/NJW, 1974, p. 1371).

It is certainly a question whether this line of ruling will continue considering that commercialization has increased already.

In any case, it is not understandable why all institutionalized parties, i.e., scientific research institutes, biobanks or the biotechnological industry, may make a commercial profit while the person whose body materials are used is excluded from benefit sharing for the very reason of avoiding commercialization.

John Moore was denied his rights to benefit sharing only because he was denied property rights of his body materials. By reverse conclusion one should think that if he does have a property right, he should also be entitled to share benefits or receive damages.

Now, what type or how much should this payment of damages be? According to the currently ruling legal opinion in Germany, a benefit sharing of the commercial success or profit would not come into question.

It will be impossible to prove material damages due to the loss of (superfluous) body materials.

One could consider a compensation for pain and suffering due to the infringement of the personal rights; however, looking at the so far restrictive legal practice at German courts, the possibility for this would be rather low. Further, one could consider an application analogous to the regulations in the copyright. This would entail requesting payment for damages amounting to the double amount of the usually paid license fee. As I described at the beginning–"an apple and an egg," so this would be two apples and two eggs!

I think it makes more sense to opt for a collective participation as compared to the rights in music via the GEMA.

Only if a right allows one to demand the omission of further use due to an infringement of property rights, higher individual participation could be negotiated. Whether a broad claim for omission would be successful, however, is doubtful.

In any case, acceptance of individual property rights is the precondition for the effective protection of the individual against the widespread practice of embezzlement or infidelity and of theft of human biological materials. In those cases, at least a penal law protection can be achieved.

Moreover, the acceptance of individual property rights seems to be suitable for increasing the sensitivity of the public for this subject.

CONCLUSION

The person from whom body materials were collected first becomes their owner in analogy to § 953 German Civil Code. This applies also if the body materials were taken within the framework of a treatment contract and remain there after the treatment/diagnosis has ended. Only if an explicit consent of the patient was given to the effect that the body materials are to be owned by the clinic/doctor is it possible to transfer ownership. A conclusive transfer of ownership does not occur.

A complete anonymization makes the body materials ownerless since an assignment to a person is not possible any longer. In this case, the former carrier of the body materials has no further chance to make claim to any rights. In that case, science and the biotechnological industry are free to utilize the body materials. The former carrier of the body materials may make claims to benefit sharing only if instead of an anonymization only a pseudonymisation takes place. However, it is common practice that the payment for damages according to existing rights to benefit sharing after unlawful use is quite small.

Only a trustee model could achieve that property rights be observed, that personal or personality rights be protected in the best possible manner and that proper benefit sharing be done.

REFERENCES

BGH/NJW. (1974).

BGH/NJW. (1994).

Breyer. (2004). *MedR,* 661.

Fluck, Jürgen. Kreislaufwirtschafts-, Abfall- und Bodenrecht, zu § 3 KrW-/AbfG Rn 79, 62. ErgL. 03/06.

V. Freier, Friedrich. (2005). Getrennte Körperteile in der Forschung zwischen leiblicher Selbstverfügung und Gemeinbesitz. MedR, 321 f.;

Freund/Weiss. (2004). *MedR,* 346.

Halàsz. (2004). Das Recht auf bio-materielle Selbstbestimmung, 13 f. (mit einem Überblick über die verschiedenen Theorien).

Laurie, G. (2005). (Intellectual) property? Let's think about taking a claim to our own genetic samples. Retrieved from www.law.ed.ac.uk/ahrb/publications/onine/GLPaper.htm.

Lippert. (2001). *MedR,* 408 (406).

Lippert (2004). *MedR,* 158.

Palandt/Heinrichs. § 90 BGB RdNr 3.

Schröder/Taupitz. (1991). Menschliches Blut verwendbar nach Belieben des Arztes?.

Schünemann. (1985). Die Rechte am menschlichen Körper.

Spranger. (2005). *NJW,* 1085 (1084).

Stellungnahme der Zentralen Ethikkommission der Bundesärztekammer vom 20.02.2003. (2003). Die Weiterverwendung von menschlichen Körpermaterialien für Zwecke der medizinischen Forschung. *DÄBl,* A-1632.

Tallacchini, Mariachiara. Rhetoric of anonymity and property rights in human biological materials (hbms). *Rev. Der. Gen H,* 22, 153-175.

Taupitz. *JZ,* 92, 1089.

V. Lersner in Von Lersner, Heinrich / Wendenburg, Helge, „Recht der Abfallbeseitigung" 0103, zu § 3 abs. 1 KrW-/AbfG, Rn. 5.

Zinkant, Kathrin. (2003). Gebt her Eure Gene. *FAZ 14.09.2003,* 65.

ENDNOTE

1 Moore, v., Regents of University of California, Cal. App. 2 Dist. (1988), e 51 Cal. 3D (1990).

Compilation of References

(1982). *President's Commission for the Study of Ethical Problems in Medicine and Biomedical and Behavioral Research*. Washington, DC: Splicing Life.

(1997). *National Bioethics Advisory Commission*. Rockville: Cloning Human Beings.

(1998). *Collins English Dictionary*. Glasgow: Harper Collins Publishers.

(1998). *The New Oxford English Dictionary*. Oxford, UK: Oxford University Press.

(2003). *National Bioethics Advisory Commission*. Rockville: Beyond Therapy.

Abdullkhodjaeva, M. (2007). *Ethical Review of Biomedical research in CIS Countries (Social and Cultural Aspects)*. Saint-Petersburg.

Academy of Medical Sciences. (2006). *Personal Data for Public Good: Using Health Information in Medical Research.*

Achrorova, Z. (2007). *Ethical Review of Biomedical Research in CIS Countries (Social and Cultural Aspects)*. Saint-Petersburg.

Adeoye, S., & Bozic, K. J. (2007). Direct to consumer advertising in healthcare: history, benefits, and concerns. *Clinical Orthopaedics and Related Research*, *457*, 96–104.

Akande, D. (2006). International organizations. In Malcolm, D. (Ed.), *Evans, International Law* (2nd ed.). Oxford, UK: Oxford University Press.

Allen, A. (1997). Genetic privacy: emerging concepts and values. In Rothstein, M. A. (Ed.), *Genetic Secrets: Protecting privacy and Confidentiality in the Genetic Era*. New Haven, CT: Yale University Press.

Almasi, E. A., Stafford, R. S., Kravitz, R. L., & Mansfield, P. R. (2006). What are the public health effects of direct-to-consumer drug advertising? *PLoS Medicine*, *3*(3), e145. doi:10.1371/journal.pmed.0030145

Congregation for the Doctrine of Faith Vatican. (1992). Instruction on respect for human life in its origin and on the dignity of procreation. In Alpern, D. K. (Ed.), *The Ethics of Reproductive Technology*. New York: Oxford University Press.

Altman, R. B. (2009). Direct-to-Consumer Genetic Testing: Failure Is Not an Option. *Clinical Pharmacology and Therapeutics*, *86*(1), 15–17. doi:10.1038/clpt.2009.63

American College of Medical Genetics. (2004). ACGM statement on direct-to-consumer genetic testing. *Genetics in Medicine*, *6*(1), 60. doi:10.1097/01.GIM.0000106164.59722.CE

Anderson, W. F., & Friedmann, T. (1995). Strategies for gene therapy. In W. T. Reich (Ed.), *Encyclopedia of Bioethics* (907-914). New York: MacMillan.

Andrews, L. B. (2001). *Future Perfect: Confronting Decisions About Genetics*. New York: Columbia University Press.

Anonymous (2008). Psychiatric genetic tests raise concerns. Home kits are now available, but they may not be testing for the right genes. *The Harvard Mental Health Letter*, *24*(11), 4.

Anonymous,. (2008). Direct-to-consumer genetic tests: flawed and unethical. *The Lancet Oncology*, *9*, 1113. doi:10.1016/S1470-2045(08)70288-2

Aplenc, R., Thompson, J., Han, P., La, M., Zhao, H., Lange, B., & Rebbeck, T. (2005). Methylenetetrahydofolate reductase polymorphism and therapy response in pediatric acute lymphoblast leukemia. *Cancer Research*, *65*(6), 2482–2487. doi:10.1158/0008-5472.CAN-04-2606

Aquinas, T. (1947). *Summa Theologica.* William Benton. Retrieved from http://www.ccel.org/ccel/aquinas/summa.html.

Arendt, H. (1999). *The Human Condition.* Sydney, Australia: B&T Publisher.

Aristotle (1936). *Metaphysics. (Tredennick Transl.).* Cambridge: University Press.

Aristotle (1999). *Nicomachean Ethics. (Terence Irwin Transl.).* Indianapolis, IN: Hackett.

Aristotle (1941). *Metaphysics.* (W. D. Ross Transl.). In McKeon, R. (Ed.), *The Basic Works of Aristotle.* New York: Random House.

Association of British Insurers. (2005). *Insurers and government reach agreement on genetic testing and insurance (*March 14, 2005). Retrieved from http://www.abi.org.uk/Media/Releases/2005/03/Insurers_and_Government_reach_agreement_on_genetic_testing_and_insurance.aspx. Accessed January 9, 2010.

Augustine,. (1887). On Marriage and Concupiscence. In Schaff, P. (Ed.), *The Nicene and post-Nicene Fathers of the Christian Church.* New York: The Christian Literature.

Australian Government First Stage Response to the Australian Law Reform Commission Report 108. (2009). *For Your Information: Australian Privacy Law and Practice.* Retrieved from http://www.dpmc.gov.au/privacy/alrc.cfm. Accessed 25 October 2009.

Australian Law Reform Commission (ALRC). 2008. *For Your Information: Privacy Law and Practice* (Report 108).

Australian Law Reform Commission and Australian Health Ethics Committee (ALRC/AHEC). (2003). *Essentially Yours, the Protection of Human Genetic Information in Australia, Report 96.*

Badano, J. L., & Katsanis, N. (2002). Beyond Mendel: an evolving view of human genetic disease transmission. *Nature Reviews. Genetics*, *3*(10), 779–789. doi:10.1038/nrg910

Bainbridge, W. S. (2003). Religious opposition to cloning. *Journal of Evolution and Technology*, *13*(2). Retrieved from http://jetpress.org/volume13/bainbridge.html accessed on 1 May 2008.

Bainbridge, D. L. (1992). *Intellectual Property.* London: Pitman Publishing.

Barboza, R., Fox, J. H., Shaffer, L. E., Opalek, F. M., & Faroki, S. (2009). Incidental findings in the cervical spine at CT for trauma evaluation. *AJR. American Journal of Roentgenology*, *192*(3), 725–729. doi:10.2214/AJR.08.1420

Bashiri, I. (2003). *From the Hymns of Zarathustra to the Songs of Borbad.* Dushanbe.

Basil. (1978). Letter to Amphilochius. In Philip Schaff & Henry Wace (Eds.), *The Nicene and Post-Nicene Fathers.* London: W M. B. Eerdmans.

Basille, C., Frydman, R., & El Aly, A. (2009). Preimplantation genetic diagnosis: state of the art. *European Journal of Obstetrics, Gynecology, and Reproductive Biology*, *145*(1), 9–13. doi:10.1016/j.ejogrb.2009.04.004

Beier, F. K. (1985). *Biotechnology and Patent Protection.* Paris: OECD.

Benn, P. A., & Chapman, A. R. (2010). Ethical challenges in providing noninvasive prenatal diagnosis. *Current Opinion in Obstetrics & Gynecology*, *22*(2), 128–134. doi:10.1097/GCO.0b013e3283372352

Bennet, A. M., Di Angelantonio, E., & Ye, Z. (2007). Association of apolipoprotein E genotypes with lipid levels and coronary risk. *Journal of the American Medical Association*, *298*(11), 1300–1311. doi:10.1001/jama.298.11.1300

Bent, S. A. (1987). *Intellectual Property Rights in Biotechnology Worldwide*. New York: Stockton Press.

Bereano, P. L. (1995) Patent pending: the race to own DNA. *The Seattle Times* (Aug. 27). Retrieved from http://weber.u.washington.edu/~radin/guaymi.htm.

Berg, C., & Fryer-Edwards, K. (2008). The ethical challenges of direct-to-consumer genetic testing. *Journal of Business Ethics*, *77*, 17–31. doi:10.1007/s10551-006-9298-8

Bergel, S. D. (2000). Aspectos Éticos y Jurídicos del Proyecto Genoma Humano: Patentamiento de Genes y Secuencias. *Medicina*, *60*, 729–730.

Bergel, S. D. (2002). Los Derechos Humanos entre la Bioética y la Genética. *Acta Bioética*, *VIII*, 315–329.

Beyleveld, D., & Townend, D. (2004). When is personal data rendered anonymous? Interpreting Recital 26 of Directive 95/46/EC. *Medical Law International*, *6*, 73.

BGH/NJW. (1974).

BGH/NJW. (1994).

Bhardwaj, M. (2007). From farm to pharma: public health challenges of nutrigenomics. *Personalised medicine*, 4(4), 423-430.

Bioethics Advisory Committee. Singapore (BAC). (2002a). *Ethical, Legal and Social Issues in Human Stem Cell Research, Reproductive and Therapeutic Cloning: A Report from the Bioethics Advisory Committee, Singapore* (June 21, 2002). Retrieved from http://www.bioethics-singapore.org/.

Bioethics Advisory Committee. Singapore (BAC). (2002b). *Human Tissue Research: A Report by the Bioethics Advisory Committee, Singapore* (November 12, 2002). Retrieved from http://www.bioethics-singapore.org/.

Bioethics Advisory Committee. Singapore (BAC). (2004). *Research Involving Human Subjects – Guidelines for IRBs: A Report by the Bioethics Advisory Committee, Singapore* (November 23, 2004). Retrieved from http://www.bioethics-singapore.org/.

Bioethics Advisory Committee. Singapore (BAC). (2005). *Genetic Testing and Genetic Research: A Report by the Bioethics Advisory Committee, Singapore* (November 25, 2005). Retrieved from http://www.bioethics-singapore.org/.

Bioethics Advisory Committee. Singapore (BAC). (2007). *Personal Information in Biomedical Research: A Report by the Bioethics Advisory Committee, Singapore* (May 7, 2007). Retrieved from http://www.bioethics-singapore.org/.

Bioethics Advisory Committee. Singapore (BAC). (2008). *Donation of Human Eggs for Research: A Report by the Bioethics Advisory Committee, Singapore* (November 3, 2008). Retrieved from http://www.bioethics-singapore.org/.

Bisang, R., Campi, M., & Cesa, V. (2009). *Biotecnología y Desarrollo." Comisión Económica para América Latina y el Caribe*. CEPAL.

Black, T. (1989). *Intellectual Property in Industry*. London, Edinburgh: Butterworths.

Blauw, H. M., Veldink, J. H., & van Es, M. A. (2008). Copy number variation in sporadic amyotrophic lateral sclerosis: a genome-wide screen. *The Lancet Neurology*, *7*(4), 319–326. doi:10.1016/S1474-4422(08)70048-6

Bloch, M., & Hayden, M. R. (1990). Opinion: predictive testing for Huntington disease in childhood: challenges and implications. *American Journal of Human Genetics*, *46*(1), 1–4.

Blow, N. (2008). DNA sequencing: generation next-next. *Nature Methods*, *5*, 267–274. doi:10.1038/nmeth0308-267

Bobbert, M. (2006). Ethical questions concerning research on human embryos: embryonic stem cells and chimeras. *Biotechnology Journal*, *1*, 352–1369. doi:10.1002/biot.200600179

Bolnick, D. A., Fullwiley, D., & Duster, T. (2007). Genetics. The science and business of genetic ancestry testing. *Science*, *318*(5849), 399–400. doi:10.1126/science.1150098

Borman, S. (2008a). Bacterial genome made from scratch. *Chemical and Engineering News*, *86*(4), 16.

Borman, S. (2008b). Synthetic genome paves the way to artificial life. *Nature*, *451*(7178), 511. doi:10.1038/451511c

Borry, P., Goffin, T., Nys, H., & Dierickx, K. (2007). Attitudes regarding carrier testing in incompetent children: a survey of European clinical geneticists. *European Journal of Human Genetics*, *15*(12), 1211–1217. doi:10.1038/sj.ejhg.5201909

Borry, P., Goffin, T., Nys, H., & Dierickx, K. (2008). Predictive genetic testing in minors for adult-onset genetic diseases. *The Mount Sinai Journal of Medicine, New York*, *75*(3), 287–296. doi:10.1002/msj.20038

Borry, P., Howard, H. C., Sénécal, K., & Avard, D. (2009). Direct-to-consumer genome scanning services. Also for children? *Nature Reviews. Genetics*, *10*(1), 8. doi:10.1038/nrg2501

Borry, P., Howard, H. C., Sénécal, K., & Avard, D. (2010). Health-related direct-to-consumer genetic testing: a review of companies' policies with regard to genetic testing in minors. *Familial Cancer*, *9*(1), 51–59. doi:10.1007/s10689-009-9253-9

Borry, P. (2008). Europe to ban direct-to-consumer genetic tests? *Nature Biotechnology*, *25*(7), 736. doi:10.1038/nbt0708-736

Bovio, S., Cataldi, A., & Reimondo, G. (2006). Prevalence of adrenal incidentaloma in a contemporary computerized tomography series. *Journal of Endocrinological Investigation*, *29*(4), 298–302.

Bray, M. B. (1990). Personalizing personalty: toward a property right in human bodies. *Texas Law Review*, *69*, 209–244.

Breck, J., & Breck, L. (2005). *Stages on Life's Way*. New York: St. Vladimir's Seminary Press.

Breitowitz, Y. (2002). What's so bad about human cloning? *Kennedy Institute of Ethics Journal*, *12*(4), 325–341. doi:10.1353/ken.2002.0023

Breyer. (2004). *MedR,* 661.

Brookes, A. J. (1999). The essence of SNPs. *Gene*, *234*(2), 177–186. doi:10.1016/S0378-1119(99)00219-X

Brose, M. S., Rebbeck, T. R., & Calzone, K. A. (2002). Cancer risk estimates for BRCA1 mutation carriers identified in a risk evaluation program. *Journal of the National Cancer Institute*, *94*(18), 1365–1372.

Brown, M. T. (2007). The potential of the human embryo. *The Journal of Medicine and Philosophy*, *32*, 585–618. doi:10.1080/03605310701680973

Business Week. (1992). March 2.

Callahan, D. (1996). Biotechnology and ethics: a blueprint for the future. Keynote: setting and communicating the limits in biotechnology. Retrieved from http://www.biotech.nwu.edu/nsf/callahan.html.

Cambon-Thomson, (2003). *Populations and Genetics: Legal Social-Ethical Perspectives. The Netherlands* (Knoppers, B. M., Ed.). Leiden, Boston.

Cameron, L. D., & Muller, C. (2009). Psychosocial aspects of genetic testing. *Current Opinion in Psychiatry*, *22*(2), 218–223. doi:10.1097/YCO.0b013e3283252d80

Cancado-Trindade, A. A. (1948). Universal Declaration of Human Rights, Paris, 10 December 1948. *Audiovisual Library of International Law.* Retrieved from http://untreaty.un.org/cod/avl/ha/udhr/udhr.html. Accessed 19 September 2010.

Cappuzzo, F., Varella-Garcia, M., & Shigematsu, H. (2005). Increased *HER2* gene copy number is associated with response to gefitinib therapy in epidermal growth factor receptor-positive non-small-cell lung cancer patients. *Journal of Clinical Oncology*, *23*(22), 5007–5018. doi:10.1200/JCO.2005.09.111

Cartagena Agreement. (1996). Retrieved from http://www.comunidadandina.org.

Catalogue of Mendelian Inheritance Diseases OMINM. http://www.ncbi.nlm.nih.gov/sites/entrez?db=omim

Caulfield, T., Ries, N. M., Ray, P. N., Shuman, C., & Wilson, B. (2010). Direct-to-consumer genetic testing: good, bad or benign? *Clinical Genetics*, *77*(2), 101–105. doi:10.1111/j.1399-0004.2009.01291.x

Caulfield, T. (2007). Biobanks and blanket consent: the proper place of the public good and public perception rationales. *King's Law Journal*, *18*(2), 209–226.

Cavoukian, A. (1995). Confidentiality issues in genetics: the need for privacy and the right 'not to know.'. *Law and the Human Genome Review*, *2*, 53.

Centers for Disease Control and Prevention (CDC). (2001). Using tandem mass spectrometry for metabolic disease screening among newborns: A report of a work group. *MMWR. Morbidity and Mortality Weekly Report*, *50*(RR03), 1–22.

Centers for Disease Control and Prevention (CDC). (2004). Genetic testing for breast and ovarian cancer susceptibility: evaluating direct-to-consumer marketing--Atlanta, Denver, Raleigh-Durham, and Seattle, 2003. *MMWR. Morbidity and Mortality Weekly Report*, *53*(27), 603–606.

Cervino, A. C., & Hill, A. V. (2000). Comparison of tests for association and linkage in incomplete families. *American Journal of Human Genetics*, *67*(1), 120–132. doi:10.1086/302992

Chadwick, R., & Berg, K. (2001). Solidarity and equity: new ethical frameworks for genetic databases. *Nature Reviews. Genetics*, *2*(4), 318–321. doi:10.1038/35066094

Chadwick, R. (2003). Genomics, publich health and identity. *Acta Bioethica*, *9*(2), 209–218. Retrieved from http://www.scielo.cl/scielo.php?script=sci_arttext&pid=S1726-569X2003000200007&lng=es Accessed 5 August 2009. doi:10.4067/S1726-569X2003000200007

Chadwick, R. (2009). Nutrigenomics. In P. Atkinson (Ed.), *Handbook of Genetics and Society* (94-104). Routledge.

Chan, C. K. (2003). Risk perception and coping responses in a SARS infectious outbreak. *Third World Resurgence*, (July/August): 28–30.

Chan, S., & Quigley, M. (2007). Frozen embryos, genetic information and reproductive rights. *Bioethics*, *21*(8), 439–448. doi:10.1111/j.1467-8519.2007.00581.x

Chapman, A. R. (1999). *Unprecedented Choices: Religious Ethics at the Frontiers of Genetic Science*. Minneapolis: Fortress Press.

Chapman, A. R. (2002). Genetic engineering and theology: exploring the interconnections. *Theology Today (Princeton, N.J.)*, *59*(1), 21–86.

Cherkas, L. F., Oelsner, E. C., Mak, Y. T., Valdes, A., & Spector, T. D. (2004). Genetic influences on female infidelity and number of sexual partners in humans: a linkage and association study of the role of the vasopressin receptor gene (AVPR1A). *Twin Research*, *7*(6), 649–658. doi:10.1375/1369052042663922

Childress: Religion has important role in bioethics debate. (2001). Speaking at a conference at Princeton University in 2001. Retrieved from the news section of the Princeton University Website http://www.princeton.edu/main/news/archive/A97/89/83Q20/index.xml?section on 10 November 2009.

Ching, L Li. (2007). *Biosafety First*. Tapir Academic Press.

Cho, M. K. (2008). Understanding incidental findings in the context of genetics and genomics. *The Journal of Law, Medicine & Ethics*, *36*(2), 280–285. doi:10.1111/j.1748-720X.2008.00270.x

Cho, M. K. (1999). Ethical considerations in synthesizing a minimal genome. *Science*, *286*(5447), 2087–2090. doi:10.1126/science.286.5447.2087

Clague, A., & Thomas, A. (2002). Neonatal biochemical screening for disease. *Clinica Chimica Acta*, *315*(1-2), 99–110. doi:10.1016/S0009-8981(01)00716-1

Clarke, A. (2007). Should families own genetic information? No. *British Medical Journal*, *335*(7609), 23. doi:10.1136/bmj.39252.392940.AD

Claude, V. (2004). Genética y Bioética en América Latina. *Acta Bioética*, *10*(2), 155–166. Retrieved from http://www.scielo.cl/scielo.php?script=sci_arttext&pid=S1726-569X2004000200004&lng=es Accessed 5 August 2009.

Claus, E. B., Schildkraut, J. M., Thompson, W. D., & Risch, N. J. (1996). The genetic attributable risk of breast and ovarian cancer. *Cancer*, *77*(11), 2318–2324. doi:10.1002/(SICI)1097-0142(19960601)77:11<2318::AID-CNCR21>3.0.CO;2-Z

Cloning Human Beings. (1997). Report and Recommendations of the National Bioethics Advisory Commission, 1997. Retrieved from http://bioethics.gov/reports/past_commissions/nbac_cloning.pdf on 15 August 2009.

Cohen, H. Floris. (1994). *The Scientific Revolution: A Historiographical Inquiry.* Chicago: University of Chicago Press.

Cohen, R. (2002, June). Neglected diseases and the health burden in poor countries. *Multinational Monitor.*

Cole-Turner, R. (1993). *The New Genesis: Theology and the Genetic Revolution*. Louisville: Westminster John Knox Press.

Colgrove, J. (2002). The McKeown thesis: a historical controversy and its enduring influence. *Am J Public Health*, 92, 725–729.

Collins, F. S., & Guttmacher, A. E. (2003). Welcome to the genomic era. *The New England Journal of Medicine*, *349*(10), 996–998. doi:10.1056/NEJMe038132

Commercialization of traditional medicine. (2010). Retrieved from http://en.wikipedia.org/wiki/Biopiracy. Accessed 4 February 2010.

Commission on Human Rights Resolution. (1993).

Convention on Biological Diversity. (1992). United Nations. June 1992.

Cooper, I. P. (1985). *Biotechnology and the Law*. New York: Clark Boardman.

Correa, C. M. (1999). Access to plant genetic resources and intellectual property rights. Commission on Genetic Resources for Food and Agriculture, Background Study Paper No. 8. FAO. Retrieved from http://www.fao.org.

Corthesy-Theulaz, I. (Ed.). (2005). Nutrigenomics: the impact of biomics technology on nutrition research. *Annals of Nutrition & Metabolism*, *49*, 355–365. doi:10.1159/000088315

Council for Responsible Genetics (CRG). (2009). Retrieved from http://www.essential.org/crg/crg.html

Council of Europe. (2003*). The Protection of the Human Embryo In Vitro. Report by the Working Party on the Protection of the Human Embryo and Foetus.* CDBI –CO- GT3, 13. Strasbourg.

Council of Europe. Additional Protocol to the Convention on Human Rights and Biomedicine, concerning Genetic Testing for Health Purposes. Strasbourg, November 27, 2008. Accessed March 4, 2010. http://conventions.coe.int/Treaty/en/Treaties/Html/203.htm

Craig Ventor, J. (2001). The sequence of the human genome. *Science*, *291*, 1304–1351..doi:10.1126/science.1058040

Cree, L. M. (2008). A reduction of mitochondrial DNA molecules during embryogenesis explains the rapid segregation of genotypes. *Nature Genetics*, *40*(2), 249–254. doi:10.1038/ng.2007.63

Criswell, L. A., Pfeiffer, K. A., & Lum, R. F. (2005). Analysis of families in the Multiple Autoimmune Disease Genetics Consortium (MADGC) Collection: the PTPN22 620W allele associates with multiple autoimmune phenotypes. *American Journal of Human Genetics*, *76*(4), 561–571. doi:10.1086/429096

Crosbie, D. (2000). *Protection of Genetic Information: An International Comparison*. London: Report to the Human Genetics Commission.

Croyle, R. T., Smith, K. R., Botkin, J. R., Baty, B., & Nash, J. (1997). Psychological responses to BRCA1 mutation testing: preliminary findings. *Health Psychology*, *16*(1), 63–72. doi:10.1037/0278-6133.16.1.63

Cummings, S. (2000). The genetic testing process: how much counseling is needed? *Journal of Clinical Oncology*, *18*(Suppl 21), 60S–62S.

Dahl, E. (2004). Attitudes towards preconception sex selection: a representative survey from Germany. *Reproductive Biomedicine Online*, *9*(6), 600–603. doi:10.1016/S1472-6483(10)61767-1

Danoff, T. M., Campbell, D. A., & McCarthy, L. C. (2004). A Gilbert's syndrome UGT1A1 variant confers susceptibility to tranilast-induced hyperbilirubinemia. *The Pharmacogenomics Journal*, *4*(1), 49–53. doi:10.1038/sj.tpj.6500221

Data Protection Working Party. (2004). *Working Document on Genetic Data, 5 and 7.*

Data Protection Working Party. (2007).

Davey Smith, G., & Lynch, J. (2004). Social capital, social epidemiology, and disease aetiology. *International Journal of Epidemiology*, *33*, 691–700. doi:10.1093/ije/dyh261

de Chaves, E. P., & Narayanaswami, V. (2008). Apolipoprotein E and cholesterol in aging and disease in the brain. *Future Lipidology*, *3*(5), 505–530. doi:10.2217/17460875.3.5.505

De Constantino, B. F. M. G., de Miranda, G. Y., & Coutinho, A. F. G. (2006). Saúde pública e ética na era da medicina genômica: rastreamentos genéticos. *Revista Brasileira de Saú de Materno Infantil*, *6*(1), 141–146. Retrieved from http://www.scielo.br/scielo.php?script=sci_arttext&pid=S1519-38292006000100017&lng=en Accessed 5 August 2009. doi:.doi:10.1590/S1519-38292006000100017

Decoding Estonia. (2010). Retrieved from http://www.geenivaramu.ee/index.php?id=237. Accessed 7 March 2010.

Demyttenaere, K. (2000). Anxiety and depression in subfertility. In Meir Steiner, Kimberley A. Yonkers and Elias Eriksson (Eds.), *Mood Disorders in Women* (371-379). London: Martin Dunitz.

Denier, Y. (2006). Need or desire? a conception and moral phenomenology of the child wish. *The International Journal of Applied Philosophy*, *20*(1), 81–95.

Deussen, P. (1908). *Philosophy of Upanishads. (A. S. Geden Trans.)*. Edinburgh: T. & T. Clark.

Devolder, K., & Harris, J. (2005). Compromise and moral complicity in the embryonic stem cell debate. In N. Athanassoulis (Ed.), *Philosophical Reflections on Medical Ethics* (88-108). Palgrave Macmillan.

Ding, Z., Harding, C. O., & Thöny, B. (2004). State-of-the-art 2003 on PKU gene therapy. *Molecular Genetics and Metabolism*, *81*(1), 3–8. doi:10.1016/j.ymgme.2003.09.010

Dinorshoev, M. (1999). *The Contribution of the Samanids Epoch to the Cultural Heritage of the Central Asia: The Proceedings of the International Colloquium*. Dushanbe: UNESCO.

Dorff, E. N. (2001). Stem cell research: Jewish perspective. In Suzanne Holland, Karen Lebacqz, & Laurie Zoloth (Eds.), *The Human Embryonic Stem Cell Debate: Science, Ethics, and Public Policy* (89-94). Cambridge, MA: MIT Press.

Döring. Ole. (2007). Grenzen der kommerziellen und medizinischenVerfügbarkeit des menschlichen Körpers. Ein Fallbeispiel aus der chinesischen Medizinethik. In J. Taupitz (Ed.), *Kommerzialisierung des menschlichen Körpers* (283-290). Heidelberg: Springer.

Dorval, M., Patenaude, A. F., & Schneider, K. A. (2000). Anticipated versus actual emotional reactions to disclosure of results of genetic tests for cancer susceptibility: Findings from *p53* and *BRCA1* testing programs. *Journal of Clinical Oncology*, *18*(10), 2135–2142.

Douglas, H. A., Hamilton, R. J., & Grubs, R. E. (2009). The Effect of BRCA Gene Testing on Family Relationships: A Thematic Analysis of Qualitative Interviews. *Journal of Genetic Counseling*, *18*(5), 418–435. doi:10.1007/s10897-009-9232-1

Dragnea, B. (2008). Unnatural life. *Nature Materials*, *8*, 102–104. doi:10.1038/nmat2108

Duncan, R. E., & Delatycki, M. B. (2006). Predictive genetic testing in young people for adult-onset conditions: where is the empirical evidence? *Clinical Genetics*, *69*(1), 8–16. doi:10.1111/j.1399-0004.2005.00505.x

Durbin, P. T. (1984). Thomism and technology: natural law theory and problems of a technological society. In Mitcham, C., & Grote, J. (Eds.), *Theology and Technology: Essays in Christian Analysis and Exegesis. Lanham: University Press.*

Dutfield, G. (2000). *Intellectual Property Rights, Trade and Biodiversity*. London: Earthscan.

Dutney, A. (2001). *Playing God*. Melbourne: Harper Collins.

Dworkin, R. (1994). *Life's Dominion: An Argument about Abortion and Euthanasia*. Vintage.

Ebbesen, M., & Jensen, T. G. (2006). Nanomedicine: techniques, potentials, and ethical implications. *Journal of Biomedicine & Biotechnology*, *51516*, 1–11. doi:10.1155/JBB/2006/51516

Elizalde, J. (1995). Confidentiality, privacy and genetic data. In *The Human Genome Project: Legal Aspects, Vol. I*. Bilbao: Fundación BBV.

Engelhardt, H. T. (1989). The context of health care: persons, possessions, and states. In Beauchamp, T., & Walters, L. (Eds.), *Contemporary Issues in Bioethics* (3rd ed.). Belmont, CA: Wadsworth.

Epel, E. S., Blackburn, E. H., & Lin, J. (2004). Accelerated telomere shortening in response to life stress. *Proceedings of the National Academy of Sciences of the United States of America*, *101*(49), 17312–17315. doi:10.1073/pnas.0407162101

Epps. (2006). Singapore's multi-billion gamble. *The Journal of Experimental Medicine*, 203, 1139-1142.

Ethical Guidelines for Biomedical Research on Human Subjects. (2000). Delhi.

European Standards on Confidentiality and Privacy in Healthcare. (2006). Paragraph 3.4.2.

Evans, J. P. (2008). Recreational genomics; what's in it for you? *Genetics in Medicine*, *10*(10), 709–710. doi:10.1097/GIM.0b013e3181859959

Evans v. The United Kingdom, 7 March 2006.

Everett, M. (2003, Spring). The gene business: the body as property in the biotech century. *Bulletin of General Anthropology Division*, *9*(2), 1–5.

Failure to define law on privacy could cost society dear. (2001). *Times of India,* August 26, 2001.

FAO. (1999). *The State of the World's Plant Genetic Resources.*

Farkas, D. H., & Holland, C. A. (2009). Direct-to-consumer genetic testing: two sides of the coin. *The Journal of Molecular Diagnostics*, *11*(4), 263–265. doi:10.2353/jmoldx.2009.090034

Farrer, L. A., Cupples, L. A., & Haines, J. L. (1997). Effects of age, sex and ethnicity on the association between apolipoprotein E genotype and Alzheimer disease: a meta-analysis. *Journal of the American Medical Association*, *278*(16), 1349–1356. doi:10.1001/jama.278.16.1349

Feinberg, J. (1974). The rights of animals and future generations (appendix: the paradoxes of potentiality). In W. T. Blackstone (Ed.), *Philosophy and Environmental Crisis* (67-68). Athens, GA: University of Georgia Press.

Feinberg, J. (1980). Abortion. In T. Regan (Ed.), *Matters of Life and Death* (183-216). Second Edition. New York: Random House.

Feuk, L., Carson, A. R., & Scherer, S. W. (2006). Structural variation in the human genome. *Nature Reviews. Genetics*, *7*(2), 85–97. doi:10.1038/nrg1767

Fields, C., Adams, M. D., White, O., & Venter, J. C. (1994). How many genes in the human genome? *Nature Genetics*, *7*(3), 345–346. doi:10.1038/ng0794-345

Final Act Embodying the Results of the Uruguay Round of Multilateral Trade Negotiations. (1994). Marrakesh, 15 April 1994.

Fletcher, J. (1970). Technological devices in medical care. In Kenneth L Vaux (Ed.), *Who Shall Live?: Medicine, Technology, Ethics*. Philadelphia: Fortress Press.

Fluck, Jürgen. Kreislaufwirtschafts-, Abfall- und Bodenrecht, zu § 3 KrW-/AbfG Rn 79, 62. ErgL. 03/06.

Ford, D., Easton, D. F., & Bishop, D. T. (1994). Risks of cancer in BRCA1-mutation carriers. Breast Cancer Linkage Consortium. *Lancet, 343*(8899), 692–695. doi:10.1016/S0140-6736(94)91578-4

Foster, C., Watson, M., & Eeeles, R. (2007). Predictive genetic testing for BRCA1/2 in a UK clinical cohort: three-year follow-up. *British Journal of Cancer, 96*(5), 718–724. doi:10.1038/sj.bjc.6603610

Fowler, C., & Mooney, P. (1990). *Shattering: Food, Politics and the Loss of Genetic Diversity*. University of Arizona Press.

Francisco, R., et al. (2003). Análisis de ADNmt de restos esqueletales del sitio arqueológico de Tiwanaku y su relación con el origen de sus constructores. *Chungará (Arica),* 35(2). Retrieved from http://www.scielo.cl/scielo.php?script=sci_arttext&pid=S0717-73562003000200006&lng=es&nrm=iso. Accessed 31 August 2009. DOI: 10.4067/S0717-73562003000200006.

Frankfurt, H. (2004). *The Importance of What We Care About*. Cambridge University Press.

Freeman, J. L., Perry, G. H., & Feuk, L. (2006). Copy number variation: new insights in genome diversity. *Genome Research, 16*(8), 949–961. doi:10.1101/gr.3677206

Freund/Weiss. (2004). *MedR,* 346.

Frith, L. (2001). Reproductive technologies. In Chadwick, R. (Ed.), *Ethics of New Technologies*. San Diego: Academic Press.

Frosch, D. L., Grande, D., Tarn, D. M., & Kravitz, R. L. (2010). A decade of controversy: balancing policy with evidence in the regulation of prescription drug advertising. *American Journal of Public Health, 100*(1), 24–32. doi:10.2105/AJPH.2008.153767

Frosch, D. L., Krueger, P. M., Hornik, R. C., Cronholm, P. F., & Barg, F. K. (2007). Creating demand for prescription drugs: a content analysis of television direct-to-consumer advertising. *Annals of Family Medicine, 5*(1), 6–13. doi:10.1370/afm.611

Frosst, P., Blom, H. J., & Milos, R. (1995). A candidate genetic risk factor for vascular disease: a common mutation in methylenetetrahydrofolate reductase. *Nature Genetics, 10*(1), 111–113. doi:10.1038/ng0595-111

Fryer, A. (1995). Genetic testing of children. *Archives of Disease in Childhood, 73*(2), 97–99. doi:10.1136/adc.73.2.97

Fryer, A. (2000). Inappropriate genetic testing of children. *Archives of Disease in Childhood, 83*(4), 283–285. doi:10.1136/adc.83.4.283

Fyrö, K., & Bodegård, G. (1987). Four-year follow-up of psychological reactions to false positive screening tests for congenital hypothyroidism. *Acta Paediatrica Scandinavica, 76*(1), 107–114. doi:10.1111/j.1651-2227.1987.tb10424.x

Gabriel, K. J. (2004). *Gurumukathuninnu.* Kottayam: Sophia Books.

Gareth Jones, D. (2005). Genetic prospects: finding a balance between choice and acceptance. *Perspectives on Science and Christian Faith, 57*(3), 202–210.

Gareth Jones, D. (2009). *Playing God: scientific, ethical and technological challenges*. Retrieved from http://www.st-edmunds.cam.ac.uk/faraday/CIS/st-edmunds/jones/pdf/jones_lecture.pdf

Gaskell, G. (2006). Europeans and biotechnology in 2005: patterns and trends. *Final Report on Eurobarometer, 64*(3), 51.

Geary, J. (1996). Battle of the bean genes. *Time, 28*(October), 46–47.

Genetic Information Non-Discrimination Act of 2008, Public Law 110-233, 12 Stat. 881, May 21, 2008.

Genetics and Public Policy Center. (2007). *Survey of direct-to-consumer testing statutes and regulations*. Berman Institute of Bioethics, Johns Hopkins University. http://www.dnapolicy.org

Geransar, R., & Einsiedel, E. (2008). Evaluating online direct-to-consumer marketing of genetic tests: informed choices or buyers beware? *Genetic Testing, 12*(1), 13–23. doi:10.1089/gte.2007.0024

Gewin, V. (2003). Genetically modified corn - environmental benefits and risks. *PLoS Biology*, *1*(1), e8. doi:10.1371/journal.pbio.0000008

Gigerenzer, G., Gaissmeier, W., Kurz-Milcke, E., Schwartz, L. M., & Woloshin, S. (2008). Helping doctors and patients make sense of health statistics. *Psychological Science in the Public Interest*, *8*(2), 53–96.

Gilbar, R. (2010). Genetic testing of children for familial cancers: a comparative legal perspective on consent, communication of information and confidentiality. *Familial Cancer*, *9*(1), 75–87. doi:10.1007/s10689-009-9268-2

Gilbody, S., Lewis, S., & Lightfoot, T. (2006). Methylenetetrahydrofolate reductase (MTHFR) genetic polymorphisms and psychiatric disorders: a HuGE review. *American Journal of Epidemiology*, *165*(1), 1–13. doi:10.1093/aje/kwj347

Glass v. The United Kingdom, 9 March 2004.

Goddard, K. A., Robitaille, J., & Dowling, N. F. (2009). Health-related direct-to-consumer genetic tests: a public health assessment and analysis of practices related to Internet-based tests for risk of thrombosis. *Public Health Genomics*, *12*(2), 92–104. doi:10.1159/000176794

Gollin, M. A. (2009). *Biopiracy: the legal perspective.* Retrieved from http://www.actionbioscience.org/biodiversity/gollin.html. Accessed 7 January 2009.

Gollust, S. E., Hull, S. C., & Wilfond, B. S. (2002). Limitations of direct-to-consumer advertising for clinical genetic testing. *Journal of the American Medical Association*, *288*(14), 1762–1767. doi:10.1001/jama.288.14.1762

Gollust, S. E., Wilfond, B. S., & Hull, S. C. (2003). Direct-to-consumer sales of genetic services on the Internet. *Genetics in Medicine*, *5*(4), 332–337. doi:10.1097/01.GIM.0000076972.83711.48

Gomez-Lobo, A. (2004). Does respect for embryos entail respect for gametes? *Theoretical Medicine*, *25*, 199–208. doi:10.1023/B:META.0000040038.52317.08

Gomez-Lobo, A. (2005). On Potentiality and Respect for Embryos, A Reply to Mary Mahowald. *Theoretical Medicine and Bioethics*, *26*(2), 105–110. doi:10.1007/s11017-005-1235-9

Gonzalez, E., Kulkarni, H., & Bolivar, H. (2005). The influence of *CCL3L1* gene-containing segmental duplications on HIV-1/AIDS susceptibility. *Science*, *307*(5714), 1434–1440. doi:10.1126/science.1101160

González, G. (1999). *Derechos humanos. La condición humana en la sociedad tecnológica.* Madrid: Tecnos.

Gonzalez, G. (2009). Alerta en Chile ante el supersalmón. http://www.tierramerica.net/2001/0812/articulo.shtml. Accessed 5 August 2009.

González. Elio A. Prieto. (2007). Deterioro genómico y manipulación genética: Dessequilibrio en la prioridad de las agendas públicas. *Acta Bioethica*, 13(2), 223-231. Retrieved from http://www.scielo.cl/scielo.php?script=sci_arttext&pid=S1726-569X2007000200010&lng=es. Accessed 5 August 2009.

Goodacre, R. (Ed.). (2004). Metabolomics by numbers: acquiring and understanding global metabolite data. *Trends in Biotechnology*, *22*, 245–252. doi:10.1016/j.tibtech.2004.03.007

Government Accountability Office. (2006) Nutrigenetic testing: tests purchased from four Web sites mislead consumers. Available from http://www.gao.gov/products/GAO-06-977T. Accessed February 21, 2010.

Government Response to Recommendations. (2009). Retrieved from http://www.alrc.gov.au/inquiries/title/alrc96/response.htm. Accessed 24 September 2009.

Graham, D. S., Akil, M., & Vyse, T. J. (2007). Association of polymorphisms across the tyrosine kinase gene, TYK2 in UK SLE families. *Rheumatology (Oxford, England)*, *46*(6), 927–930. doi:10.1093/rheumatology/kel449

Graham, R. (1971). Foreword. In Frazier, C. A. (Ed.), *Should Doctors Play God?*Nashville: Broadman.

GRAIN. (2003). ¿Un sistema mundial de patentes? El Tratado sobre el Derecho Sustantivo de Patentes de la OMPI. Documento de Análisis. http://www.grain.org

Gray, S. W., O'Grady, C., Karp, L., Smith, D., Schwartz, J. S., Hornik, R. C., & Armstrong, K. (2009). Risk information exposure and direct-to-consumer genetic testing for BRCA mutations among women with a personal or family history of breast or ovarian cancer. *Cancer Epidemiology, Biomarkers & Prevention*, *18*(4), 1303–1311. doi:10.1158/1055-9965.EPI-08-0825

Green, M. R. (Ed.). (2003). Nutrigenetics: where next for food industry? *The Pharmacogenomics Journal*, *3*, 191–193. doi:10.1038/sj.tpj.6500180

Green, M., & Botkin, J. (2003). "Genetic exceptionalism" in medicine: clarifying the differences between genetic and nongenetic tests. *Annals of Internal Medicine*, *138*, 571–575.

Gregorios, P. (1980). *Cosmic Man*. New Delhi: Sophia Pub.

Gregorios, P. M. (1977). *The Human Presence: An Orthodox View of Nature*. Geneva: World Council of Churches.

Gregorios, P. M. (2007). *Cosmic Man*. New Delhi: Sophia.

Gregorios, P. (2007). *An Eastern Orthodox perspective of God, Man, Nature.* Retrieved from http://www.paulosmargregorios.info. Accessed on 15 November 2007.

Gregorios, Paulos Mar. (1992). *Human God.* Kottayam: Mar Gregorios Fondation.

Gros-Espiell, H. (1997). El proyecto de Declaración Universal sobre el Genoma Humano y los Derechos de la Persona Humana de la UNESCO. *Revista de Derecho y Genoma Humano. Law and the Human Genome Review*, 7.

Grubb, P. W. (1986). *Patents in Chemistry and Biotechnology*. Oxford: Clarendon Press.

Grünhage, F., Nattermann, J., & Gressner, O. A. (2010). Lower copy numbers of the chemokine *CCL3L1* gene in patients with chronic hepatitis C. *Journal of Hepatology*, *52*(2), 153–159. doi:10.1016/j.jhep.2009.11.001

Guidelines for the Protection of Privacy and Transborder Flows of Personal Data. *(*1980*)*.

Gurian, E. A., Kinnamon, D. D., Henry, J. J., & Waisbren, S. E. (2006). Expanded newborn screening for biochemical disorders: the effect of a false-positive result. *Pediatrics*, *117*(6), 1915–1921. doi:10.1542/peds.2005-2294

Gurwitz, D., & Bregman-Eschet, Y. (2009). Personal genomics services: whose genomes? *European Journal of Human Genetics*, *17*(7), 883–889. doi:10.1038/ejhg.2008.254

Guyton, A. C., & Hall, J. E. (1999). *Textbook of Medical Physiology.* 10th Ed.

Halàsz. (2004). Das Recht auf bio-materielle Selbstbestimmung, 13 f. (mit einem Überblick über die verschiedenen Theorien).

Hamilton, J. G., Lobel, M., & Moyer, A. (2009). Emotional distress following genetic testing for hereditary breast and ovarian cancer: a meta-analytic review. *Health Psychology*, *28*(4), 510–518. doi:10.1037/a0014778

Hansen, B., & Schotsman, P. (2001). Cloning: the human as created co-creator. *Ethical Perspectives*, *8*(2), 84. doi:10.2143/EP.8.2.503828

Hanson, S. S. (2006). More on Respect for Embryos and Potentiality: Does respect for Embryos Entail Respect for In Vitro Embryos? *Theoretical Medicine and Bioethics*, *27*, 215–226. doi:10.1007/s11017-006-9001-1

Hansson, M. G., Dillner, J., Bartram, C. R. J., Carlson, A., & Helgesson, G. (2006). Should donors be allowed to give broad consent to future biobank research? *The Lancet Oncology*, *7*(3), 266–269. doi:10.1016/S1470-2045(06)70618-0

Harakas, S. (1999). *Wholeness of Faith and Life: Orthodox Christian Ethics* (*Vol. III*). Massachusetts: Holy Cross Orthodox Press.

Harris, J. (1985). *The Value of Life*. London: Routledge.

Harry, D. (1995). Patenting of life and its implications for indigenous peoples. Kellogg Foundation. http://user.uni-frankfurt.de/~ecstein/gen/iatp/ipr-info7.html. Accessed 17 August 2009.

Hartmann, R. G. (2005). Face value: challenges of transplant technology. *American Journal of Law & Medicine*, *31*(1), 7.

Hefner, P. (1993). *The Human Factor: Evolution, Culture and Religion*. Philadelphia: Fortress Press.

Heidegger, M. (1996). *Being and Time. (J. Stambaugh Transl)*. Albany, NY: State University of New York Press.

Helgadottir, A., Thorleifsson, G., & Manolescu, A. (2007). A common variant on chromosome 9p21 affects the risk of myocardial infarction. *Science*, *316*(5830), 1491–1493. doi:10.1126/science.1142842

Hellegers, A. (1978). Fetal development. In T. L. Beauchamp (Ed.), *Contemporary Issues in Bioethics* (194-199). Encino, Calif.: Dickenson.

Hemmings, S. M., Kinnear, C. J., & Lochner, C. (2006). Genetic correlates in trichotillomania. A case- control association study in the South African Caucasian population. *The Israel Journal of Psychiatry and Related Sciences*, *43*(2), 93–101.

Hertzberg, M., Neville, S., Favaloro, E., & McDonald, D. (2005). External quality assurance of DNA testing for thrombophilia mutations. *American Journal of Clinical Pathology*, *123*(2), 189–193. doi:10.1309/09F827BC91M3D91C

Hertzberg, M., Neville, S., & McDonald, D. (2006). External quality assurance of molecular analysis of haemochromatosis gene mutations. *Journal of Clinical Pathology*, *59*(7), 744–747. doi:10.1136/jcp.2005.026005

Hewlett, J., & Waisbren, S. E. (2006). A review of the psychosocial effects of false-positive results on parents and current communication practices in newborn screening. *Journal of Inherited Metabolic Disease*, *29*(5), 677–682. doi:10.1007/s10545-006-0381-1

Hill, Richard L. (1994). OSU study finds genetic altering of bacterium upsets natural order. *The Oregonian,* August 8.

Hira, D. G. (2000). Beberapa kasus paten atas kehidupan dan biopiracy. Email discussion, July 9.

Hirtzlin, I. (2003). An empirical survey on biobanking of human genetic material and data in six EU countries. *European Journal of Human Genetics*, *11*, 475–488. doi:10.1038/sj.ejhg.5201007

HM Government and the Association of British Insurers. (2005). *Concordat and Moratorium on Genetics and Insurance* (March 2005). Retrieved from http://www.abi.org.uk/Information/Codes_and_Guidance_Notes/528.pdf.

Hoffmaster, B. (1991). Between the sacred and the profane: bodies, property, and patents in Moore case. *Intellectual Property Journal*, *7*, 115–148.

Hofgärtner, W. T., & Tait, J. F. (1999). Frequency of problems during clinical molecular-genetic testing. *American Journal of Clinical Pathology*, *112*(1), 14–21.

Hogarth, S., Javitt, G., & Melzer, D. (2008). The current landscape for direct-to-consumer genetic testing: legal, ethical, and policy issues. *Annual Review of Genomics and Human Genetics*, *9*, 161–182. doi:10.1146/annurev.genom.9.081307.164319

Hogarth, S., & Melzer, D. (2007). The IVD Directive and genetic testing problems and proposals. Retrieved from http://www.eshg.org/documents/BriefingIVDDirective-FINAL_july_20071.pdf.

Hogarth, S., Javitt, G. and Melzer, D. (2008). The current landscape for direct-to-consumer genetic testing: legal, ethical, and policy issues. *Annual Review of Genomics and Human Genetics,* 9, 165 ff.

Hollon, M. F. (2005). Direct-to-consumer advertising. A haphazard approach to health promotion. *Journal of the American Medical Association*, *293*(16), 2030–2033. doi:10.1001/jama.293.16.2030

Hollowell, K. (1999). *Cloning: Redefining when Life Begins. Exposing Flaws in the Pre embryo-Embryo Distinction.*

Hollox, E. J., Armour, J. A. L., & Barber, J. C. K. (2003). Extensive normal copy number variation of a beta-defensin antimicrobial-gene cluster. *American Journal of Human Genetics*, *73*(3), 591–600. doi:10.1086/378157

Holm, S. (1996). The moral status of the pre-personal human being: the argument from potential reconsidered. In *Conceiving the Embryo, Ethics, Law and Practice in Human Embryology* (193-220). Martinus Nijhoff.

Holtzman, N. A. (1999). Are genetic tests adequately regulated? *Science*, *286*(5439), 409. doi:10.1126/science.286.5439.409

Holtzman, N. A., & Marteau, T. M. (2000). Will genetics revolutionize medicine? *The New England Journal of Medicine*, *343*(2), 141. doi:10.1056/NEJM200007133430213

Honeywell, C. R., Gollob, M. H., Rutberg, J., Gow, R. M., & Geraghty, M. T. (2008). Discrepant DNA analysis in three patients with inherited arrhythmia: molecular genetic test results deserve a second glance. *American Journal of Medical Genetics. Part A*, *146A*(11), 1466–1469. doi:10.1002/ajmg.a.32336

Hongladarom, S. (2009). Privacy, the individual and genetic information: A Buddhist perspective. *Bioethics*, *23*(7), 403–412. doi:10.1111/j.1467-8519.2009.01716.x

House of Lords Science and Technology Committee. (2009). *Genomic Medicine. 2nd Report of Session 2008-09*.

Hudson, K., Javitt, G., Burke, W., & Byers, P., & American Society of Human Genetics Social Issues Committee. (2007). ASHG Statement on direct-to-consumer genetic testing in the United States. *Obstetrics and Gynecology*, *110*(6), 1392–1395.

Hughes, J. J., & Keown, D. (1995). Buddhism and medical ethics: a bibliographic introduction. *Journal of Buddhist Ethics*, 2 (Online Journal retrieved from http://www.buddhistethics.org/2/dkhughes.html).

HUGO Ethics Committee. (2002). Statement on Human Genomic Databases. Retrieved from http://www.hugo-international.org/img/genomic_2002.pdf. Accessed 15 February 2010.

HUGO Ethics Committee. (2002). *Statement on Benefit Sharing*. Retrieved from http://www.gene.ucl.ac.uk/hugo/benefit.html. Accessed 23 October 2002.

Hull, S. C., & Prasad, K. (2001). Reading between the lines: direct-to-consumer advertising of genetic testing in the USA. *Reproductive Health Matters*, *9*(18), 44–48. doi:10.1016/S0968-8080(01)90089-8

Human and Fluss. (2001). *The World Medical Association's Declaration of Helsinki: historical and contemporary perspectives* (World Medical Association: 24 July 2001). Retrieved from http://www.wma.net/en/20activities/10ethics/10helsinki/draft_historical_contemporary_perspectives.pdf. Accessed 20 November 2009.

Human and Fluss. (2003). Dismantling the Helsinki Declaration. *Editorial Canadian Medical Association Journal,* Nov. 11, 169.10.

Human Genetic Commission. (2009). Common framework of principles for direct-to-consumer geentic testing services. Principles and consultation questions. Retrieved from http://www.tga.gov.au/ivd/forthcoming.htm. Accessed 29 September 2009.

Human Genetic Examination Act of Germany. (Gesetz über genetische Untersuchungen bei Menschen (Gendiagnostikgesetz - GenDG). Enactment No. 374/09 of the German Federal Parliament (Bundestag), April 24, 2009, original German text and a parallel English translation available from EuroGentest at http://www.eurogentest.org/uploads/1247230263295/GenDG_German_English.pdf, and accessed January 10, 2010.

Human Genetics Commission. (2003). Genes Direct: Ensuring the effective oversight of genetic tests supplied directly to the public. London: Department of Health. Available at: http://www.dh.gov.uk/en/index.htm, Accessed March 07, 2010.

Human Genome Diversity Project, North American Regional Committee. (1995). *Model Ethical Protocol for Collecting DNA Samples,* in HGDP. Retrieved from http://www-leland.stanford.edu/group/morrinst/Protocol.html#Q0.

Huthwaite, J. S., Martin, R. C., Griffith, H. R., Anderson, B., Harrell, L. E., & Marson, D. C. (2006). Declining medical decision-making capacity in mild AD: a two-year longitudinal study. *Behavioral Sciences & the Law*, *24*(4), 453–463. doi:10.1002/bsl.701

Ida, R. (2009). Should we improve human nature? an interrogation from an Asian perspective. In Savulescu, J., & Bostrom, N. (Eds.), *Human Enhancement*. Oxford University Press.

IDRC Canada. (2010). *Viewpoint – what is biopiracy?* Retrieved from http://www.idrc.ca/idrcbulletin/ev-64405-201-1-DO_TOPIC.html. Accessed 4 February 2010.

Iervolino, A. C., Perroud, N., Fullana, M. A., Guipponi, M., Cherkas, L., Collier, D. A., & Mataix-Cols, D. (2009). Prevalence and heritability of compulsive hoarding: a twin study. *The American Journal of Psychiatry*, *166*(10), 1156–1161. doi:10.1176/appi.ajp.2009.08121789

Imelfort, I., Batley, J., Grimmond, S., & Edwards, D. (2009). Genome sequencing approaches and successes. *Methods in Molecular Biology (Clifton, N.J.)*, *513*, 345–358. doi:10.1007/978-1-59745-427-8_18

Indian Genome Variation Consortium, Brahmachari, Samir K. et al. (2005). The Indian Genome Variation database (IGVdb): a project overview. *Human Genetics*, *118*, 1–11. doi:10.1007/s00439-005-0009-9

Indigenous Peoples Coalition Against Biopiracy (IPCB). (2009). Retrieved from http://www.niec.net/ipcb/

Instituto de Estudios Ecologistas del Tercer Mundo. (2006). *Encuentro taller internacional 'servicios ambientales': la naturaleza como mercancía*. Quito: Mayo.

International Covenant on Civil and Political Rights. *(*1966*)*.

International Declaration on Human Genetic Data. (2003). Records of the General Conference 32nd Session, 29 September - 17 October 2003, adopted 16 October 2003.

International Disease Classification (IDC). (2010). Retrieved from http://apps.who.int/classifications/apps/icd/icd10online/. Accessed 15 March 2010.

International Human Genome Sequencing Consortium. (2004). Finishing the euchromatic sequence of the human genome. *Nature*, *431*(7011), 931–945. doi:10.1038/nature03001

International Human Genome Sequencing Consortium et al. (2001). Initial sequencing and analysis of human genome. *Nature*, *409*, 860–921..doi:10.1038/35057062

International Review of Industrial Property and Copyright Law. (1970).

Interview with Victoria M. Sopelak, Ph.D. (2002). University of Mississippi Medical Centre (UMC), June, 11, 2002.

Ioannidis, J. P. (2009). Personalized genetic prediction: too limited, too expensive, or too soon? *Annals of Internal Medicine*, *150*(2), 139–141.

Ionita-Laza, I., Rogers, A. J., Lange, C., Raby, B. A., & Lee, C. (2009). Genetic association analysis of copy-number variation (CNV) in human disease pathogenesis. *Genomics*, *93*(1), 22–26. doi:10.1016/j.ygeno.2008.08.012

Israel extends its human reproductive cloning moratorium for another 7 years till 2016. (2009). Retrieved from *Jerusalem Post.* Retrieved from http://www.jpost.com/servlet/Satellite?cid=1255450644033&pagename=JPArticle%2FShowFull on 1 November 2009.

Iwantoro, S. (2007). *Issues and concerns affecting biosecurity in Indonesia. PowerPoint presentation.* Director General of Agriculture Quarantine Agency of Indonesia.

Iwantoro, S. (2006). The role of agriculture quarantine agency of Indonesia (AQAI) as a trade instrument in global trade liberalization. PowerPoint presentation in Nanning, China, November 2, 2006. Director General of Agriculture Quarantine Agency of Indonesia.

Jackson, R. (2001). Expression of mouse interleukin-4 by a recombinant ectromelia virus suppresses cytolytic lymphocyte responses and overcomes genetic resistance to mousepox. *Journal of Virology*, *75*, 1205–1210. doi:10.1128/JVI.75.3.1205-1210.2001

Jameson, J. L. (1998). *Principles of Molecular Medicine* (1st ed.). Totowa, NJ: Humana Press.

Janson-Smith, D. (2002). *The genetics and insurance committee*, July 19, 2002. Retrieved from http://genome.wellcome.ac.uk/doc_WTD021010.html. Accessed January 10, 2010.

Janssens, A. C., Gwinn, M., Bradley, L. A., Oostra, B. A., van Duijn, C. M., & Khoury, M. J. (2008). A critical appraisal of the scientific basis of commercial genomic profiles used to assess health risks and personalize health interventions. *American Journal of Human Genetics*, *82*(3), 593–599. doi:10.1016/j.ajhg.2007.12.020

Johnstone, C., & Kaye, J. (2004). Does the UK biobank have a legal obligation to feedback individual findings to participants? *Medical Law Review*, *12*, 239. doi:10.1093/medlaw/12.3.239

Jonas, H. (1995). *El Principio Responsabilidad*. Barcelona: Círculo de Lectores/Herder.

Jonathan, R. (1999). Ethics and transgenic crops: a review. *Electronic Journal of Biotechnology*, *2*, 5–6. Retrieved from http://www.scielo.cl/scielo.php?script=sci_arttext&pid=S0717-34581999000200003&lng=es Accessed 5 August 2009.

Ambrose of Milan. (1995). Explanation of David the prophet. In Jurgens, W. (Ed.), *The Faith of the Early Fathers* (*Vol. II*). Bangalore Theological Publication.

Kaan, T. (2009). *Walking the tightrope: the regulation of biomedical research in Singapore*. Paper presented at the 2009 ILST Conference on Innovation, Competition and Regulation, Taiwan on December 3-4, 2009, organized by the Institute of Law for Science and Technology, National Tsing Hua University, Hsinchu, Taiwan.

Kaklamani, V. G., Wisinski, K. B., & Sadmi, M. (2008). Variants of the adiponectin (ADIPOQ) and adiponectin receptor 1 (ADIPOR1) genes and colorectal cancer risk. *Journal of the American Medical Association*, *300*(13), 1523–1531. doi:10.1001/jama.300.13.1523

Kalfoglou, A. L. (2005). Opinions about new reproductive genetic technologies: hopes and fears for our genetic future. *Fertility and Sterility*, *83*(6), 1612–1621. doi:10.1016/j.fertnstert.2005.01.090

Kaput, J. (2008). Nutrigenomics research for personalized nutrition and medicine. *Current Opinion in Biotechnology*, *19*(2), 110–120. doi:10.1016/j.copbio.2008.02.005

Kass, L. (2003). Ageless bodies, happy souls: biotechnology and the pursuit of perfection. *New Atlantis (Washington, D.C.)*, *1*, 9–28.

Katsanis, S. H., Javitt, G., & Hudson, K. (2008). Public health. A case study of personalized medicine. *Science*, *320*(5872), 53–54. doi:10.1126/science.1156604

Kawanishi, C., Lundrgen, S., Agren, H., & Bertilsson, L. (2004). Increased incidence of *CYP2D6* gene duplication in patients with persistent mood disorders: ultrarapid metabolism of antidepressants as a cause of nonresponse. A pilot study. *European Journal of Clinical Pharmacology*, *9*(11), 803–807.

Kaye, J. (2008). The regulation of direct-to-consumer genetic tests. *Human Molecular Genetics*, *17*(R2), R180–R183. doi:10.1093/hmg/ddn253

Kaye, J. (2008). The regulation of direct-to-consumer genetic tests. *Human Molecular Genetics*, *17*, 182. doi:10.1093/hmg/ddn253

Kehrer-Sawatzki, H. (2007). What a difference copy number variation makes. *BioEssays*, *29*(4), 311–313. doi:10.1002/bies.20554

Kennedy, S. (2004). Women health biobank in India. *The Indian Journal of Medical Research*, *120*, 131–132.

Keogh, L. (2009). Is uptake of genetic testing for colorectal cancer influenced by knowledge of insurance implications? *The Medical Journal of Australia*, *191*, 255.

Keown, D. (2001a). *Buddhism and Bioethics*. Palgrave Macmillan.

Keown, D. (2001b). *The Nature of Buddhist Ethics*. Palgrave Macmillan.

Kevorkian, J. (1992). A controlled auction market is a practical solution to the shortage of transplantable organs. *Medicine and Law*, 11.1-2, 47-55.

Khandanpour, N., Willis, G., & Meyer, F. J. (2009). Peripheral arterial disease and methylenetetrahydrofolate reductase (MTHFR) C677T mutations: a case-control study and meta-analysis. *Journal of Vascular Surgery*, *49*(3), 711–718. doi:10.1016/j.jvs.2008.10.004

Khor, M. (2007). *Biosafety First*. Tapir Academic Press.

Kilner, J. F. (2000). Human cloning. In Kilner, F., Cunningham, P., & David Hager, W. (Eds.), *The Reproduction Revolution: A Christian Appraisal of Sexuality, Reproductive Technologies, and the Family*. Cambridge: W. B. Eerdmans.

Kilpivaara, O., & Mukherjee, S., S., Schram, A.M., et al. (2009). A germline JAK2 SNP is associated with predisposition to the development of JAK2 (V617F)-positive myeloproliferative neoplasms. *Nature Genetics*, *41*(4), 455–459. doi:10.1038/ng.342

Kohane, I. S., Masys, D. R., & Altman, R. B. (2006). The incidentalome: a threat to genomic medicine. *Journal of the American Medical Association*, *296*(2), 212–215. doi:10.1001/jama.296.2.212

Koropatrick, S. (Ed.). (1993). Infertility: a non-event transition. *Fertility and Sterility*, *59*, 163–171.

Kottow, M. H. (2002). Public health, genetics and ethics. *Revista de Saude Publica*, *36*(5), 537–544. Retrieved from http://www.scielo.br/scielo.php?script=sci_arttext&pid=S0034-89102002000600001&lng=en Accessed 5 August 2009. doi:.doi:10.1590/S0034-89102002000600001

Kremer, Michael &Glannerster, Rachel. (2001). Creating a market for vaccines. *New York Times*. June 1.

Kriari-Catranis, I. (2003). Genetic data and confidentiality: the Estonian experiment. *Law and the Human Genome Review*, *19*, 147–157.

Krishnan, K. J. (2008). What causes mitochondrial DNA deletions in human cells? *Nature Genetics*, *40*(3), 275–279. doi:10.1038/ng.f.94

Kruuv, K., & Nomper, A. (2002). The Estonian Genome Project. Presented at the International Workshop "Law in Genetic Era" at the Central European University, Budapest, 7-9 June.

Kubar, O., & ten Have, H. (2007). *Ethical Review of Biomedical research in CIS Countries (Social and Cultural Aspects)*. Saint-Petersburg.

Kuhn, L., Schramm, D. B., & Donninger, S. (2007). African infants' *CCL3* gene copies influence perinatal HIV transmission in the absence of maternal nevirapine. *AIDS (London, England)*, *21*(13), 1753–1761. doi:10.1097/QAD.0b013e3282ba553a

Kuppuswamy, C. (2006). The role of international institutions in the formation of international bioethical law: UNESCO and the United Nations General Assembly attempt to govern human cloning. *Journal International de Bioethique*, *18*(1-2), 139–170.

Kyogoku, C., Morinobu, A., & Nishimura, K. (2009). Lack of association between tyrosine kinase 2 (TYK2) gene polymorphisms and susceptibility to SLE in a Japanese population. *Modern Rheumatology*, *19*(4), 401–406. doi:10.1007/s10165-009-0173-1

Lakusta, L., Dessalegn, B., & Landau, B. (2010). Impaired geometric reorientation caused by genetic defect. *Proceedings of the National Academy of Sciences of the United States of America*, *107*(7), 2813–2817. doi:10.1073/pnas.0909155107

Lander, E. S., Linton, L. M., & Birren, B. (2001). Initial sequencing and analysis of the human genome. *Nature*, *409*(6822), 860–921. doi:10.1038/35057062

Langley, L. S., & Blackston, J. W. (2006). Sperm, egg and a petri dish: unveiling the underlying property issues surrounding cryopreserved embryos. *Journal of Legal Medicine*, 167–206. doi:10.1080/01947640600716408

Laurie, G. (2005). (Intellectual) property? Let's think about taking a claim to our own genetic samples. Retrieved from www.law.ed.ac.uk/ahrb/publications/onine/GLPaper.htm.

Leacock, S. (1924). *The garden of folly* (pp. 122–131). New York: Dodd Mead.

Leder, D. (1984). Medicine and paradigms of embodiment. *The Journal of Medicine and Philosophy*, *9*(1), 29–43.

Lee, H. S., Korman, B. D., & Le, J. M. (2009). Genetic risk factors for rheumatoid arthritis differ in Caucasian and Korean populations. *Arthritis and Rheumatism*, *60*(2), 364–371. doi:10.1002/art.24245

Lee, S. S., & Crawley, L. (2009). Research 2.0: social networking and direct-to-consumer (DTC) genomics. *The American Journal of Bioethics*, *9*(6-7), 35–44. doi:10.1080/15265160902874452

Lee, C. K. (Ed.). (1999). Gene expression profile of aging and its retardation by caloric restriction. *Science*, *285*, 1390–1393. doi:10.1126/science.285.5432.1390

Lerman, C., Hughes, C., & Lemon, S. J. (1998). What you don't know can hurt you: adverse psychologic effects in members of BRCA1-linked and BRCA2-linked families who decline genetic testing. *Journal of Clinical Oncology*, *16*(5), 1650–1654.

Levitt, M. (2003). Public consultation in bioethics: what's the point of asking the public when they have neither scientific nor ethical expertise? *Health Care Analysis*, *11*(1), 15–25. doi:10.1023/A:1025381828650

Levy, H. L., & Albers, S. (2000). Genetic screening of newborns. *Annual Review of Genomics and Human Genetics*, *1*, 139–177. doi:10.1146/annurev.genom.1.1.139

Libby, E. N., Booker, J. K., Gulley, M. L., Garcia, D., & Moll, S. (2006). False-negative factor V Leiden genetic testing in a patient with recurrent deep venous thrombosis. *American Journal of Hematology*, *81*(4), 284–289. doi:10.1002/ajh.20543

Lippert (2004). *MedR,* 158.

Lippert. (2001). *MedR,* 408 (406).

Lipworth, W., Ankeny, R., & Kerridge, I. (2006). Consent in crisis: the need to reconceptualize consent to tissue banking research. *Internal Medicine Journal*, *36*(2), 124–128. doi:10.1111/j.1445-5994.2006.01020.x

Lissak, A., Sharon, A., Fruchter, O., Kassel, A., Sanderovitz, J., & Abramovici, H. (1999). Polymorphism for mutation of cytosine to thymine at location 677 in the methylenetetrahydrofolate reductase gene is associated with recurrent early fetal loss. *American Journal of Obstetrics and Gynecology*, *181*(1), 126–130. doi:10.1016/S0002-9378(99)70447-3

Lister, F. (1990). The role of international organizations in the 1990s and beyond. *International Relations*, *10*, 101–116. doi:10.1177/004711789001000201

Litaay, T. (2007). *Hak Kekayaan Intelektual*. Salatiga: Widya Sari.

Litaay, T., & Karetji, P. (2008). *Unity in (bio)diversity? Indonesian traditional knowledge protection in research activities*. Salatiga: PSKTI UKSW.

Litaay, T. (2007). Kemendesakan Perlindungan Pengetahuan Tradisional di Indonesia. In Louhenapessy, D. (Eds.), *Biosecurity and Indigenous Knowledge – Workshop Proceeding. PSKTI UKSW-CDU-CRCNPB-BaKTI*. Salatiga.

Litaay, T., & Karetji, P. (2009*). Public policy and protection of indigenous knowledge*. Paper presented in Asia Pacific Sociological Association Conference, 2009.

Little, J., Higgins, J. P., & Ioannidis, J. P. (2009). STrengthening the REporting of Genetic Association studies (STREGA): an extension of the STROBE Statement. *Annals of Internal Medicine*, *150*(3), 206–215.

Liu, Y., & Pearson, Y. E. (2008). Direct-to-consumer marketing of predictive medical genetic tests: assessment of current practices and policy recommendations. *Journal of Public Policy & Marketing*, *27*(2), 131–148. doi:10.1509/jppm.27.2.131

Llancaqueo, V. T. (2006). El nuevo regimen internacional de derechos de propiedad intelectual y los ederechos de los pueblos indigenas. In Berraondo, M. (Ed.), *Pueblos Indígenas y Derechos Humanos*. Centro de Políticas Públicas y Derechos Indígenas, Instituto de Derechos Humanos, Universidad de Deusto.

Llewelyn, M. (1997). The legal protection of biotechnological inventions: an alternative approach. *European Intellectual Property Review*, *19*(3), 115–127.

Lobo, D. S., & Kennedy, J. L. (2009). Genetic aspects of pathological gambling: a complex disorder with shared genetic vulnerabilities. *Addiction (Abingdon, England)*, *104*(9), 1454–1465. doi:10.1111/j.1360-0443.2009.02671.x

Lockwood, M. (1988). Warnock vs Powell (and Harradine): when does potentiality count? *Bioethics*, *2*, 188–213. doi:10.1111/j.1467-8519.1988.tb00048.x

Lodder, L., Frets, P. G., & Trijsburg, R. W. (2001). Psychological impact of receiving a BRCA1/BRCA2 test result. *American Journal of Medical Genetics*, *98*(1), 15–24. doi:10.1002/1096-8628(20010101)98:1<15::AID-AJMG1014>3.0.CO;2-0

Lolas, F., Rodríguez, E., & Valdebenito, C. (2004). El Proyecto del Genoma Humano en la Literatura Biomédica en cuatro países Latinoamericanos. *Acta Bioética*, *X*, 167–180.

Long, A. A. (1974). *Hellenistic Philosophy*. London: Duckworth.

Lowery, J. T., Byers, T., Axell, L., Ku, L., & Jacobellis, J. (2008). The impact of direct-to-consumer marketing of cancer genetic testing on women according to their genetic risk. *Genetics in Medicine*, *10*(12), 888–894. doi:10.1097/GIM.0b013e31818de6d7

Lucassen, A. (2007). Should families own genetic information? Yes. *British Medical Journal*, *335*(7609), 22. doi:10.1136/bmj.39252.386030.AD

Lucassen, A., & Parker, M. (2001). Revealing false paternity: some ethical considerations. *Lancet*, *357*(9261), 1033–1035. doi:10.1016/S0140-6736(00)04240-9

Lunshof, J. E., Chadwick, R., Vorhaus, D. B., & Church, G. M. (2008). From genetic privacy to open consent. *Nature Reviews. Genetics*, *9*(5), 406–411. doi:10.1038/nrg2360

Lynch, J., DaveySmith, G., Kaplan, G., & House, J. (2000). Income inequality and mortality: importance to health of individual income, psychosocial environment, or material conditions. *BMJ (Clinical Research Ed.)*, *320*, 1200–1204. doi:10.1136/bmj.320.7243.1200

Lysaught, M. T. (1995). Body: social theories. In T. W. Reich (Ed.), *Encyclopedia of Bioethics, 1,* 300-305.New York: Simon & Schuster Macmillan.

M.S. v.Sweden, 27 August 1997.

Maccioni, R. B., Muñoz, J. P., & Maccioni, C. (2004). Dimensiones bioéticas de la investigación sobre el genoma humano. [revista en la Internet]. *Acta Bioethica*, *10*(1), 75–80. Retrieved from http://www.scielo.cl/scielo.php?script=sci_arttext&pid=S1726-569X2004000100010&lng=es Accessed 5 August 2009. doi:10.4067/S1726-569X2004000100010

Machiola, J. I. (2009). Banco de genes indígenas alimenta sospechas en Argentina y el Mundo. Retrieved from http://www.biodiversidadla.org/content/view/full/16097. Accessed 17 August 2009.

Mackenbach, J. P. (2007). Sanitation: pragmatism works. *BMJ (Clinical Research Ed.)*, *334*, s17. doi:10.1136/bmj.39044.508646.94

Maddox, B. (2003). The double helix and the 'wronged heroine'. *Nature*, *421*(6921), 407–408. doi:10.1038/nature01399

Madrid, R. (1999). Cuestiones Jurídicas en el Proyecto del Genoma Humano: Presente y Perspectivas Futuras. *Humanitas*, *15*, 20–24.

Mahowald, M. B. (2004). Respect for embryos and the potentiality argument. *Theoretical Medicine and Bioethics*, *25*(3), 209–214. doi:10.1023/B:META.0000040065.84498.4c

Maienschein, J. (2003). *Whose View of Life?: Embryos, Cloning, and Stem Cells*. Cambridge, MA: Harvard University Press.Mallia, P., & ten Have, H. (2005). Pragmatic approaches to genetic screening. *Medicine, Health Care, and Philosophy*, *8*(1), 69–77. doi:10.1007/s11019-004-6752-1

Malo, P. E. (2000). Deciding custody of frozen embryos: many eggs are frozen but who is chosen? In Care Law, H. (Ed.), *J. DePaul* (pp. 307–312).

Manolio, T. A., Brooks, L. D., & Collins, F. S. (2008). A HapMap harvest of insights into the genetics of common disease. *The Journal of Clinical Investigation, 118*(5), 1590–1605. doi:10.1172/JCI34772

Marco, S., & Miazato, I. E. S. (2001). Bioethics, Intellectual Property and Genomics. *Rev. Hosp. Clin. Fac. Med. S. Paulo,* 56(4), 97-102. Accessed 5 August 2009. Retrieved from http://www.scielo.br/scielo.php?script=sci_arttext&pid=S0041-87812001000400001&lng=en. DOI: 10.1590/S0041-87812001000400001.

Marietta, C., & McGuire, A. L. (2009). Currents in contemporary ethics. Direct-to-consumer genetic testing: is it the practice of medicine? *The Journal of Law, Medicine & Ethics*, *37*(2), 369–374. doi:10.1111/j.1748-720X.2009.00380.x

Mariman, C. M. E. (2006). Nutrigenomics and nutrigenetics: the 'omic' revolution in nutritional sciences. *Biotechnology and Applied Biochemistry*, *44*, 119–128. doi:10.1042/BA20050112

Marks, Stephen P. (2002). Human rights assumptions of restrictive and permissive approaches to human reproductive cloning. *Health and Human Rights*, 6(1), 80-100, p. 87.

Marmot, M. (2004). *Status Syndrome: How Your Social Standing Directly Affects your Health and Life Expectancy.* London: Bloomsbury.

Marmot, M., & Wilkinson, R. G. (2001). Psychosocial and material pathways in the relation between income and health: a response to Lynch *et al. BMJ (Clinical Research Ed.)*, *322*, 1233–1236. doi:10.1136/bmj.322.7296.1233

Mattick, J. S. (2003). The human genome and the future of medicine. *The Medical Journal of Australia*, *179*, 212–216.

Maxam, A. M., & Gilbert, W. (1977). A new method for sequencing DNA. *Proceedings of the National Academy of Sciences of the United States of America*, *74*(2), 560–564. doi:10.1073/pnas.74.2.560

McCarthy, M. I., Abencasis, G. R., Cardon, L. R., Goldstein, D. B., Little, J., Ioannidis, J. P., & Hirschhorn, J. N. (2008). Genome-wide association studies for complex traits: consensus, uncertainty and challenges. *Nature Reviews. Genetics*, *9*(5), 356–369. doi:10.1038/nrg2344

McGoldrick, D. (2005). Multiculturalism and its discontents. *H.R.L. Re*, *5*(1), 27–56.

Mclnerney. Peter K. and Rainbolt, George W. Rainbolt. (1994). *Ethics.* New York: Harper Perennial.

McKeown, T., Brown, R. G., & Record, R. G. (1972). An interpretation of the modern rise of population in Europe. *Population Studies*, *26*, 345–382. doi:10.2307/2173815

McKeown, T. (1971). A historical appraisal of the medical task. In McLachlan, G., & McKeown, T. (Eds.), *Medical History and Medical Care.* London: Oxford University Press.

McKinlay, J. B., & McKinlay, S. M. (1977). The questionable contribution of medical measures to the decline of mortality in the United States in the Twentieth Century. *The Milbank Quarterly*, *55*, 405–429. Retrieved from http://www.manhattan-institute.org/html/medical_progress.htm accessed July 19, 2006.

McNeill Jr. Donald G. (2007). Indonesia may sell, not give, bird flu virus to scientists. *The New York Times*, February 7, 2007. Retrieved from http://www.nytimes.com/2007/02/07/world/asia/07birdflu.html. Accessed 21 November 2009.

Merges, R. P. (1988). Intellectual property in higher life forms: the patent system and controversial technologies. *Maryland Law Review (Baltimore, Md.)*, *47*, 1051–1075.

Metspalu, A. (2002). Biotechnology as an instrument of politics: the example of Estonia. In *Sinclair House Debates: Who Owns the Human Genome?* Bad Homburg v. d. Höhe: Herbert Quandt Foundation, The Foundation of Altana AG.

Metspalu, A. (2003). Workshop on "Biobanks for Health" in Oslo from January 28 - 31, 2003.

Metzker, M. L. (2010). Sequencing technologies - the next generation. *Nature Reviews. Genetics*, *11*(1), 31–46. doi:10.1038/nrg2626

Mezer, E., & Wygnanski-Jaffe, T. (2009). Ethical issues in ocular genetics. *Current Opinion in Ophthalmology*, *20*(5), 382–386. doi:10.1097/ICU.0b013e32832f7feb

Mikirtichian, G., Nikitina, A., Sozinov, A., Guryleva, M., & Malaysheva, E. (2007). *Ethical Review of Biomedical Research in CIS Countries (Social and Cultural Aspects)*. Saint-Petersburg.

Ministry of Health Singapore. (2003) Public consultation on the draft regulation of Biomedical Research Bill - summary of feedback received 10 November 2003 - 30 November 2003. Retrieved from http://www.moh.gov.sg/mohcorp/econsultationpast.aspx?ecid=81. Accessed January 10, 2010.

Mitchell, P. B., Meiser, B., Wilde, A., Fullerton, J., Donald, J., Wilhelm, K., & Schofield, P. R. (2010). Predictive and diagnostic genetic testing in psychiatry. *The Psychiatric Clinics of North America*, *33*(1), 225–243. doi:10.1016/j.psc.2009.10.001

Model Criminal Law Officers' Committee of the Standing Committee of Attorneys-General. (2008). Non-Consensual Genetic Testing. Discussion Paper.

Molster, C., Charles, T., Samanek, A., & O'Leary, P. (2009). Australian study on public knowledge of human genetics and health. *Public Health Genomics*, *12*(2), 84–91. doi:10.1159/000164684

Moniz, H. (2004). Privacy and intra-family communication of genetic information. *Revista de Derecho y Genoma Humano. Law and the Human Genome Review*, *21*, 111.

Morozova, O., & Marra, M. A. (2008). Applications of next-generation sequencing technologies in functional genomics. *Genomics*, *92*(5), 255–264. doi:10.1016/j.ygeno.2008.07.001

Moser, K. L., Kelly, J. A., Lessard, C. J., & Harley, J. B. (2009). Recent insights into the genetic basis of systemic lupus erythematosus. *Genes and Immunity*, *10*(5), 373–379. doi:10.1038/gene.2009.39

Mouchawar, J., Hensley-Alford, S., & Laurion, S. (2005). Impact of direct-to-consumer advertising for hereditary breast cancer testing on genetic services at a managed care organization: a naturally-occurring experiment. *Genetics in Medicine*, *7*(3), 191–197. doi:10.1097/01.GIM.0000156526.16967.7A

Muchtadi, Deddy. (N.d.) *Keamanan pangan produk hasil rekayasa genetik*. Powerpoint presentation. Departemen Ilmu dan Teknologi Pangan FATETA. Institut Pertanian Bogor.

Muntaner, C. (2004). Social capital, social class, and the slow progress of psychosocial epidemiology. *International Journal of Epidemiology*, *33*, 674–680. doi:10.1093/ije/dyh200

Murphy, W. J., Larkin, D. M., & Everts-van der Wind, A. (2005). Dynamics of mammalian chromosome evolution inferred from multispecies comparative maps. *Science*, *309*(5734), 613–617. doi:10.1126/science.1111387

Mutua, Makau wa. (1996). The ideology of human rights. *Virginia Journal of International Law*, 36. 593. Retrieved from SSRN: http://ssrn.com/abstract=1525598, p.

Myhr, A. I., & Traavik, T. (2007). *Biosafety First*. Tapir Academic Press.

Nakajima, T., Kaur, G., Mehra, N., & Kimura, A. (2008). HIV-1/AIDS susceptibility and copy number variation in CCL3L1, a gene encoding a natural ligand for HIV-1 co-receptor CCR5. *Cytogenetic and Genome Research*, *123*(1-4), 156–160. doi:10.1159/000184703

National Biosafety Framework. 2004. Dushanbe. Available at http://www.biodiv.tojikiston.com.

National Health and Medical Research Council (NHMRC). (1999). *Guidelines for Genetic Registers and Associated Genetic Material*. Issued by the National Health and Medical Research Council in accordance with the National Health and Medical Research Act, 1992 (Cth). Retrieved from http://www.nhmrc.gov.au/_files_nhmrc/file/publications/synopses/e14.pdf.

National Health and Medical Research Council. (2007). *National Statement on Ethical Conduct in Human Research*. Canberra: NHMRC.

National Health and Medical Research Council (NHMRC). (2004). *The Impact of Privacy Legislation on NHMRC Stakeholders: Comparative Stakeholder Analysis.*

National Human Genome Research Institute. (n.d.). Retrieved from www.genome.gov

Nelkin, D., & Lindee, M. S. (1995). *The DNA Mystique: The Gene as a Cultural Icon*. New York: Freeman.

Neri, D. (2009). On the concept of eugenics: preliminaries to a critical appraisal. *Cadernos de Saude Publica*. Retrieved from http://www.scielo.br/scielo.php?script=sci_arttext&pid=S0102-311X1999000500004&lng=en Accessed 5 August 2009. doi:.doi:10.1590/S0102-311X1999000500004

Newton, P., Brown, D., & Clover, C. (1999). Alarm over "Frankenstein" foods. *Electronic Telegraph,* Issue 1358, 12 February 1999. http://www.telegraph.co.uk.

Ng, T. W., Turinici, G., & Danchin, A. (2003). A double epidemic model for the SARS propagation. *BMC Infectious Diseases*, *3*, 19. doi:10.1186/1471-2334-3-19

NHMRC. (2009). *Use and Disclosure of Genetic Information to a Patient's Genetic Relatives under section 95AA of the Privacy Act 1988 (Cth)*. Guidelines for Health Practitioners in the Private Sector.

Nicol, D., Otlowski, M., & Chalmers, D. (2001). Consent, commercialisation and benefit-sharing. *Journal of Law and Medicine*, *9*, 80.

Nicoletto, M. O., Donach, M., De Nicolo, A., Artioli, G., Banna, G., & Monfardini, S. (2001). BRCA-1 and BRCA-2 mutations as prognostic factors in clinical practice and genetic counselling. *Cancer Treatment Reviews*, *27*(5), 295–304. doi:10.1053/ctrv.2001.0233

Nicoll, J. A. R., Roberts, G. W., & Graham, D. I. (1995). Apolipoprotein E4 allele is associated with deposition of amyloid beta-protein following head injury. *Nature Medicine*, *1*(2), 135–137. doi:10.1038/nm0295-135

Normile. (2007). An Asian tiger's bold experiment. *Science,* 316(5821), 38-41.

Norwegian Biobank Act. (2003). *Ministry of Health*, Law 2003-02-21, Oslo.

Novaes, S. B., & Salem, T. (2003). Embedding the embryo. In J. Harris and S. Holm (Eds.), *The Future of Human Reproduction, Ethics, Choice, and Regulation* (101-126). Oxford.

Nys, H., Romeo-Casabona, C. M., & Desmet, C. (2002). Legal aspects of prenatal testing for late-onset neurological diseases. In G. Evers-Kiebooms, M. W. Zoetewelj and P. S. Harper (Eds.), *Prenatal Testing for Late-onset Neurological Diseases* (84-109). Oxford: BIOS Scientific Publishers.

O'Keefe, M. (2004). Gender choice: is it playing God? *Christian Century (Chicago, Ill.)*, *121*(8), 12–13.

Ohata, T., Tsuchiya, A., Watanabe, M., Sumida, T., & Takada, F. (2009). Physicians' opinion for 'new' genetic testing in Japan. *Journal of Human Genetics*, *54*(4), 203–208. doi:10.1038/jhg.2009.11

Okada, Y., Mori, M., & Yamada, R. (2008). SLC22A4 polymorphism and rheumatoid arthritis susceptibility: a replication study in a Japanese population and a meta-analysis. *The Journal of Rheumatology*, *35*(9), 1723–1728.

Okonkwo, O., Griffith, H. R., & Belue, K. (2007). Medical decision-making capacity in patients with mild cognitive impairment. *Neurology*, *69*(15), 1528–1535. doi:10.1212/01.wnl.0000277639.90611.d9

OMPI. (2001). Panorama general sobre las cuestiones relativas a la propiedad intelectual y los recursos genéticos, los conocimientos tradicionales y el folclore. Documento preparado por la Secretaría, Comité Intergubernamental sobre Propiedad Intelectual y Recursos Genéticos, Conocimientos Tradicionales y Folclore, Primera sesión, Ginebra 30 de abril a 3 de mayo de 2001.

Opinion of the European Group on Ethics in Science and New Technologies to the European Commission. (2003). Ethical aspects of genetic testing in the workplace.

Otlowski, M. (2001). Protecting genetic privacy: an overview. In *Regulating the New Frontiers: Legal issues in Biotechnology*. Hobart, Melbourne: Centre for Law and Genetics.

Otlowski, M. (2002a). Employers' use of genetic test information: is there a need for regulation? *Australian Journal of Labour Law*, *15*, 1.

Otlowski, M. (2002b). Protecting genetic privacy in the research context: where to from here? *Macquarie Law Journal*, *2*, 91.

Otlowski, M. (2007). Disclosure of genetic information to at risk relatives: recent amendments to the *Privacy Act* 1988 (Cth). *The Medical Journal of Australia*, *187*, 398.

Otlowski, M. (2007). The use of legal remedies in Australia for pursuing allegations of genetic discrimination: findings from an empirical study. *International Journal of Discrimination and the Law*, *9*, 3.

Otlowski, M., & Williamson, R. (2003). Ethical and Legal Issues and the "New Genetics.". *The Medical Journal of Australia*, *178*, 4.

Overwalle, G. v. (1999). Patent protection for plants: a comparison of American and European approaches. *IDEA the Journal of Law and Technology*, *39*, 143–194.

Palandt/Heinrichs. § 90 BGB RdNr 3.

Palomaki, G. E., Bradley, L. A., Richards, C. S., Richards, C. S., & Haddow, J. E. (2003). Analytic validity of cystic fibrosis testing: a preliminary estimate. *Genetics in Medicine*, *5*(1), 15–20. doi:10.1097/00125817-200301000-00003

Panas, M., Karadima, G., Markianos, M., Kalfakis, N., & Vassilopoulos, D. (2008). Phenotypic discordance in a pair of monozygotic twins with Huntington's disease. *Clinical Genetics*, *74*(3), 291–292. doi:10.1111/j.1399-0004.2008.01036.x

Panchen, K. E. (1991). Patentability in the field of therapy and diagnosis. *International Review of Industrial Property and Copyright Law*, *22*, 879–880.

Paola, R. P., et al. (2002). Composición genética de la población chilena: Distribución de polimorfismos de DNA mitocondrial en grupos originarios y en la población mixta de Santiago. *Rev. méd. Chile* [revista en la Internet], Feb, 130(2), 125-131. Accessed 31 August 2009. Retrieved from: http://www.scielo.cl/scielo.php?script=sci_arttext&pid=S0034-98872002000200001&lng=es. DOI: 10.4067/S0034-98872002000200001.

Paparini, A., & Romano-Soica, V. (2004). Public health issues related with the consumption of food obtained from genetically modified organisms. *Biotechnology Annual Review*, *10*(1), 85–122. doi:10.1016/S1387-2656(04)10004-5

Pasacreta, J. V. (2003). Psychosocial issues associated with genetic testing for breast and ovarian cancer risk: an integrative review. *Cancer Investigation*, *21*(4), 588–623. doi:10.1081/CNV-120022380

Patents, I. P., & Diversity, H. G. (1993). RAFI Communique, Rural Advancement Foundation International, Ottawa, Canada. (May 1993).

Paul, D. B. (2008). Patient advocacy in newborn screening: continuities and discontinuities. *American Journal of Medical Genetics. Part C, Seminars in Medical Genetics*, *148C*(1), 8–14. doi:10.1002/ajmg.c.30166

Paulsen, J. S., Ferneyhough Hoth, K., Nehl, C., & Stierman, L. (2005). Critical periods of suicide risk in Huntington's Disease. *The American Journal of Psychiatry*, *162*, 725–731. Retrieved from http://www.huntington-assoc.com/Critical%20ab05.pdf. doi:10.1176/appi.ajp.162.4.725

Pearson, P. (2008). Genetic testing for everyone. *Nature*, *453*, 570–571. doi:10.1038/453570a

Peiris, J. S. M., Lai, S. T., & Poon, L. L. M. (2003). Coronavirus as a possible cause of Severe Acute Respiratory Syndrome. *Lancet*, *361*, 1319–1325. doi:10.1016/S0140-6736(03)13077-2

Pena, S. D., & Chakraborty, R. (1994). Paternity testing in the DNA era. *Trends in Genetics*, *10*(6), 204–209. doi:10.1016/0168-9525(94)90257-7

Pennisi, E. (2008). *Scientist synthesize a genome from scratch*. Retrieved from http://Sciencenow.sciencemag.org/cgi/content/short/2008/124/3.

Perry, M. (2008). Channel NewsAsia report, "Sexually active women advised to use the Pill to reduce abortion rate", posted 21 May 2008 1950 hrs, from http://www.channelnewsasia.com/stories/singaporelocalnews/view/349189/1/.html, re-accessed 10 January 2010.

Peters, T. (1995). Playing God and germline intervention. *The Journal of Medicine and Philosophy*, *20*, 365–385.

Peters, T. (2003). *Science, Theology, and Ethics*. Burlington, UK: Ashgate.

Pettersson, E., Lundeberg, J., & Ahmadian, A. (2009). Generations of sequencing technologies. *Genomics*, *93*(2), 105–111. doi:10.1016/j.ygeno.2008.10.003

Pfister, E. L., Kennington, L., & Straubhaar, J. (2009). Five siRNAs targeting three SNPs may provide therapy for three-quarters of Huntington's disease patients. *Current Biology*, *19*(9), 774–778. doi:10.1016/j.cub.2009.03.030

Phillips, J., & Firth, A. (1990). *Introduction to Intellectual Property Law*. London: Butterworths.

Plass, A. M., Baars, M. J., & Cornel, M. C. (2009). Testing the children: do non-genetic health-care providers differ in their decision to advise genetic presymptomatic testing on minors? A cross-sectional study in five countries in the European Union. *Genetic Testing and Molecular Biomarkers*, *13*(3), 367–376. doi:10.1089/gtmb.2008.0119

Pogge, T. W. (2005). World poverty and human rights. *Ethics & International Affairs*, *19*(1), 1–7. doi:10.1111/j.1747-7093.2005.tb00484.x

Poirier, J. (1994). Apolipoprotein E in animal models of CNS injury and in Alzheimer's disease. *Trends in Neurosciences*, *17*(12), 525–530. doi:10.1016/0166-2236(94)90156-2

Poulsen, J. B. (1999). Danish consumers attitudes towards functional foods. *MAPP working paper no. 62.2.*

PP No. 27 tahun (2004). tentang Tata Cara Paten oleh Pemerintah.

Preston, F. E., Kitchen, S., Jennings, I., & Woods, T. A. (1999). A UK National External Quality Assessment scheme (UK Neqas) for molecular genetic testing for the diagnosis of familial thrombophilia. *Thrombosis and Haemostasis*, *82*(5), 1556–1557.

Prodanov, V. (1996). Cultural pro-attitudes, reproductive ethics and embryo protection. In Evans (Ed.), *Conceiving the Embryo, Ethics, Law and Practice in Human Embryology*. The Hague/London/Boston: Martinus Nijhoff Publishers.

Przyluska-Fiszer, A. (1996). Human embryology and the criterion of moral standing. In *Conceiving the Embryo; Ethics, Law and Practice in Human Embryology.* Martinus Nijhoff: 165-172.

Qiu, C., De Ronchi, D., & Fratiglioni, L. (2007). The epidemiology of the dementias: an update. *Current Opinion in Psychiatry*, *20*(4), 380–385. doi:10.1097/YCO.0b013e32816ebc7b

Qiu, C., Kivipelto, M., & von Strauss, E. (2009). Epidemiology of Alzheimer's disease: occurrence, determinants, and strategies toward intervention. *Dialogues in Clinical NeuroSciences*, *11*(2), 111–128.

Raats, M. M. (Ed.). (2008). *Food for the Aging Population*. Woodhead Publishing.

Raber, J., Huang, Y., & Ashford, J. W. (2004). ApoE genotype accounts for the vast majority of the AD risk and AD pathology. *Neurobiology of Aging*, *25*(5), 641–650. doi:10.1016/j.neurobiolaging.2003.12.023

Rachels, J. (1986). *The End of Life*. Oxford: The Oxford University Press.

Rachels, J., & Ruddick, W. (1989). Lives and liberty. In Christman, J. (Ed.), *The Inner Citadel: Essays on Individual Autonomy*. New York: Oxford University Press.

Radcliffe-Richards, J., Daar, A. S., Guttmann, R. D., Hoffenberg, R., Kennedy, I., & Lock, M. (1998). The case for allowing kidney sales. *Lancet*, *352*, 1950–1952. doi:10.1016/S0140-6736(97)08211-1

Radin, M. J. (1982). Property and personhood. *Stanford Law Review*, *34*, 957–1015. doi:10.2307/1228541

Radin, M. J. (1987). Market-inalienability. *Harvard Law Review*, *100*(8), 1849–1937. doi:10.2307/1341192

Ramsay, P. (1970). *Fabricated Man: The Ethics of Genetic Control*. New York: Yale University Press.

Ramus, S. J., & Gayther, S. A. (2009). The contribution of BRCA1 and BRCA2 to ovarian cancer. *Molecular Oncology*, *3*(2), 138–150. doi:10.1016/j.molonc.2009.02.001

Redecker, N. v. (2003). The Estonian gene bank la: an example for Germany? In German-Estonian Legal Questions. Frankfurt: N. v. Redecker.

Regulation (EC) 45/2001 of the European Parliament and of the Council of 18 December 2000 European Union Directive on the Protection of Individual with Regard to the Processesing of Personal Data and on the Free Movement of such Data. (2001).

Report, G. (1998). Diez razones por las que la UPOV es un mal negocio. Conflictos entre comercio global y biodiversidad. Retrieved from http://www.grain.org/briefings/?id=76 Accessed 19 August 2009.

Resnik, D. B., & Langer, P. J. (2001). Human germline gene therapy reconsidered. *Human Gene Therapy*, *12*(11), 1449–1458. doi:10.1089/104303401750298607

Revel, M. (2003). Human reproductive cloning, embryo stem cells and germline gene intervention: an Israeli perspective. *Medicine and Law*, *22*, 701–732.

Reyes, M. S., & Rozowski, J. N. (2003). Alimentos transgénicos. *Revista Chilena de Nutricion*, *30*(1), 21–26.

Rhodes, R. (2006). Why test children for adult-onset genetic diseases? *The Mount Sinai Journal of Medicine, New York*, *73*(3), 609–616.

Ribeiro, S. (2009). Monsanto y la soya Argentina. http://www.etcgroup.org/es/materiales/publicaciones.html?pub_id=64. Accessed 5 August 2009.

Ries, N. M., & Castle, D. (2008). Nutrigenomics and ethics interface: direct-to-consumer services and commercial aspects. *OMICS: A Journal of Integrative Biology*, *12*(4), 245–250. doi:10.1089/omi.2008.0049

Rifkin, J., & Perlas, N. (1983). *Algeny*. New York: Viking.

Risch, N., Spiker, D., & Lotspeich, L. (1999). A genomic screen of autism: evidence for a multilocus etiology. *American Journal of Human Genetics*, *65*(2), 493–507. doi:10.1086/302497

Robert, J. S., & Baylis, F. (2005). Crossing species boundaries. In Thomas A. Shanno, (Ed.), *Genetics: Science, Ethics, and Public Policy* (11-32). Lanham, MD: Rowman & Littlefield Publishers.

Robertson, J. A. (2003). The $1000 genome: ethical and legal issues in whole genome sequencing of individuals. *The American Journal of Bioethics*, *3*(3), W35–W42. doi:10.1162/152651603322874762

Rodríguez, E., Valdebenito, C., Kanner, E., & Lolas, F. (2004). El Proyecto del Genoma Humano y las Regulaciones jurídicas en cuatro países latinoamericanos. *Jurisprudencia Argentina*, *IV*(5), 42–50.

Rodríguez, E., Valdebenito, C., Misseroni, A., Fernández, L., Outomuro, D., Schiattino, I., & Lolas, F. (2004). Percepciones Sociales sobre Genómica en Cuatro Países Latinoamericanos. Implicaciones Ético Legales. *Derecho y Genoma*, *21*, 141–164.

Rodríguez, E., et al. (2005). Social, ethical and legal attitudes towards genomic research in four Latin American countries. *Electronic Journal of Biotechnology*, 8(3). Available from http://www.ejbiotechnology.info/content/vol8/issue3/full/9/index.html. ISSN 0717-3458.

Rodríguez. Silvia. (2006). TLCs: el conocimiento tradicional en venta. *GRAIN, Abril 2006.* http://www.grain.org/briefings/?id=198. Accessed 19 August 2009.

Rodríguez-Guillén Mdel, R., Torres-Sánchez, L., & Chen, J. (2009). Maternal MTHFR polymorphisms and risk of spontaneous abortion. *Salud Pública de México*, *51*(1), 19–25.

Roffe, P. (2006). América Latina y la nueva arquitectura internacional de la propiedad intelectual: de los ADPIC-TRIPS a los nuevos tratados de libre comercio. In *Diálogo Regional sobre Propiedad Intelectual, Innovación y Desarrollo Sostenible*. Costa Rica, 10-12 Mayo: UNCTAD/ICTSD.

Roig, J. L. D. &Gómez, Mercedes. (2000). Riesgos sobre la Salud de los Alimentos Modificados Genéticamente: Una revisión bibliográfica. *Revista Espanola de Salud Publica*, *74*(3), 255–261. doi:10.1590/S1135-57272000000300003

Romeo-Casabona, C. M. (2002). *Los genes y sus leyes. El derecho ante el genoma humano*. Bilbao, Granada: Comares.

Romeo-Casabona, C. M. (2004). Anonymization and pseudonymization: the legal framework at a European level. In Beyleveld, D., Townend, D., Rouillé-Mirza, S., & Wright, J. (Eds.), *The Data Protection Directive and Medical Research Across Europe*. England: Ashgate.

Romeo-Malanda, S., & Nicol, D. (2007). Protection of genetic data in medical genetics: a legal analysis in the European context. *Revista de Derecho y Genoma Humano. Law and the Human Genome Review*, *27*, 97–134.

Roscam Abbing, H. D. C. (1995). Genetic Information and third party interests. How to find the right balance? *Law and the Human Genome Review*, *2*, 39.

Rosen, G. (1971). Historical trends and future prospects in public health. In McLachlan, G., & McKeown, T. (Eds.), *Medical History and Medical Care*. London: Oxford University Press.

Roses, A. D. (1996). Apolipoprotein E and Alzheimer's disease. A rapidly expanding field with medical and epidemiological consequences. *Annals of the New York Academy of Sciences*, *802*, 50–57. doi:10.1111/j.1749-6632.1996.tb32598.x

Rothenberg, M. B., & Sills, E. M. (1967). Iatrogenesis-PKU anxiety syndrome. *The American Journal of Psychiatry*, *124*(1), 109.

Rothstein, M. A. (2004). *Genetics and Life Insurance: Medical Underwriting and Social Policy*. Boston: MIT Press.

Rothstein, M. A. (1997). Genetic secret: a policy framework. In Rothstein, M. A. (Ed.), *Genetic Secrets: Protecting Privacy and Confidentiality in the Genetic Era*.

Ruan, Y. J., Lin, W. C., & Ling, A. E. (2003). Comparative full-length genome sequence analysis of 14 SARS coronavirus isolates and common mutations associated with putative origins of infection. *Lancet*, *361*, 1779–1785. doi:10.1016/S0140-6736(03)13414-9

Rural Advancement Foundation International (RAFI). (1995). Gene hunters in search of "disease genes" collect human DNA from remote island populations. In *RAFI Communiqué* (May/June). Retrieved from http://www.rafi.ca/communique/19953.html.

Rural Advancement Foundation International (RAFI). (2009). Retrieved from http://www.rafi.ca/

Rural Advancement Foundation Internacional (RAFI). (1998). Quinoa patent dropped: Andean farmers defeat U.S. University. *RAFI Genotype* – May 22, 1998. http://www.rafi.org. Accessed 5 August 2009.

Russia extends human cloning ban. (2009). Retrieved from *Ria Novosti*, http://en.rian.ru/russia/20091002/156329425.html on 28 October 2009.

Safrin, S. (2004). Hyperownership in a time of biotechnological promise: the international conflict to control the building blocks of life. *Rutgers Law School (Newark) Faculty Paper 15*. Retrieved from http://law.bepress.com/rutgersnewarklwps/fp/art15. Accessed Dec 12, 2006.

Saito, K. (Ed.). (2005). A Nutrigenomic database-integrated repository for publications and associated microarray data in nutrigenomics research. *The British Journal of Nutrition*, *94*, 493–495. doi:10.1079/BJN20051536

Saletan, W. (2007). Tinkering with humans. *New York Times*. Retrieved from http://www.nytimes.com/2007/07/08/books/review/Saletan.html?pagewanted=1&_r= 1 on 2 December 2009.

Samatovicz, R. A. (2000). Genetics and brain injury: apolipoprotein E. *The Journal of Head Trauma Rehabilitation*, *15*(3), 869–874. doi:10.1097/00001199-200006000-00002

Sanchez, G. J. (2003). Developing a platform for genomic medicine in Mexico. *Science*, *300*, 295–296. doi:10.1126/science.1084059

Sands, P., & Klein, P. (2001). *Bowett's Law of International Institutions*. London: Sweet and Maxwell.

Sanger, F., & Coulson, A. R. (1975). A rapid method for determining sequences in DNA by primed synthesis with DNA polymerase. *Journal of Molecular Biology*, *94*(3), 441–448. doi:10.1016/0022-2836(75)90213-2

Sanger, F., Nicklen, S., & Coulson, A. R. (1977). DNA sequencing with chain-terminating inhibitors. *Proceedings of the National Academy of Sciences of the United States of America*, *74*(12), 5463–5467. doi:10.1073/pnas.74.12.5463

Santos, M. A. (1997). Aspectos científicos de los principales avances de la genética humana. El Impacto Social de la Manipulación Genética. *Humanistas*, *9*, 16–27.

Saunders, A. M., Strittmatter, W. J., & Schmechel, D. (1993). Association of apolipoprotein E allele epsilon 4 with late-onset familial and sporadic Alzheimer's disease. *Neurology*, *43*(8), 1467–1472.

Savulescu, J. (2005). How will history judge cloning? Retrieved from the *Times Higher Education*, 6 May 2005 at http://www.timeshighereducation.co.uk/story.asp?storycode=195874.

Savulescu, J. Compulsory genetic testing for APOE Epsilon 4 and Boxing. InTamburrini CM, Tännsjö T. (Ed.) *Genetic Technology and Sport: Ethical Questions (Ethics and Sport)*. 1st Edition, Routledge.

Schatz, U. (1998). Patentability of genetic engineering inventions in European patent office practice. *International Review of Industrial Property and Copyright Law,* 29, 2-16.

Schiattino, I., Silva, C., Lolas, F., Valdebenito, C., & Rodríguez, E. (2005). Percepciones y estados emocionales sobre el proyecto genoma humano en actores sociales seleccionados en la Región Metropolitana, Chile. *Revista Chilena de Salud Pública*, *9*, 154–161.

Schiattino, I., Silva, C., Lolas, F., Valdebenito, C., Rodríguez, E. (2005). Descripción de las percepciones sobre el proyecto genoma humana en Chile, Perú, Argentina y México, *Revista Quirón,* 36(1/3).

Schickedanz, A. D., & Herdman, R. C. (2009). Direct-to-consumer genetic testing: the need to get retail genomics right. *Clinical Pharmacology and Therapeutics*, *86*(1), 17–20. doi:10.1038/clpt.2009.56

Schröder/Taupitz. (1991). Menschliches Blut verwendbar nach Belieben des Arztes?.

Schummer, J. (2001). Aristotle on technology and nature. *Philosophia Naturalis*, *38*, 105.

Schünemann. (1985). Die Rechte am menschlichen Körper.

Scriver, C. R., & Waters, P. J. (1999). Monogenic traits are not simple: lessons from phenylketonuria. *Trends in Genetics*, *15*(7), 267–272. doi:10.1016/S0168-9525(99)01761-8

Seeds of discontent. (1999). *The Economist,* 20 February 1999, 93-95.

Service, R. F. (2006). Gene sequencing: The Race for the $1000 Genome. *Science*, *311*, 1544–1546. doi:10.1126/science.311.5767.1544

Setoyama, K. (2005). Privacy of genetic information. *Osaka University Law Review* 52,81 ff.

Shastry, B. S. (2002). SNP alleles in human disease and evolution. *Journal of Human Genetics*, *47*(11), 561–566. doi:10.1007/s100380200086

Shastry, B. S. (2007). SNPs in disease gene mapping, medicinal drug development and evolution. *Journal of Human Genetics*, *52*(11), 871–880. doi:10.1007/s10038-007-0200-z

Sheets-Johnstone, M. (Ed.). (1992). The materialization of the body: a history of Western medicine, a history in progress. In *Giving the Body Its Due* (132-158). Albany, NY: State University of New York Press.

Shendure, J., Mitra, R. D., Varma, C., & Church, G. M. (2004). Advanced sequencing technologies: methods and goals. *Nature Reviews. Genetics*, *5*(5), 335–344. doi:10.1038/nrg1325

Shestack, J. J. (1998). The philosophic foundations of human rights. *Human Rights Quarterly*, *20*(2), 201–234. doi:10.1353/hrq.1998.0020

Siegel, S., Dittrich, R., & Vollmann, J. (2008). Ethical opinions and personal attitudes of young adults conceived by in vitro fertilisation. *Journal of Medical Ethics*, *34*(4), 236–240. doi:10.1136/jme.2007.020487

Sigurdsson, S., Nordmark, G., & Göring, H. H. (2005). Polymorphisms in the tyrosine kinase 2 and interferon regulatory factor 5 genes are associated with systemic lupus erythematosus. *American Journal of Human Genetics*, *76*(3), 528–537. doi:10.1086/428480

SIMAS Comunicación para el desarrollo rural. (2007). Terminator: Las semillas suicidas son semillas homicidas. Retrieved from http://www.simas.org.ni/simasnoticia/359. Accessed 5 August 2009.

Simpson, A. J. (2000). The genome sequence of the plant pathogen *Xylella* fastdiosa: the *Xylella* fastidiosa consortium of the organization for nucleotide sequencing and analysis. *Nature*, *6792*, 151–157.

Simpson, A. J. G. (2001). Genomics in Brazil. *Research Coordination, Pan American Health Organization* (online). Retrieved from http://www.paho.org/english/hdp/HDR/ACHR-02-Simpson.PDF. Accessed 31 August, 2009.

Singapore Department of Statistics. (2009). *Yearbook of Statistics Singapore 2009.* Retrieved from http://www.singstat.gov.sg/pubn/reference.html. Accessed 10 January 2010.

Skene, L. (1998). Patients' rights or family responsibilities? two approaches to genetic testing. *Medical Law Review*, *6*, 1. doi:10.1093/medlaw/6.1.1

Smith, A., Moran, A., & Boyd, M. C. (2007). Phenocopies in BRCA1 and BRCA2 families: evidence for modifier genes and implications for screening. *Journal of Medical Genetics*, *44*(1), 10–15. doi:10.1136/jmg.2006.043091

Smith, D. H., & Cohen, C. B. (2003). *A Christian Response to the New Genetics*. Lanham, MD: Rowman & Littlefield Publishers.

Snead, O. C. (2009). Bioethics and self-governance: the lessons of the Universal Declaration on Bioethics and Human Rights. *The Journal of Medicine and Philosophy*, *34*, 204–222. doi:10.1093/jmp/jhp024

Soatov, I. (2003). *From the Hymns of Zarathustra to the Songs of Borbad.* Dushanbe.

Sobel, S., & Cowan, C. B. (2003). Ambiguous loss and disenfrenchized grief: the impact of DNA predictive testing on the family as a system. *Family Process*, *42*(1), 47–57. doi:10.1111/j.1545-5300.2003.00047.x

Sobrino, B., Brión, M., & Carracedo, A. (2005). SNPs in forensic genetics: a review on SNP typing methodologies. *Forensic Science International*, *154*(2-3), 181–194. doi:10.1016/j.forsciint.2004.10.020

Sobrino, B., & Carracedo, A. (2005). SNP typing in forensic genetics: a review. *Methods in Molecular Biology (Clifton, N.J.)*, *297*, 107–126.

Sommer, A., & Davidson, F. R. (2002). Assessment and control of vitamin A deficiency: the annecy accords. *The Journal of Nutrition*, *132*, 2845S–2850S.

Spranger. (2005). *NJW*, 1085 (1084).

Spriggs, M. (2004). Compulsory brain scans and genetic tests for boxers-or should boxing be banned? *Journal of Medical Ethics*, *30*, 515–516. doi:10.1136/jme.2003.003541

Statute of the International Bioethics Committee. (1998). Retrieved from the UNESCO IBC portal on the 10 July 2009 http://portal.unesco.org/shs/en/ev.php-URL_ID=2026&URL_DO=DO_TOPIC&URL_SECTION=201.html.

Steiner, H. (1994). *An Essay on Rights*. Oxford: Blackwell.

Steiner, H. (2003). Persons of lesser value: Moral argument and the 'final solution. In Garrard, E., & Scarre, G. (Eds.), *Moral Philosophy and the Holocaust*. Burlington, UK: Ashgate.

Stellungnahme der Zentralen Ethikkommission der Bundesärztekammer vom 20.02.2003. (2003). Die Weiterverwendung von menschlichen Körpermaterialien für Zwecke der medizinischen Forschung. *DÄBl*, A-1632.

Sterckx, S. (1998). Some ethically problematic aspects of the proposal for a directive on the legal protection of biotechnological Inventions. *European Intellectual Property Review*, *20*(4), 123–128.

Stewart, A. D. (1989). Screening for cystic fibrosis. *Nature*, *341*(6244), 696. doi:10.1038/341696b0

Sudarsky, L., Myers, R. H., & Walshe, T. M. (1983). Huntington's disease in monozygotic twins reared apart. *Journal of Medical Genetics*, *20*(6), 408–411. doi:10.1136/jmg.20.6.408

Suharto. (2007a). *Tugas dan fungsi pusat informasi dan keamanan hayati dalam pelaksanaan pengawasan keamanan pangan*. PowerPoint presentation 30 October 2007. Pusat Informasi dan Keamanan Hayati / Centre of Information and Biosafety, Agriculture Quarantine Agency of Indonesia.

Suharto. (2007b). *Peran karantina dalam rangka perlindungan sumberdaya genetik dan keanekaragaman hayati*. PowerPoint presentation 11 December 2007 in Pontianak, West Kalimantan. Pusat Informasi dan Keamanan Hayati / Centre of Information and Biosafety, Agriculture Quarantine Agency of Indonesia.

Sullivan, S. D., & Salladay, S. A. (2007). Gene therapy: restoring health or playing God? *Journal of Christian Nursing*, *24*(4), 199–205.

Svenaeus, F. (2007). A Heideggerian defence of therapeutic cloning. *Theoretical Medicine and Bioethics*, *28*, 31–62. doi:10.1007/s11017-007-9025-1

Sweeney, James A. (2005). Margins of appreciation: cultural relativity and the European Court of Human Rights in the post-Cold War era. *I.C.L.Q.*, 54(2), 459-474.

Szreter, S. (2002). Rethinking McKeown: the relationship between public health and social change. *American Journal of Public Health*, *92*, 722–725. doi:10.2105/AJPH.92.5.722

Szreter, S. (2004). Debating mortality trends in 19th century Britain. *International Journal of Epidemiology*, *33*, 705–709. doi:10.1093/ije/dyh143

Takala, T. (2004). The (im)morality of (un)naturalness. *Cambridge Quarterly of Healthcare Ethics*, *13*(1), 17. doi:10.1017/S0963180104131046

Tallacchini, Mariachiara. Rhetoric of anonymity and property rights in human biological materials (hbms). *Rev. Der. Gen H*, 22, 153-175.

Taubenberger, J. K. (2005). Characterization of the 1918 influenza virus polymerase genes. *Nature*, *437*, 889–893. doi:10.1038/nature04230

Taupitz. *JZ*, 92, 1089.

Teasdale, G. M., Murray, G. D., & Nicoll, J. A. (2005). The association between APOE epsilon4, age and outcome after head injury: a prospective cohort study. *Brain*, *128*(Pt 11), 2556–2561. doi:10.1093/brain/awh595

Teasdale, G. M., Nicoll, J. A., Murray, G., & Fiddes, M. (1997). Association of apolipoprotein E polymorphism with outcome after head injury. *Lancet*, *350*(9084), 1069–1071. doi:10.1016/S0140-6736(97)04318-3

Thakur, M., Grossman, I., & McCrory, D. C. (2007). Review of evidence for genetic testing for CYP450 polymorphisms in management of patients with nonpsychotic depression with selective serotonin reuptake inhibitors. *Genetics in Medicine*, *9*(12), 826–835. doi:10.1097/GIM.0b013e31815bf98f

Thangadurai, S. (2004). The Human Genome Project: the role of analytical chemists. *Analytical Sciences*, *20*(4), 595–601. doi:10.2116/analsci.20.595

The Agency for Science, Technology and Research, with the Ministry of Health. (2008). Joint Media Release: "13th BMS IAC meeting announces key achievements in Translational & Clinical Research efforts in Singapore," 17 October 2008. Retrieved from https://www.nmrc.gov.sg/corp/uploadedFiles/NMRC/News,_Views_And_Events/News/IAC%20Media%20Release.pdf. Accessed 20 November 2009.

The American Fertility Society Ethics Committee. (1988). Ethical considerations of the new reproductive technologies: biomedical research and respect for the pre-embryo. In *'Fertility and Sterility: 'Official Journal of the American Fertility Society,* 49(2) suppl. 1, The Protection of the Human Embryo In Vitro. (2003). Report by the Working Party on the Protection of the Human Embryo and Foetus. CDBI –CO- GT3, 13, Council of Europe, Strasbourg, 19 June 2003.

The Council for International Organizations of Medical Sciences (CIOMS) in collaboration with the World Health Organization. (WHO). (2002). *International Ethical Guidelines for Biomedical Research Involving Human Subjects*. Current edition, Geneva. Retrieved from http://www.cioms.ch.

The Human Genetics Society of Australasia (HGSA). (1990). *Guidelines for Human DNA Banking*. July, 1990.

The Huntington's Disease Collaborative Research Group. (1993). A novel gene containing a trinucleotide repeat that is expanded and unstable on Huntington's disease chromosomes. *Cell*, *72*(6), 971–983. doi:10.1016/0092-8674(93)90585-E

The International Conference on Harmonization of Technical Requirements for Registration of Pharmaceuticals for Human Use. (1996). *Guideline for Good Clinical Practice E6 (R1), 10 June 1996.*

The International HapMap Consortium. (2003). The International HapMap Project. *Nature*, *426*, 789–796. doi:10.1038/nature02168

The International HapMap Consortium. (2007). A second generation human haplotype map of over 3.1 million SNPs. *Nature*, *449*, 851–861. doi:10.1038/nature06258

The International Schizophrenia Consortium. (2008). Rare chromosomal deletions and duplications increase risk of schizophrenia. *Nature*, *455*(7210), 237–241. doi:10.1038/nature07239

The National Medical Ethics Committee (NMEC). Singapore. (1998). *National Medical Ethics Committee: A Review of Activities, 1994-1997.* Ministry of Health Singapore, 18 March 1998. Retrieved from http://www.moh.gov.sg/mohcorp/publicationsreports.aspx?id=2406.

The Nuremberg Code: Directives for Human Experimentation. (1949). Trials of War Criminals before the Nuremberg Military Tribunals under Control Council Law No. 10, Vol. 2. Washington, D.C.: U.S. Government Printing Office. Retrieved from http://ohsr.od.nih.gov/guidelines/nuremberg.html.

The World Medical Association. (2008). *The Declaration of Helsinki (Ethical Principles for Medical Research Involving Human Subjects).* Current edition by the 59th General Assembly of the WMA, Seoul, 22 October 2008. Retrieved from http://www.wma.net.

Thiele, F. (2003). Genetic tests in the insurance system: criteria for a moral evaluation, *Poiesis Prax*, pp. 193 f.

Thorsteinsdóttir, H., Daar, A. S., Sáenz, T. W., & Singer, P. A. (2009). Building a biopharmaceutical innovation system in Cuba: growth through linkages. In Mytelka (Ed.), *Pharmaceutical Innovation (Bio)Building K. L.: Pharmaceutical Innovation Systems in Developing Countries*. Maastricht: Institute for New Technologies (INTECH), United Nations University.

Timmons, M. (2006). *Moral Theory*. Oxford, UK: Rowman & Littlefield.

Timuragaoglu, A., Dizlek, S., Uysalgil, N., Tosun, O., & Yamac, K. (2006). Methylenetetrahydrofolate reductase C677T polymorphism in adult patients with lymphoproliferative disorders and its effect on chemotherapy. *Annals of Hematology*, *8*(12), 863–868. doi:10.1007/s00277-006-0175-4

Townsend, P., & Davidson, N. (Eds.). (1988). *Inequalities in Health: The Black Report and the Health Divide*. Harmondsworth: Penguin.

Tripodi, A., Chantarangkul, V., Menegatti, M., Tagliabue, L., & Peyvandi, F. (2005). Performance of clinical laboratories for DNA analyses to detect thrombophilia mutations. *Clinical Chemistry*, *517*, 1310–1311. doi:10.1373/clinchem.2005.049981

Trouiller, P., Olliaro, P., Torreele, E., Orbinski, J., Laing, R., & Ford, N. (2002). Drug development for neglected diseases: a deficient market and a public health policy failure. *Lancet*, *359*, 2188–2194. doi:10.1016/S0140-6736(02)09096-7

Tsoutsman, T., Bagnall, R. D., & Semsarian, C. (2008). Impact of multiple gene mutations in determining the severity of cardiomyopathy and heart failure. *Clinical and Experimental Pharmacology & Physiology*, *35*(11), 1349–1357. doi:10.1111/j.1440-1681.2008.05037.x

Turner, L. (2004). Bioethics needs to rethink its agenda. *BMJ (Clinical Research Ed.)*, *328*(7432), 175. doi:10.1136/bmj.328.7432.175

Tymstra, T. (1986). False positive results in screening tests: experiences of parents of children screened for congenital hypothyroidism. *Family Practice*, *3*(2), 92–96. doi:10.1093/fampra/3.2.92

UK Human Genetic Commission. (2007). *More Genes Direct: A Report on Developments in the Availability, Marketing and Regulation of Genetic Tests Supplied Directly to the Public*. (UK Human Genetics Commission, London; http://www.phgfoundation.org/news/3933/)

UNESCO. (2006). *Avicenna and the Ethics of Science and Technology Today*. Paris: UNESCO.

UNESCO. (1997). Universal Declaration on the Human Genome and Human Rights. Retrieved from http://portal.unesco.org/en/ev.phpURL_ID=13177&URL_DO=DO_TOPIC&URL_ SECTION=201.html. Accessed 15 February 2010.

UNESCO. (2003). Article 22 International Declaration on Human Genetic Data. Retrieved from http://portal.unesco.org/en/ev.phpURL_ID=17720&URL_DO=DO_TOPIC&URL_SECT ION=201.html. Accessed 15 February 2010.

Universal Declaration on Bioethics and Human Rights. (2005). Records of the General Conference, 33rd session Paris, 3-21 October 2005, adopted 19 October 2005.

Universal Declaration on the Human Genome and Human Rights. (1997). Records of the General Conference, Twenty-ninth Session Paris, 21 October to 12, adopted 11 November 1997.

Untung, K. (2008). National policy on biological diversity: community management of biosecurity. *Special Co-Publication between Kritis-Journal of Interdisciplinary Development Studies (Indonesia) and Learning Communities-International Journal of Learning in Social Contexts (Australia)*, 228-238.

UU No. 14 tahun (2001). tentang Paten.

UU No. 28 tahun (2000). tentang Perlindungan Varietas Tanaman.

V. Freier, Friedrich. (2005). Getrennte Körperteile in der Forschung zwischen leiblicher Selbstverfügung und Gemeinbesitz. MedR, 321 f.;

V. Lersner in Von Lersner, Heinrich / Wendenburg, Helge, „Recht der Abfallbeseitigung" 0103, zu § 3 abs. 1 KrW-/AbfG, Rn. 5.

van Dellen, A., & Hannan, A. J. (2004). Genetic and environmental factors in the pathogenesis of Huntington's disease. *Neurogenetics*, *5*(1), 9–17. doi:10.1007/s10048-003-0169-5

van der Ploeg, C. P., Lanting, C. I., Kauffman-de Boer, M. A., Uilenburg, N. N., de Ridder-Sluiter, J. G., & Verkerk, P. H. (2008). Examination of long-lasting parental concern after false-positive results of neonatal hearing screening. *Archives of Disease in Childhood*, *93*(6), 508–511. doi:10.1136/adc.2007.129320 van der Werf, M. J. (Ed.). (2001). Nutrigenomics: application of genomics technologies in nutritional sciences and food technology. *Journal of Food Science*, *66*(6), 772–780. doi:10.1111/j.1365-2621.2001.tb15171.x

Varghese, P. (1974). *Freedom and Authority*. Madras: CLS-ISPCK-LPH.

Venter, J. C., Adams, M. D., & Myers, E. W. (2001). The sequence of the human genome. *Science*, *291*(5507), 1304–1351. doi:10.1126/science.1058040

Verhey, A. (1995). Playing God and invoking a perspective. *The Journal of Medicine and Philosophy*, *20*, 347–364.

Views of the Holy See on human embryonic cloning, July 17, 2003. (2003). Retrieved from http://www.holyseemission.org/cloning2003eng.html, retrieved 7 December 2009.

Visudha Kurbana Thaksa. (2003). *Kottayam: M.O.* C Publications.

Vitrano, G. (2009). Perœ al servicio de transnacionales y biopiratas. www.selvas.org. Accessed 17 August 2009.

von Harnack, A. (2009). *History of Dogma – Volume V*. Retrieved from http://www.ccel.org/ccel/harnack/dogma5.ii.ii.i.iv.ii.html.

Wadman, M. (1996). Genetic resistance spreads to consumers. *Nature*, *383*, 564. doi:10.1038/383564a0

Wang, X. P., Lin, Q. D., Ma, Z. W., & Zhao, A. M. (2004). [C677T and A1298C mutation of the methylenetetrahydrofolate reductase gene in unexplained recurrent spontaneous abortion]. *Zhonghua Fu Chan Ke Za Zhi*, *39*(4), 238–241.

Warnock, M. (1985). *A Question of Life: The Warnock Report on Human Fertilisation and Embryology*. Oxford: Blackwell.

Wasson, K. (2008). Consumer alert: ethical issues raised by the sale of genetic tests directly to consumers. *The American Journal of Bioethics*, *8*(6), 16–18. doi:10.1080/15265160802248351

Wasson, K., Cook, E. D., & Helzlsouer, K. (2006). Direct-to-consumer online genetic testing and the four principles: an analysis of the ethical issues. *Ethics & Medicine*, *22*(2), 83–91.

Waters, M. D., & Fostel, J. M. (2004). Toxicogenomics and systems toxicology: aims and prospects. *Nature Reviews. Genetics*, *5*, 936–948. doi:10.1038/nrg1493

Waters, M. D., Selkirk, J. K., & Olden, K. (2003). The impact of new technologies on human population studies. *Mutation Research*, *544*, 349–360. doi:10.1016/j.mrrev.2003.06.022

Watson, J. D., & Crick, F. H. (1953). Molecular structure of nucleic acids; a structure for deoxyribose nucleid acid. *Nature*, *171*(4356), 737–738. doi:10.1038/171737a0

Watson, M., Foster, C., & Eeles, R. (2004). Psychosocial impact of breast/ovarian (BRCA1/2) cancer-predictive genetic testing in a UK multi-centre clinical cohort. *British Journal of Cancer*, *91*(10), 1787–1794. doi:10.1038/sj.bjc.6602207

Webster's Third New International Dictionary. (1976). G. & C. Merriam Company.

Weckworth, W., & Fiehn, O. (2002). Can we discover novel pathways using metabolomics analysis? *Current Opinion in Biotechnology*, *12*, 156–160. doi:10.1016/S0958-1669(02)00299-9

Weil, V. (1996). Biotechnology and ethics: a blueprint for the future. Biotechnology: social impact and quandaries. http://www.biotech.nwu.edu/nsf/weil.html.

Weisgraber, K. H., Rall, S. C. Jr, & Mahley, R. W. (1981). Human E apoprotein heterogeneity. Cysteine–arginine interchanges in the amino acid sequence of the apo-E isoforms. *The Journal of Biological Chemistry*, *256*(17), 9077–9083.

Weiss, T. G., & Carayannis, T. (2001). Whither United Nations economic and social ideas?: a research agenda. *Global Social Policy*, *1*(1), 25–47. doi:10.1177/146801810100100103

Wendler, D. (2006). One-time consent for research on biological samples. *British Medical Journal*, *332*, 544–547. doi:10.1136/bmj.332.7540.544

Wertz, D. C., Fanos, J. H., & Reilly, P. R. (1994). Genetic testing for children and adolescents. Who decides? *Journal of the American Medical Association*, *272*(11), 875–881. doi:10.1001/jama.272.11.875

Wheeler, D. A., Srinivasan, M., & Egholm, M. (2008). The complete genome of an individual by massively parallel DNA sequencing. *Nature*, *452*(7189), 872–876. doi:10.1038/nature06884

Whiteman, D. C., Clutton, C., & Hill, D. (2006). Australian public's views on privacy and health research. *British Medical Journal*, *332*, 1274. doi:10.1136/bmj.332.7552.1274-a

Whitfield, P. D. (Ed.). (2004). Metabolomics: an emerging post genomic tool for nutrition. *The British Journal of Nutrition*, *92*, 549–555. doi:10.1079/BJN20041243

Wilde, A., Meisner, B., Mitchell, P. B., & Schofield, P. R. (2010). Public interest in predictive genetic testing, including direct-to-consumer testing, for susceptibility to major depression: preliminary findings. *European Journal of Human Genetics*, *18*(1), 47–51. doi:10.1038/ejhg.2009.138

Wilkins, M. R. (Ed.). (1996). Progress with proteome projects: why all proteins expressed by genomes should be identified and how to do it. *Biotechnology & Genetic Engineering Reviews*, *13*, 19–50.

Williams, R. A., Mamotte, C. D., & Burnett, J. R. (2008). Phenylketonuria: an inborn error of phenylalanine metabolism. *The Clinical Biochemist. Reviews / Australian Association of Clinical Biochemists*, *29*(1), 31–41.

Williams. 2008. The Declaration of Helsinki and public health. *Bulletin of the World Health Organization*, 86(8).

Williams-Jones, B. (1999). Concepts of personhood and the commodification of the body. *Health Law Review*, *7*(3), 11–13.

Williams-Jones, B. (2006, October). Bioethics and patent law: the cases of Moore and the Hagahai people. *Wipo Magazine*, *6*, 17–18.

Wilmut, (1997). Viable offspring derived from foetal and adult mammalian cells. *Nature*, *385*(6619), 810–813. doi:10.1038/385810a0

Wilmut, I., Campbell, K., & Tudge, C. (2000). *The Second Creation: Dolly and the Age of Biotechnology*. London: Headline.

Winston, R. (2003). Playing God? *Nature*, *426*(6967), 603. doi:10.1038/426603a

WIPO. (1984). *Industrial Property Protection of Biotechnological Inventions. WIPO Pub. Biot/CE/I/2. Bull, A. T., Holt, G. & Lilly, M. D. (1982). Biotechnology: International Trends and Perspectives*.Paris: OECD.

Wolf, S. M. (2008). Introduction: the challenge of incidental findings. *The Journal of Law, Medicine & Ethics*, *36*(2), 216–218. doi:10.1111/j.1748-720X.2008.00265.x

Wolfe, S. M. (2002). Direct-to-consumer advertising--education or emotion promotion? *The New England Journal of Medicine*, *346*(7), 524–526. doi:10.1056/NEJM200202143460713

Woloshin, S., Schwartz, L. M., Tremmel, J., & Welch, H. G. (2001). Direct-to-consumer advertisements for prescription drugs: what are Americans being sold? *Lancet*, *358*(9288), 1141–1146. doi:10.1016/S0140-6736(01)06254-7

World Health Organization. (2003). A multicentre collaboration to investigate the cause of severe acute respiratory syndrome. *Lancet*, *361*, 1730–1733. doi:10.1016/S0140-6736(03)13376-4

Xiong, T., Richardson, M., Woodroffe, R., Halligan, S., Morton, D., & Lilford, R. J. (2005). Incidental lesions found on CT colonography: their nature and frequency. *The British Journal of Radiology*, *78*(925), 22–29. doi:10.1259/bjr/67998962

Yang, Y., Chung, E. K., & Wu, Y. L. (2007). Gene copy-number variation and associated polymorphisms of complement component C4 in human systemic lupus erythematosus (SLE): low copy number is a risk factor for and high copy number is a protective factor against SLE susceptibility in European Americans. *American Journal of Human Genetics*, *80*(6), 1037–1054. doi:10.1086/518257

Yeoh, K. C. (2008). Singapore's biomedical sciences landscape. *Journal of Commercial Biotechnology*, *14*(2), 141–148. doi:10.1057/palgrave.jcb.3050083

Young, J. (2003). *The Death of God and the Meaning of Life*. London: Routledge.

Z. v.Finland, 25 February 1997.

Zeggini, E., & Ioannidis, J. P. (2009). Meta-analysis in genome-wide association studies. *Pharmacogenomics*, *10*(2), 191–201. doi:10.2217/14622416.10.2.191

Zhang, X. (Ed.). (2008). Novel omics technologies in nutrition research. *Biotechnology Advances*, *26*, 169–176. doi:10.1016/j.biotechadv.2007.11.002

Zimmern, R. L., & Kroese, M. (2007). The evaluation of genetic tests. *Journal of Public Health (Oxford, England)*, *29*(3), 246–250. doi:10.1093/pubmed/fdm028

Zinkant, Kathrin. (2003). Gebt her Eure Gene. *FAZ 14.09.2003*, 65.

Zoloth, L. (2001). Jordan's Banks: A view from the first years of human embryonic stem cell research. In Holland, S., Lebacqz, K., & Zoloth, L. (Eds.), *The Human Embryonic Stem Cell Debate: Science, Ethics, and Public Policy*. Cambridge, MA: MIT Press.

About the Contributors

Soraj Hongladarom is an associate professor of philosophy and Director of the Center for Ethics of Science and Technology at Chulalongkorn University in Bangkok, Thailand. He has published books and articles on such diverse issues as bioethics, computer ethics, and the roles that science and technology play in the culture of developing countries. His concern is mainly on how science and technology can be integrated into the life-world of the people in the so-called Third World countries, and what kind of ethical considerations can be obtained from such relation. A large part of this question concerns how information technology is integrated in the lifeworld of the Thai people, and especially how such integration is expressed in the use of information technology in education. He is the editor, together with Charles Ess, of Information Technology Ethics: Cultural Perspectives, also published by IGI Global. His works have also appeared in Bioethics, The Information Society, AI & Society, Philosophy in the Contemporary World, and Social Epistemology, among others.

* * *

Minakshi Bhardwaj is a research associate at the ESRC Centre for Economic and Social Aspects of Genomics (Cesagen) at Cardiff University, UK. She has a background in biological sciences and bioethics. Her main interest and expertise is in the ethical and governance issues raised by new technologies; in particular genetic technologies and comparative studies between developing and rich countries. Mina has worked as researcher at the United Nations Food and Agriculture Organisation (FAO) and as a consultant to UNESCO on the ethical issues raised by new energy technologies. Mina is secretary to the Human Genome Organisation (HUGO) ethics committee since 2005. Mina held grants for international projects from Daiwa Foundation and British Academy looking into ethical issues arising out of biobanking and genetic databases in developing countries. Mina is currently involved in the nutritional genomics project at Cesagen, looking into the diet-gene association and health.

Leonardo D. de Castro is Professor of Philosophy at the University of the Philippines and currently Senior Research Fellow at the National University of Singapore's Centre for Biomedical Ethics. A member of the UNESCO Advisory Expert Committee for the Teaching of Ethics, he is also President of the Asian Bioethics Association and Secretary of the International Association of Bioethics. He was previously Vice Chair of the UNESCO International Bioethics Committee. Prof. de Castro was primarily responsible for setting up a regional Research Ethics Training Program at the University of the Philippines. As consultant to the European Commission, he helped establish and train the National Ethics Committees for Health Research of Cambodia and Lao PDR. He has served as consultant to WHO on the ethics of organ transplantation, member of the Custodian Group for the Declaration of Istanbul, and member of the Asian Working Group against Organ Trafficking. He has been awarded the Takashi Fujii Prize of the International Federation of Social Science Organizations, a National Book Award (Manila Critics Circle), several International Publication Awards, and an Outstanding Monograph Award (National Academy of Science and Technology). He is Editor-in-Chief of the Asian Bioethics Review and serves on the Board of several bioethics journals.

Chan Chee Khoon is a professor of health and social policy at the Center for Policy Research & International Studies (CenPRIS), Universiti Sains Malaysia. He graduated from the Massachusetts Institute of Technology with Bachelor and Master's degrees in Life Sciences, and has a Doctor of Science degree in Epidemiology from Harvard University. He has served as consultant and technical adviser to the World Bank, Malaysian Institute of Economic Research, European Commission (EC-INCO expert evaluator for health sector research), World Health Organisation (global consultations on Genomics & Health; IDEAHealth advisory board), United Nations Economic and Social Commission for Asia and Pacific (UNESCAP Expert Group Meeting on Health & Development, 2004), United Nations Educational, Scientific and Cultural Organisation (World Commission on Ethics of Scientific Knowledge and Technology, Bangkok regional consultation, 2006), Japan International Cooperation Agency (third-party evaluator), Penang State Government (Task Force on Viral Encephalitis, Task Force on SARS Epidemic), Ministry of Health (Malaysia), Socioeconomic & Environmental Research Institute, Penang (senior fellow), Malaysian Medical Association, Federation of Malaysian Consumer Associations (health policy adviser), and the Department of Environment in Malaysia. Elected to a two-year term on the inaugural Executive Board of the International Society for Equity in Health, he also serves on the editorial advisory boards of the International Journal for Equity in Health, Global Health Promotion and Global Social Policy (2003-2005). His current research interests include emerging and re-emerging infectious diseases, ethical and policy issues in science and technology, environment and development, and health systems in transition.

Varghese M. Daniel is a (Adjunct) Research Associate in the Faculty of Arts at Monash University, Melbourne, Australia. His PhD was on 'the moral status of human person' from Melbourne, Australia. He studied Master of Bioethics at the Katholic University, Belgium. He also completed Master of Science (Psychology) from Madras University. He had his Bachelors degrees in Theology and Arts from Serampur University and Osmania Universities respectively. He has been awarded Erasmus Mundus European Fellowship, Australian Post-graduation Award and Research Training Scheme scholarship. His research interests include Religion and Bioethics, Human Rights, Eastern Philosophy & Theology and Social Psychology. He also serves as a minister of the Indian Orthodox Church. He has married Dr. Smitha (Biomedical Scientist, CSIRO, Australia) and have two children.

Ole Döring (PhD and MA phil) is a Philosopher and Sinologist with a specialty in cross-cultural ethics. He has been conducting a research program on bioethics in China at Bochum University and the Institute for Asian Studies in Hamburg. Between 2006 and 2009, he was instrumental in the EC6 consortium project BIONET: Ethical Governance of Biological and Biomedical Research: Chinese - European Co-operation, which, among others, fosued on biobanking. He is currently teaching and developing a related research infrastructure at the Horst-Görtz-Institute for the Theory, History and Ethics of Chinese

Life Sciences (Charité) in Berlin. His extensive publications include "China's struggle for practical regulations in medical ethics", Nature Reviews Genetics 4, 233 –239; Advances in Chinese Medical Ethics. Chinese and International Perspectives, (ed., with RB Chen) Hamburg (IfA): 2002; Chinese Scientists and Responsibility: Ethical issues of Human Genetics in Chinese and International Contexts, Ole Döring (ed.) MIA 314, Hamburg, 1999; and Life Sciences in Translation – A Sino-European Dialogue on Ethical Governance of the Life Sciences. The BIONET's Textbook (ed.), (online publication November 2009at URL:http://www.bionet-china.org/pdfs/BIONET%20Textbook%20-%202nd%20Edition%20-%20 Dec%2009.pdf); Chinas Bioethik verstehen; Hamburg (Abera): 2004. Kultur und Bioethik. Eigentum am eigenen Körper, Baden-Baden (Nomos), Ed. (with Christian Steineck), 2. and revised edition: 2009. In 2009, Dr. Döring was rewarded with the "Outstanding Contributions Award" by the Chinese Journal Yixue yu Zhexue (Medicine and Philosophy).

Elena Ignovska works at the Institute of Civil Law in Macedonia. She graduated at the Faculty of Law "Iustinianus Primus" in Skopje in 2005, among the first three students in her generation. In February 2006 she was employed as a teaching assistant at the Institute of Civil Law, Faculty of Law, on the subjects Family Law and Law on Inheritance. Additionaly, she was engaged on the subjects Contemporary Systems of Inheritance, Sociology of Family and European Family Law. Currently, she is working on developing new course on The New Reproductive Technologies and the Law. In 2008 she joined the Erasmus Mundus Master Studies of Bioethics at three Universities: Katholieke Universiteit Leuven, Radboud Universiteit Nijmegen and Universita degli Studi di Padova. Elena Ignovska is a member of National Transplant Foundation of Republic of Macedonia.

Brigitte E.S. Jansen studied Educational Science, Philosophy, Psychology and Sociology at the Universities of Bielefeld and Lüneburg, Germany. She is doctor of philosophy and doctor rerum publicarum (public administration). Beside this she also has a practical agricultural degree. In 1996 she developed the first multi-media work report in management cybernetic (Heidelberg 1996). From 1991 to 1996 she was project leader of FOKUS, the Research Group of Cybernetic Company Strategy at the Institute of Business Administration (Organization and Decision Making) of the Leuphana University Lüneburg. Together with Prof. Dr. Jürgen Simon she has built up the Research Centre BioEthicsLaw e.V. and worked on a wide variety of projects on an international level in leading positions. She is the director of this institution and also visiting Professor University of Madras since 2005 till now (Bioethics, Bioethics-Law, Public Health). She is and was also member of the board of examiner of T.N. Medical University (Chennai), of the University of Madras and the Mother Theresa Women University (Kodaikanal). Since 1999 till now she is EU-expert in the evaluation of project proposals (Bioethics, Medical ethics, Gender aspects) and also reviewer of several international journals in the field. She was co-investigator of the International workshop "Technology and Culture: Genetics and its Social and Ethical Implications in Asia and Europe", Bangkok 2007 and responsible for several panels on international conferences. Her main research interest focused on Biomedical Ethics, Biolaw, Governance and Intercultural Aspects of Research Ethics, questions of justice and Gender studies. She is author and editor of some books and articles concerning these subjects in the last years.

Terry S.H. Kaan is Associate Professor of Law at the Faculty of Law of the National University of Singapore, where he teaches the Law of Torts, and Biomedical Law & Ethics. He has served as a member of Singapore's Bioethics Advisory Committee, its National Medical Ethics Committee, and the

Ministry of Health's Organ Transplantation Committee, among others. He is currently a board member of the Singapore Nursing Board.

Yakub Adi Krisanto is a lecturer in Faculty of Law and researcher at the Centre of Eastern Indonesia Studies at Satya Wacana Christian University, Salatiga, Indonesia. He graduated from Satya Wacana Christian University (undergraduate, Law) and finished his master of laws degree at Pelita Harapan University Jakarta. His works cover the topic of Public Information, Public utilities, and Good Governance. He is also an active civil society member to combat corruption in Indonesia. He is now preparing the establishment of the Center of Anti-Corruption Studies in Satya Wacana Christian University, Salatiga, Indonesia.

Jakkrit Kuanpoth is currently a Senior Lecturer of University of Wollongong in Australia where he teaches Property and Trusts, Intellectual Property Law, and International Trade Law. He holds a Thai law degree and was admitted to the Thai Bar as a barrister-at-law. He also holds an LLM in International Economic Law from the University of Warwick, England. He wrote his PhD thesis at the University of Aberdeen, Scotland.

Chamu Kuppuswamy is a legal academic, obtained her undergraduate and postgraduate degrees at the University of Madras, and continued her education in the UK. As a Chevening Scholar she studied for a Masters Degree in security and defence at Hull, and went on to do a PhD in law at Sheffield. She has obtained numerous awards including the University of Sheffield Centenary Achievement Award. She is a member of the Board of the International Association of Bioethics (IAB). She has consulted for UNESCO and is one of the coordinators of the Arts-Bioethics Network of IAB. She is on the organising team of Sheffield Cafe Scientifique.

Teoh Chin Leong is Research Fellow at the National University of Singapore's Centre for Biomedical Ethics, assisting with the development of outreach programs with a special focus on public education and bioethics education in schools. With interests in ethical, political and educational theory and practice, he is currently pursuing his doctoral studies at the Yong Loo Lin School of Medicine at the National University of Singapore.

Theofransus Litaay is a lecturer of the faculty of law of Satya Wacana Christian University (SWCU) and a member of the Centre of Eastern Indonesia Studies of SWCU, Salatiga, Indonesia. His research interests related to the issues of law and public policy, especially on the areas of indigenous knowledge protection, biosecurity, and conflict resolution. He has conducted research activities in three provinces of Indonesia, namely Papua province, Papua Barat province and East Nusa Tenggara province. He graduated from Satya Wacana Christian University Faculty of Law in 1996 and finished his master of law degree (LL.M) at Vrije Universiteit, Amsterdam in 2002. In 2006-2007 he was granted with the United Board Fellowship at Ateneo School of Law in Ateneo de Manila University, Manila – Philippines. In 2008, he became the United Board Fellow at Valparaiso University School of Law at Valparaiso, Indiana, US. Since 2007, he took active part in the Australian-Indonesian Biosecurity Community Management Project (Ausindo-Biocom), a project of CRC-Plant Biosecurity and Charles Darwin University together with a group of Indonesian universities. He is a PhD candidate at Charles Darwin University in Darwin, Australia.

Fernando Lolas is a psychiatrist at the University of Chile. He is a member of the Chilean Language Academy and Spanish Royal Academy. He is also Professor Faculty of Medicine and Social Sciences, University of Chile, Director Bioethics Program PAHIO/WHO, Chile.

Firuza Nasyrova obtained her PhD in biochemistry from Moscow, Russia and a Doctor of Science degree from Tajikistan. Currently she is Principal Investigator, Institute of Plant Physiology and Genetics of the Tajik Academy of Sciences, Dushanbe, Tajikistan. Before that she used to be Project manager of ISTC project #T-1105 "Genome Analyses of Cereals in Tajikistan," National Expert of the Biodiversity Enabling Activities/Development of National Biosafety Framework for Tajikistan, and Chief Scientific Secretary of Tajik Department of International Higher Education Academy of Sciences (TD IHEAS), Dushanbe, Tajikistan.

Dianne Nicol is a Professor at the Law Faculty, University of Tasmania. The broad theme of Dianne's research in the law discipline is the regulation of biotechnology and human genetics. She is particularly interested in the commercialisation of genetic knowledge and patenting of genetic inventions as well as the specific regulatory issues associated with genetic, cloning and stem cell technology and biobanking. She teaches in the areas of intellectual property law, information technology and the law, biotechnology and the law, media law and equity. Dianne started her academic career as a scientist, receiving a PhD from Dalhousie University in Canada in 1987. In 1996 she obtained a first class honours degree in law from the University of Tasmania and in 1997 she completed an LLM on patenting of genetic technologies in Australia. She was admitted as a barrister and solicitor to the Supreme Court of Tasmania and the High Court in 1998. She commenced work as an academic at the University of Tasmania in 2000.

Margaret Otlowski is Professor of Law at University of Tasmania, Deputy Director of the Centre for Law and Genetics, and Dean of the Law Faculty and Head of School. She has longstanding experience in health law and bioethics with particular reference to the interface between law, genetics and ethics. She has published extensively in the field, and has been engaged by Commonwealth and State governments and agencies as consultant and member for various committees, working parties and Tribunals, including current membership with the Tasmanian Anti-Discrimination Tribunal. She has been involved with a number of funded collaborative research projects through the Centre for Law and Genetics and funded by the Australian Research Council where her work has focussed on issues of regulation, privacy, discrimination and law reform.

Dyah Hapsari Prananingrum is a Lecturer in Faculty of Law at the Satya Wacana Christian University, Salatiga, Indonesia. As a lecture, she was teaching several subject materials, which are relating with business law, such as investment law, commercial law and intellectual property law. Besides teaching, she was involved in academic activity covering articles writing and other academic papers. Some of results in writing academic she has been published in internal and national journals. Now, she is a PhD candidate at the post graduate program Gadjah Mada University Indonesia.

Somparn Promta is Professor of Buddhist philosophy at philosophy department, Chulalongkorn University, Thailand. He has written a number of books on Buddhist ethics, Buddhist philosophy, Buddhism, Indian philosophy, philosophy, novel, short stories, and poetry. He also serves as the editor of the Chulalongkorn Journal of Buddhist Studies.

Eduardo Rodríguez obtained his PhD Genetics from New York University in the USA. He also has a Masters in theology degree from Saint John's University, also in the USA. Currently he is a Consultant Bioethics Program PAHIO/WHO, and Coordinator in the Ethics of Research Program Interdisciplinary Center for Studies on Bioethics, University of Chile.

Jürgen Robienski studied law in Hannover (Germany) and Örebro (Sweden). Since 1996 he is working as a lawyer with offices in Hannover and Müden / Aller. From June 2003 until June 2009 he worked as assistant of Prof. Dr. Jürgen Simon at the Leuphana University of Lüneburg. He also worked in the developing of a master program of the Friedrich-Schiller-University of Jena (Prof. Dr. Nicolaus Knöpffler) and in several national and international research programs, for example the European and Latin American project "Latinbanks" (Prof. Dr. mult. Carlos Romeo Casabona). Additionally, he worked several times as a researcher for the Büro für Technikfolgen-Abschätzung beim Deutschen Bundestag (Office of Technology Assessment at the German Federal Parliament). His research concerns medical and economic law in general, esp.legal questions of bio banking, human tissue, gene diagnosis, gene doping, enhancement and synthetic biology. About these questions he has published several articles and books and held several lectures on international conferences. His thesis, he finished in 2009, also concerns the legal changes of bio banking, gene diagnosis and gene doping in regard of the new German laws about tissue and gene diagnosis.

Sergio Romeo-Malanda is a Lecturer in Criminal Law at the Law Faculty, University of Las Palmas de Gran Canaria (Spain), and a research fellow of the Inter-University Chair BBVA Foundation – Provincial Government of Biscay in Law and the Human Genome (University of Deusto, University of the Basque Country, Bilbao, Spain). He is also an Honorary Professor of the Catholic University of Santa Maria de Arequipa, National University of St Agustin (both in Arequipa, Peru) and of the University José Carlos Mariátegui (Moquegua, also in Peru). He holds a Diploma in Advanced Studies in Criminal Law (LLM) and a PhD in Law. His research and teaching interests focus on Biomedical Law, Biotechnology Law and Criminal Law, with a particular emphasis on crimes related to biotechnology, such as the offences of genetic manipulation and cloning, an the legal status of minors in the Health System, fields in which he has published and lectured widely in several countries.

Juergen Walter Simon, First and Second State Examen in Law, also Master of Business Administration. He had written his thesis in history of law and his habilitation in Civil Law, Economic Law and Law of Informatics. He has been the head of the Institute of Law at the University of Lueneburg since 1989, and has been the head of the Research Centre of Biotechnology and Law at the same university. He is a member of the Advisory Committee of the Biotechnology Federal Association, Braunschweig. In 1998 he receved the BAUM Environment Award. He has participated in and has organised, both nationally and internationally, numerous congresses, symposiums and workshops in Germany, Europe, Asia and America. He has also directed several research projects, some of which were for the Office of Technology Assessment of the German Federal Council, for the VW-Foundation, the DFG (German Research Association), the Federal Ministry of Research, some other Ministries, private Enterprises and Academies. He has collaborated as an expert or scientific advisor for the European Commission, the VW-Foundation, Konrad-Adenauer-Foundation, Friedrich-Ebert-Foundation in Germany, Hungaria, Croatia and Vietnam. He had directed the EU project on "Xenotransplantation: Legal, ethical, social ad economic aspects" and was member of some other European projects up to now. He was a participant of three Leonardo projects, had organised an European symposium on 'Genetherapy' in 1997 and one on 'Cloning' in 2000. In the last few years, his research has been mainly focused on themes relating to Law and Biotechnology, Environment, Economics and new information technologies. He is the author of many books and articles in these areas.

Richard A. Stein received his MD from the "Iuliu Hatieganu" University of Medicine and Pharmacy, Cluj-Napoca, Romania (1996) and his PhD in Biochemistry from the Department of Biochemistry and Molecular Genetics, University of Alabama at Birmingham (2005), where he was the recipient of the 2005 "Outstanding Graduate Student" Award. His research articles on the genetics and molecular biology of microorganisms were published in journals that include Proceedings of the National Academy of Sciences of the USA, Molecular Microbiology, Journal of Structural Biology, Protein Science, and Journal of Biological Chemistry. Dr. Stein is currently a postdoctoral research associate in the Department of Molecular Biology at Princeton University. In addition to biomedical research, Dr. Stein is interested in science communication and in exploring anthropogenic factors that shape emerging and re-remerging infectious diseases worldwide. His articles, commentaries, and editorials on topics that include infectious disease outbreaks, antibiotic resistance, and public health, were published in Annals of Internal Medicine, The International Journal of Infectious Diseases, Annals of Family Medicine, and International Journal of Clinical Practice. Dr. Stein authored several invited book reviews for The Journal of The American Medical Association (JAMA), Trends in Endocrinology and Metabolism, American Journal of Physical Anthropology, New England Journal of Medicine (NEJM), and Journal of Infection. He served a 4-year appointment on the Editorial Board of the American Journal of Infection Control (2006-2009), and currently serves as an Editorial Board member of the European Journal of Internal Medicine and as an Associate Editor at the International Journal of Clinical Practice.

Index

W

Breinigsville, PA USA
15 November 2010
249270BV00004B/6/P